Communications in Computer and Information Science 2030

Rationale

The CCIS series is devoted to the publication of proceedings of computer science conferences. Its aim is to efficiently disseminate original research results in informatics in printed and electronic form. While the focus is on publication of peer-reviewed full papers presenting mature work, inclusion of reviewed short papers reporting on work in progress is welcome, too. Besides globally relevant meetings with internationally representative program committees guaranteeing a strict peer-reviewing and paper selection process, conferences run by societies or of high regional or national relevance are also considered for publication.

Topics

The topical scope of CCIS spans the entire spectrum of informatics ranging from foundational topics in the theory of computing to information and communications science and technology and a broad variety of interdisciplinary application fields.

Information for Volume Editors and Authors

Publication in CCIS is free of charge. No royalties are paid, however, we offer registered conference participants temporary free access to the online version of the conference proceedings on SpringerLink (http://link.springer.com) by means of an http referrer from the conference website and/or a number of complimentary printed copies, as specified in the official acceptance email of the event.

CCIS proceedings can be published in time for distribution at conferences or as post-proceedings, and delivered in the form of printed books and/or electronically as USBs and/or e-content licenses for accessing proceedings at SpringerLink. Furthermore, CCIS proceedings are included in the CCIS electronic book series hosted in the SpringerLink digital library at http://link.springer.com/bookseries/7899. Conferences publishing in CCIS are allowed to use Online Conference Service (OCS) for managing the whole proceedings lifecycle (from submission and reviewing to preparing for publication) free of charge.

Publication process

The language of publication is exclusively English. Authors publishing in CCIS have to sign the Springer CCIS copyright transfer form, however, they are free to use their material published in CCIS for substantially changed, more elaborate subsequent publications elsewhere. For the preparation of the camera-ready papers/files, authors have to strictly adhere to the Springer CCIS Authors' Instructions and are strongly encouraged to use the CCIS LaTeX style files or templates.

Abstracting/Indexing

CCIS is abstracted/indexed in DBLP, Google Scholar, EI-Compendex, Mathematical Reviews, SCImago, Scopus. CCIS volumes are also submitted for the inclusion in ISI Proceedings.

How to start

To start the evaluation of your proposal for inclusion in the CCIS series, please send an e-mail to ccis@springer.com.

Kanubhai K. Patel · KC Santosh · Atul Patel ·
Ashish Ghosh

Editors

Soft Computing and Its Engineering Applications

5th International Conference, icSoftComp 2023
Changa, Anand, India, December 7–9, 2023
Revised Selected Papers, Part I

 Springer

Editors
Kanubhai K. Patel (iD)
Charotar University of Science
and Technology
Changa, India

Atul Patel (iD)
Charotar University of Science
and Technology
Changa, India

KC Santosh (iD)
University of South Dakota
Vermillion, SD, USA

Ashish Ghosh (iD)
Indian Statistical Institute
Kolkata, India

ISSN 1865-0929   ISSN 1865-0937 (electronic)
Communications in Computer and Information Science
ISBN 978-3-031-53730-1   ISBN 978-3-031-53731-8 (eBook)
https://doi.org/10.1007/978-3-031-53731-8

This Springer imprint is published by the registered company Springer Nature Switzerland AG
The registered company address is: Gewerbestrasse 11, 6330 Cham, Switzerland

Paper in this product is recyclable.

Preface

It is a matter of great privilege to have been tasked with the writing of this preface for the proceedings of The Fifth International Conference on Soft Computing and its Engineering Applications (icSoftComp 2023). The conference aimed to provide an excellent international forum for emerging and accomplished research scholars, academicians, students, and professionals in the areas of computer science and engineering to present their research, knowledge, new ideas, and innovations. The conference was held during 07–09 December 2023, at Charotar University of Science & Technology (CHARUSAT), Changa, India, and organized by the Faculty of Computer Science and Applications, CHARUSAT. There are three pillars of Soft Computing viz., i) Fuzzy computing, ii) Neuro computing, and iii) Evolutionary computing. Research submissions in these three areas were received. The Program Committee of icSoftComp 2023 is extremely grateful to the authors from 17 different countries including UK, USA, France, Germany, Portugal, North Macedonia, Tunisia, Lithuania, United Arab Emirates, Sharjah, Saudi Arabia, Bangladesh, Philippines, Finland, Malaysia, South Africa, and India who showed an overwhelming response to the call for papers, submitting over 351 papers. The entire review team (Technical Program Committee members along with 14 additional reviewers) expended tremendous effort to ensure fairness and consistency during the selection process, resulting in the best-quality papers being selected for presentation and publication. It was ensured that every paper received at least three, and in most cases four, reviews. Checking of similarities was also done based on international norms and standards. After a rigorous peer review 44 papers were accepted with an acceptance ratio of 12.54%. The papers are organised according to the following topics: Theory & Methods, Systems & Applications, and Hybrid Techniques. The proceedings of the conference are published as two volumes in the Communications in Computer and Information Science (CCIS) series by Springer, and are also indexed by ISI Proceedings, DBLP, Ulrich's, EI-Compendex, SCOPUS, Zentralblatt Math, MetaPress, and SpringerLink. We, in our capacity as volume editors, convey our sincere gratitude to Springer for providing the opportunity to publish the proceedings of icSoftComp 2023 in their CCIS series.

icSoftComp 2023 exhibited an exciting technical program. It also featured high-quality workshops, two keynotes and six expert talks from prominent research and industry leaders. Keynote speeches were given by Theofanis P. Raptis (CNR, Italy), Sardar Islam (Victoria University, Australia), and Massimiliano Cannata (SUPSI, Switzerland). Expert talks were given by Xun Shao (Toyohashi University of Technology, Japan), Ashis Jalote Parmar (NTNU, Norway), Chang Yoong Choon (Universiti Tunku Abdul Rahman, Malaysia), Biplab Banerjee (IIT Bombay, India), Sonal Jain (SPU, India), and Sharnil Pandya (Linnaeus University, Sweden). We are grateful to them for sharing their insights on their latest research with us.

The Organizing Committee of icSoftComp 2023 is indebted to R.V. Upadhyay, Provost of Charotar University of Science and Technology and Patron, for the confidence that he invested in us in organizing this international conference. We would also like to take this opportunity to extend our heartfelt thanks to the honorary chairs of this conference, Kalyanmoy Deb (Michigan State University, MI, USA), Witold Pedrycz (University of Alberta, Alberta, Canada), Leszek Rutkowski (IEEE Fellow) (Czesto-chowa University of Technology, Czestochowa, Poland), and Janusz Kacprzyk (Polish Academy of Sciences, Warsaw, Poland) for their active involvement from the very beginning until the end of the conference. The quality of a refereed volume primarily depends on the expertise and dedication of the reviewers, who volunteer with a smiling face. The editors are further indebted to the Technical Program Committee members and external reviewers who not only produced excellent reviews but also did so in a short time frame, in spite of their very busy schedules. Because of their quality work it was possible to maintain the high academic standard of the proceedings. Without their support, this conference could never have assumed such a successful shape. Special words of appreciation are due to note the enthusiasm of all the faculty, staff, and students of the Faculty of Computer Science and Applications of CHARUSAT, who organized the conference in a professional manner.

It is needless to mention the role of the contributors. The editors would like to take this opportunity to thank the authors of all submitted papers not only for their hard work but also for considering the conference a viable platform to showcase some of their latest findings, not to mention their adherence to the deadlines and patience with the tedious review process. Special thanks to the team of EquinOCS, whose paper submission platform was used to organize reviews and collate the files for these proceedings. We also wish to express our thanks to Amin Mobasheri (Editor, Computer Science Proceedings, Springer Heidelberg) for his help and cooperation. We gratefully acknowledge the financial (partial) support received from Department of Science & Technology, Government of India and Gujarat Council on Science & Technology (GUJCOST), Government of Gujarat, Gandhinagar, India for organizing the conference. Last but not least, the editors profusely thank all who directly or indirectly helped us in making icSoftComp 2023 a grand success and allowed the conference to achieve its goals, academic or otherwise.

December 2023
                                                        Kanubhai K. Patel
                                                        KC Santosh
                                                        Atul Patel
                                                        Ashish Ghosh

Organization

Patron

R. V. Upadhyay   Charotar University of Science and Technology, India

Honorary Chairs

Kalyanmoy Deb   Michigan State University, USA
Witold Pedrycz   University of Alberta, Canada
Leszek Rutkowski   Czestochowa University of Technology, Poland
Janusz Kacprzyk   Polish Academy of Sciences, Poland

General Chairs

Atul Patel   Charotar University of Science and Technology, India
George Ghinea   Brunel University London, UK
Dilip Kumar Pratihar   Indian Institute of Technology Kharagpur, India
Pawan Lingras   Saint Mary's University, Canada

Technical Program Committee Chair

Kanubhai K. Patel   Charotar University of Science and Technology, India

Technical Program Committee Co-chairs

Ashish Ghosh   Indian Statistical Institute, Kolkata, India
KC Santosh   University of South Dakota, USA
Deepak Garg   Bennett University, India
Gayatri Doctor   CEPT University, India

| Maryam Kaveshgar | Ahmedabad University, India |
| Ashis Jalote-Parmar | Norwegian University of Science and Technology, Norway |

Advisory Committee

Arup Dasgupta	Geospatial Media and Communications, India
Valentina E. Balas	University of Arad, Romania
Bhuvan Unhelkar	University of South Florida Sarasota-Manatee, USA
Dharmendra T. Patel	Charotar University of Science and Technology, India
Indrakshi Ray	Colorado State University, USA
J. C. Bansal	Soft Computing Research Society, India
Narendra S. Chaudhari	Indian Institute of Technology Indore, India
Rajendra Akerkar	Vestlandsforsking, Norway
Sudhir Kumar Barai	BITS Pilani, India
S. P. Kosta	Charotar University of Science and Technology, India

Technical Program Committee Members

Abhijit Datta Banik	IIT Bhubaneswar, India
Abdulla Omeer	Dr. Babasaheb Ambedkar Marathwada University, India
Abhineet Anand	Chitkara University, India
Aditya Patel	Kamdhenu University, India
Adrijan Božinovski	University American College Skopje, North Macedonia
Aji S.	University of Kerala, India
Akhil Meerja	Vardhaman College of Engineering, India
Aman Sharma	Jaypee University of Information Technology, India
Ami Choksi	C.K. Pithawala College of Engg. and Technology, India
Amit Joshi	Malaviya National Institute of Technology, India
Amit Thakkar	Charotar University of Science and Technology, India
Amol Vibhute	Symbiosis Institute of Computer Studies and Research, India
Anand Nayyar	Duy Tan University, Vietnam

Angshuman Jana   IIIT Guwahati, India
Ansuman Bhattacharya   IIT (ISM) Dhanbad, India
Anurag Singh   IIIT-Naya Raipur, India
Aravind Rajam   Washington State University, Pullman, USA
Arjun Mane   Government Institute of Forensic Science, India
Arpankumar Raval   Charotar University of Science and Technology,
                                India
Arti Jain   Jaypee Institute of Information Technology, India
Arunima Jaiswal   Indira Gandhi Delhi Technical University for
                                Women, India
Asha Manek   RVITM Engineering College, India
Ashok Patel   Florida Polytechnic University, USA
Ashok Sharma   Lovely Professional University, India
Ashraf Elnagar   University of Sharjah, UAE
Ashutosh Kumar Dubey   Chitkara University, India
Ashwin Makwana   Charotar University of Science and Technology,
                                India
Avimanyou Vatsa   Fairleigh Dickinson University - Teaneck, USA
Avinash Kadam   Dr. Babasaheb Ambedkar Marathwada
                                University, India
Ayad Mousa   University of Kerbala, Iraq
Bhaskar Karn   BIT Mesra, India
Bhavik Pandya   Navgujarat College of Computer Applications,
                                India
Bhogeswar Borah   Tezpur University, India
Bhuvaneswari Amma   IIIT Una, India
Chaman Sabharwal   Missouri University of Science and Technology,
                                USA
Charu Gandhi   Jaypee University of Information Technology,
                                India
Chirag Patel   Innovate Tax, UK
Chirag Paunwala   SCET, India
Costas Vassilakis   University of the Peloponnese, Greece
Darshana Patel   Rai University, India
Dattatraya Kodavade   DKTE Society's Textile and Engineering
                                Institute, India
Dayashankar Singh   Madan Mohan Malaviya University of
                                Technology, India
Deepa Thilak   SRM University, India
Deepak N. A.   RV Institute of Technology and Management,
                                India
Deepak Singh   IIIT, Lucknow, India
Delampady Narasimha   IIT Dharwad, India

Dharmendra Bhatti	Uka Tarsadia University, India
Digvijaysinh Rathod	National Forensic Sciences University, India
Dinesh Acharya	Manipal Institute of Technology, India
Divyansh Thakur	IIIT Una, India
Dushyantsinh Rathod	Alpha College of Engineering and Technology, India
E. Rajesh	Galgotias University, India
Gururaj Mukarambi	Central University of Karnataka, India
Gururaj H. L.	Vidyavardhaka College of Engineering, India
Hardik Joshi	Gujarat University, India
Harshal Arolkar	GLS University, India
Himanshu Jindal	Jaypee University of Information Technology, India
Hiren Joshi	Gujarat University, India
Hiren Mewada	Prince Mohammad Bin Fahd University, Saudi Arabia
Irene Govender	University of KwaZulu-Natal, South Africa
Jagadeesha Bhatt	IIIT Dharwad, India
Jaimin Undavia	Charotar University of Science and Technology, India
Jaishree Tailor	Uka Tarsadia University, India
Janmenjoy Nayak	AITAM, India
Jaspher Kathrine	Karunya Institute of Technology and Sciences, India
Jimitkumar Patel	Charotar University of Science and Technology, India
Joydip Dhar	ABV-IIITM, India
József Dombi	University of Szeged, Hungary
Kamlendu Pandey	VNSGU, India
Kamlesh Dutta	NIT Hamirpur, India
Kiran Trivedi	Northeastern University, USA
KiranSree Pokkuluri	Shri Vishnu Engineering College for Women, India
Krishan Kumar	National Institute of Technology Uttarakhand, India
Kuldip Singh Patel	IIIT Naya Raipur, India
Kuntal Patel	Ahmedabad University, India
Latika Singh	Sushant University, India
M. Srinivas	National Institute of Technology-Warangal, India
M. A. Jabbar	Vardhaman College of Engineering, India
Maciej Ławrynczuk	Warsaw University of Technology, Poland
Mahmoud Elish	Gulf University for Science and Technology, Kuwait

Mandeep Kaur	Sharda University, India
Manoj Majumder	IIIT Naya Raipur, India
Meera Kansara	Gujarat Vidyapith, India
Michał Chlebiej	Nicolaus Copernicus University, Poland
Mittal Desai	Charotar University of Science and Technology, India
Mohamad Ijab	National University of Malaysia, Malaysia
Mohini Agarwal	Amity University Noida, India
Monika Patel	NVP College of Pure and Applied Sciences, India
Mukti Jadhav	Marathwada Institute of Technology, India
Neetu Sardana	Jaypee University of Information Technology, India
Nidhi Arora	Solusoft Technologies Pvt. Ltd., India
Nilay Vaidya	Charotar University of Science and Technology, India
Nitin Kumar	National Institute of Technology Uttarakhand, India
Parag Rughani	GFSU, India
Parul Patel	VNSGU, India
Prashant Pittalia	Sardar Patel University, India
Priti Sajja	Sardar Patel University, India
Pritpal Singh	Jagiellonian University, Poland
Punya Paltani	IIIT Naya Raipur, India
Rajeev Kumar	NIT Hamirpur, India
Rajesh Thakker	Vishwakarma Govt Engg College, India
Ramesh Prajapati	LJ Institute of Engineering and Technology, India
Ramzi Guetari	University of Tunis El Manar, Tunisia
Rana Mukherji	ICFAI University, Jaipur, India
Rashmi Saini	GB Pant Institute of Engineering and Technology, India
Rathinaraja Jeyaraj	National Institute of Technology Karnataka, India
Rekha A. G.	State Bank of India, India
Rohini Rao	Manipal Academy of Higher Education, India
S. Shanmugam	Concordia University Chicago, USA
S. Srinivasulu Raju	VR Siddhartha Engineering College, India
Sailesh Iyer	Rai University, India
Saman Chaeikar	Iranians University e-Institute of Higher Education, Iran
Sameerchand Pudaruth	University of Mauritius, Mauritius
Samir Patel	PDPU, India
Sandeep Gaikwad	Symbiosis Institute of Computer Studies and Research, India
Sandhya Dubey	Manipal Academy of Higher Education, India

Sanjay Moulik	IIIT Guwahati, India
Sannidhan M. S.	NMAM Institute of Technology, India
Sanskruti Patel	Charotar University of Science and Technology, India
Saurabh Das	University of Calcutta, India
S. B. Goyal	City University of Malaysia, Malaysia
Shachi Sharma	South Asian University, India
Shailesh Khant	Charotar University of Science and Technology, India
Shefali Naik	Ahmedabad University, India
Shilpa Gite	Symbiosis Institute of Technology, India
Shravan Kumar Garg	Swami Vivekanand Subharti University, India
Sohil Pandya	Charotar University of Science and Technology, India
Spiros Skiadopoulos	University of the Peloponnese, Greece
Srinibas Swain	IIIT Guwahati, India
Srinivasan Sriramulu	Galgotias University, India
Subhasish Dhal	IIIT Guwahati, India
Sudhanshu Maurya	Graphic Era Hill University, Malaysia
Sujit Das	National Institute of Technology-Warangal, India
Sumegh Tharewal	Dr. Babasaheb Ambedkar Marathwada University, India
Sunil Bajeja	Marwadi University, India
Swati Gupta	Jaypee University of Information Technology, India
Tanima Dutta	Indian Institute of Technology (BHU), India
Tanuja S. Dhope	Rajarshi Shahu College of Engineering, India
Thoudam Singh	NIT Silchar, India
Tzung-Pei Hong	National University of Kaohsiung, Taiwan
Vana Kalogeraki	Athens University of Economics and Business, Greece
Vasudha M. P.	Jain University, India
Vatsal Shah	BVM Engineering, India
Veena Jokhakar	VNSGU, India
Vibhakar Pathak	Arya College of Engg. and IT, India
Vijaya Rajanala	SR Engineering College, India
Vinay Vachharajani	Ahmedabad University, India
Vinod Kumar	IIIT Lucknow, India
Vishnu Pendyala	San José State University, USA
Yogesh Rode	Jijamata Mahavidhyalaya Buldana, India
Zina Miled	Indiana University, USA

Additional Reviewers

Anjali Mahavar
Falguni Parsana
Himanshu Patel
Kamalesh Salunke
Krishna Kant
Lokesh Sharma
Mihir Mehta

Parag Shukla
Prashant Dolia
Shanti Verma
Tejasvi Koti
Rachana Parikh
Ramesh Chandra Goswami
Rayeesa Tasneem

Contents – Part I

Theory and Methods

System and Applications

Contents – Part II

Hybrid Techniques

Theory and Methods

A Simple Difference Based Inter Frame Video Forgery Detection and Localization

B. H. Shekar[1] , Wincy Abraham[2](✉) , and Bharathi Pilar[3]

[1] Mangalore University, Mangalagangothri, Mangalore, India
[2] Assumption College, Changanacherry, Kottayam, India
wincya@gmail.com
[3] University College, Mangalore, Mangalore, India

Abstract. A video becomes forged, if it is altered by changing the information contained within a frame or by changing the original sequencing of frames by deleting some frames or adding some frames in between, referred to as intra-frame forgery and inter-frame forgery respectively. This paper proposes an effective method for inter-frame video forgery detection which is capable of detecting duplication of frames, deletion of frames and also insertion of frames in the video. The method proposed is also capable of locating the forgery. There are many other existing methods which detect video forgery using features such as correlation coefficient between adjacent frames, optical flow, Zernike moment and so on. The proposed method detects forgery in a simple method compared to the existing ones. It consumes less computational power and time. The fact that manipulation done on the video alters the original sequencing of frames, which can be detected by examining the difference in pixel intensities of adjacent frames is made use of by this method. This method separates the frames of the video and uses the difference in pixel intensities of adjacent frames in two different ways to detect forgery. The original sequence of frames in the video follows a smooth pattern of adjacent frame differences, but any change occurring to the sequencing causes spikes. By checking the presence of these spikes, forgery along with the location of forgery can be detected. This method is found to have better accuracy compared to state-of-the-art methods and experimentation is done using the publicly available datasets.

Keywords: Video forgery · Pixel intensity · adjacent frames · outlier detection

1 Introduction

Nowadays video evidence plays a vital role in the detection of crimes, as video capturing of events is a common practice now due to the availability of surveillance cameras and other digital cameras for common use. Video forgery has become a common phenomenon as video manipulation tools are easily available and can be used even without much expertise. To hide evidence of crimes from

K. K. Patel et al. (Eds.): icSoftComp 2023, CCIS 2030, pp. 3–15, 2024.
https://doi.org/10.1007/978-3-031-53731-8_1

the video, it is modified by deleting some frames or removing some objects within the frame of the video. Sometimes video is altered by inserting some other frames into it or by doing any other type of forgery. This means that manipulation of the video is a common practice in every society. The detection of it is highly demanded especially by the court of law as it is used as evidence and also by the common man because people want to see what is real and not something fake. Many methods exist for video forgery detection but none is foolproof. So better methods are still in demand.

Video forgery can be classified as inter-frame forgery or intra-frame forgery. We know that a video is made up of a sequence of frames. In video compression, 3 types of frames are used, I-frame, P-frame and B-frame. I-frame(Intra coded picture) is the least compressed and is a complete image. P-frames(Predicted picture) contain only the difference from the previous I-frame to save space. B-frames(Bidirectional predicted picture) store the difference from both the previous and the following frames to specify its content, thus saving more space. In inter-frame forgery, the forgery takes place between the two adjacent frames of the video. It may be by inserting a new set of frames or by copy-pasting frames in between any two frames, or by deleting a set of frames from the video. Some video forgery detection algorithms use the changes happening to the different types of compressed frames of the video while others consider only the decompressed frames. In intra-frame video forgery, manipulation takes place within the frames of the video. Some objects in the frame may be removed or new objects added and so on. Various methods exist to detect inter-frame forgery and intra-frame forgery and forgery which is a combination of both types. They differ in the approach used, the data set considered and the level of detection accuracy.

This paper proposes a simple and efficient method for inter-frame video forgery detection which considers the intensity variations in the frames of the video and yields results better than some of the state-of-the-art methods. The following section discusses the works carried out by other researchers on this topic. Section 3 describes the methodology mean while Sect. 4 discusses various other methods we tried for forgery detection. Section 5 deals with the experimentation carried out and the results obtained for the various datasets and concluding remarks are expressed in Sect. 6.

2   Review of Related Literature

Video forgery occurs in various forms like object duplication, frame duplication, object deletion, frame deletion, object splicing, frame shuffling, frame insertion and so on. Various methods exist which perform the detection of these types of forgeries. Many of the detection methods which are capable of detecting one type of forgery do not work well for forgery of another type. A single technique which can detect video forgery irrespective of the type of forgery is highly demanded.

J. A. Aghamaleki et al. [1] detect inter frame video forgery by the analysis of quantization effect of DCT coefficients of I and P frames. Soumya et al. [2] in their method localizes inter-frame forgery due to frame replication and the potential tampered frames are identified through frame similarity analysis. Statistical measures of entropy and contrast of residual frames along with similarity measures serve as the feature set for detection. They have done manipulation on surveillance video and self-captured smartphone video to create a dataset for the experimentation. Wang et al. [3] use correlation coefficient as a measure of similarity to find frame duplication. The similarity in the spatial and temporal correlations is used as an indication of frame duplication. Shanableh T et al. [4] use prediction residuals, percentage of intra-coded macroblocks, quantization scales and an estimate of the PSNR values using the I, P and B frames in the video for detecting frame deletion. K. Sitara et al. [5] proposes an inter-frame forgery detection algorithm based on tamper traces from spatiotemporal and compressed domains. They experimented using a dataset containing 23,586 videos which comprised inter-frame video forgeries like insertion, deletion, duplication, and shuffling. Evaluation results demonstrate that the model outperforms other methods, especially the inter-frame shuffling detection. Zhao et al. [6] propose a method which performs detection by HSV colour histogram difference between adjacent frames as a similarity measure and matching recheck based on tampering positions. The method is capable of detecting frame insertion, copying and deletion. Experimental results demonstrated that the model achieved precision, recall, and accuracy of 98.07%, 100%, and 99.01%, respectively. The model, therefore, outperformed the existing state-of-the-art methods of that time. Akumba et al. [7] use the mean and standard deviation of the correlation coefficient between adjacent frames as the feature to perform classification as authentic and forged. An accuracy of 100% is obtained for the VIFFD dataset and locally manipulated video dataset. In the method by Gurvinder Singh et al. [8] different algorithms are proposed to detect frame duplication and region duplication forgeries in videos. The authors claim higher detection accuracy and execution efficiency compared to the existing methods. The dataset used consists of some videos from SULFA [13] and some downloaded from the internet. In the method by Vinay Kumar et al. [9] the inter-frame correlation coefficient and correlation distance measure are used for the detection and localization of forgery and has 83% accuracy at the video level. They use the publicly available VIFFD [11] dataset for experimentation. In the work by Nitin et al. [10] entropy coded (DistrEn2D and MSE2D) frames are used for forgery detection using the correlation consistency between them. Two-dimensional distribution entropy (DistrEn2D) and bi-dimensional multiscale entropy (MSE2D), are used for the detection. Experimentation is done on original data collected from SULFA [13], REWIND [14], and VTL [18] and then applying various forgery operations. The overall detection accuracy is claimed to be 96.6%. Sondos et al. detect [16] inter-frame forgery operations like frame shuffling, frame insertion and deletion and are localized using the temporal average and the universal image quality index. The method was tested using 15 tampered videos which are made publicly available. In [17] the authors propose a two-stage inter-frame forgery detection technique with low computational cost for HEVC-coded videos. In the first stage, abnormal points are detected based on compression domain features, and in the second stage, the abnormal points and their locations are validated. Li et al. [21] in their method use camera sensor pattern noise to detect inter-frame splicing forgery detection. The method first estimates reference SPN, and then calculates signed peak-to-correlation energy (SPCE) at the block level for classification. We concentrate our attention here on inter-frame forgery. The proposed method is capable of handling any kind of inter-frame forgeries like frame insertion, frame duplication and frame deletion.

3 Proposed Method

Frames of the video can be separated and each frame can be treated as an image. Each image is represented using the intensity at each pixel. The consecutive frames in the video will have small differences in some but not all of the corresponding pixel values. This fact is made use of in finding the inter-frame forgery. Our proposed method makes use of the difference in pixel intensities of the adjacent frames as the feature for the detection of forgery.

There are two differences considered in this approach for more accurate detection. Pixel-wise difference of adjacent frames and difference of adjacent frame averages. The video is first divided into constituent frames. Then frames are converted to gray scale. We then focus on the difference between the intensity values of adjacent frames for the clue for forgery as the difference will be high if any frame under consideration is a newly inserted one or which became adjacent due to the deletion of some frames in between. If there are scene changes in the video, it will result in frames with varying content. The variation of content in adjacent frames may be due to camera motion, movement of objects within the scene, noise and change in lighting [15]. In the case of videos captured using surveillance cameras, camera motion problems will not be present. The difference caused by noise and change in lighting can be neglected as it will be small compared to the difference caused by moving objects. Thus the main cause of scene change when the video is captured using surveillance cameras is the movement of objects. But in videos with high frame rate, the change in adjacent frame intensities due to forgery outweigh the change due to object motion.

The proposed method uses two difference arrays for video forgery detection each of which contains the difference between adjacent frames in two different ways. The presence of outliers in these arrays indicates forgery. They are computed as follows.

1. differenceArray DA is found by finding the difference in intensity of each pixel in the adjacent frames as

$$DA[i,j] = f_k[i,j] - f_{k+1}[i,j] \tag{1}$$

where k varies from 1 to one less than the total number of frames n of the video, and i,j take values as per the size of each frame.

DiffAvg, the average of differenceArray DA is found by

$$DiffAvg = \frac{\sum_{i=0}^{M-1} \sum_{j=0}^{N-1} DA(i,j)}{M \times N} \tag{2}$$

2. AvgDiff is found by finding the average intensity of each frame, as

$$AvgFrame_k = \frac{\sum_{i=0}^{M-1} \sum_{j=0}^{N-1} f_k(i,j)}{M \times N} \tag{3}$$

and then taking the difference between adjacent frame averages.

$$AvgDiff = AvgFrame_k - AvgFrame_{k+1} \tag{4}$$

where k varies from 1 to one less than the total number of frames n of the video.

Both the differences hold the difference between the frame intensities and are found for the entire video. Thus there are two difference arrays created for each video. One to contain each DiffAvg corresponding to each adjacent frames as the element and the other to hold each AvgDiff as the element. These difference arrays are called differenceArray1 and differenceArray2 respectively. The location of the maximum element of the array is a suspected location of forgery. It is confirmed by checking whether the value found as the maximum of the array under consideration is a threshold time larger than the average value of its two neighbours at each side of the array. Otherwise, it is not a forgery location. It is formulated as

$$m > threshold \times \frac{differenceArrayx[n-1] + differenceArrayx[n+1]}{2} \tag{5}$$

where m is the value at the location n of the $differenceArrayx$, x is either 1 or 2. The threshold is set as 20 for $differenceArray1$ and 8 for $differenceArray2$ empirically. If m is found to satisfy the above inequality, n is considered as a valid location of forgery and it is deleted from the array and the next maximum is found and the process is repeated. Thus if two such locations are found for a video it is considered as the location of insertion and copy forgeries while for deletion forgery only one value must appear as deviating from the normal values in the difference array. Since the inequality above checks for values which very much deviate from the normal values, many of them that come up during the process due to scene changes do not emerge as the locations of forgery.

   The proposed method first checks differenceArray1 for outliers after which outlier detection is done on the differenceArray2 as well for which a threshold of 8 is chosen empirically instead of 20 used previously. Two types of arrays are used by the method since the forgery may remain undetected if only one array is used. Forgery undetected in one array is found to get detected in the other array. In case there is an outlier in any one of the arrays, the video is considered as forged. Otherwise, it is treated as authentic. It is seen that in most of the cases, the difference caused due to scene change or any other reason does not emerge as an outlier as the difference caused due to the manipulation of video is much larger compared to the differences due to the other reasons. The index of the outlier in the array is taken as the location of the forgery. The algorithm used in the proposed method is listed below.

Input: Video snippet
Output: Decision whether the video is forged or not
Method:

1. Separate all frames in the video.
2. Convert frames into gray colour format.
3. Calculate the cumulative difference in pixel intensities of adjacent pairs of frames.
4. Find the average intensity of each frame and then find the difference between adjacent frame pairs.
5. Use the difference array in step 3 for calculation and set threshold = 20.
6. From the difference array find the maximum value.
7. Confirm whether the maximum value is an outlier by checking whether it is more than threshold times the average of its neighbours at both sides in the array.
8. If any value emerges as an outlier in the above step keep it as a possible location of forgery, remove it from the array and repeat step 6 once more, go to step 10 otherwise.
9. If the threshold is not equal to 8, go to step 6 using the difference array of step 4 and by setting threshold = 8.
10. If forgery is detected in step 8 mark the video as forged and as authentic otherwise.
11. Exit

The block diagram depicting the method is shown in Fig. 1.

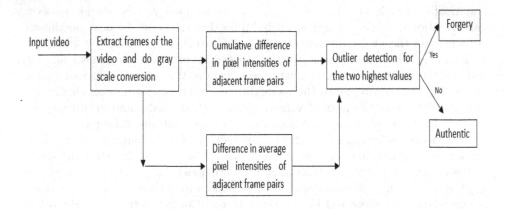

Fig. 1. Block diagram of the proposed classifier

The variation occurring to the frame content on forgery or scene change for various forged videos and original videos are illustrated in the following figures. Figures 2 and 3 depict the distribution of these differences for video forged by frame duplication, the average value of adjacent frame differences (as mentioned in step 4 of the algorithm) and the difference of adjacent frame averages (as mentioned in step 6 of the algorithm) respectively. The spikes occur at the locations where the frame duplication occurs. It occurs at multiple places in this video. Both the differences indicate forgery.

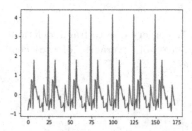

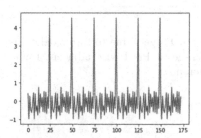

Fig. 2. Forgery due to frame duplication-Average of Pixel wise adjacent frame differences

Fig. 3. Forgery due to frame duplication-Difference of averages of adjacent frames

Figures 4 and 5 illustrate the spike caused due to frame insertion at the beginning of the video. A set of frames have been inserted at the beginning. The spike occurs at the point of insertion and at the end of inserted frames. Here frames have been inserted at the beginning of the video. So the spike is seen only at one location. The difference graphs in Fig. 4 and Fig. 5 are shown in the same way as the above for a video forged due to frame insertion.

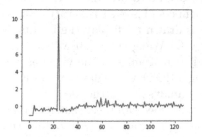

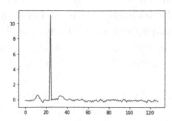

Fig. 4. Forgery due to frame insertion-Average of Pixel wise adjacent frame differences

Fig. 5. Forgery due to frame insertion-Difference of averages of adjacent frames

Figures 6 and 7 depicts the differences as before and the spikes occur at the locations where the frame deletion occurs. Since it is deletion forgery spike occurs only at one place if there is no sudden scene change. Figures 8 and 9 show the differences for video with no forgery and the graph appears not to have any spike indicating no forgery.

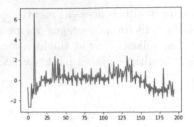

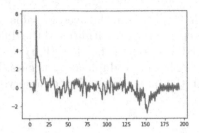

Fig. 6. Forgery due to frame deletion-Average of Pixel wise adjacent frame differences

Fig. 7. Forgery due to frame deletion-Difference of averages of adjacent frames

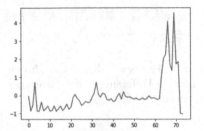

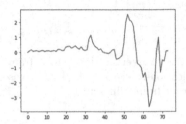

Fig. 8. Original Video-Average of Pixel wise adjacent frame differences

Fig. 9. Original video-Difference of averages of adjacent frames

4   Other Methods Considered

We borrowed some ideas from the state-of-the-art literature and tried various other methods for forgery detection. None of them could give satisfactory results, two of which are discussed below and a table is drawn to display the results of the comparison. One of them is the method by Qi Wang et al. [20] where they used the correlation coefficient between the adjacent frames of the video as the feature vector. We experimented the same using VIFFD as the dataset. For each video, the correlation coefficients obtained were analysed and the idea used was that if any value is below a certain lower limit, the video can be considered to be forged. Various lower limits were tried and the appropriate value to obtain the best result was found to be 0.95. However, it was found that only insert forgery it was able to identify well and in all other cases, it showed low performance. We ourselves considered another method which takes the standard deviation of the first 100 high-frequency coefficients of the DCT (Discrete Cosine Transform) of each frame of the video as the feature vector. Feature vectors extracted from all the videos in the dataset are supplied to SVM (Support Vector Machine) classifier for classification as authentic or forged. 80% of the feature vectors were used for training the classifier and the rest 20% were used for testing. However, the method could not achieve promising results. Table 1 shows the results obtained for both the methods considered in this section. The first method considered was designed for forgery detection and localization while the second could do only forgery detection.

Table 1. Accuracy of the two other methods for VIFFD dataset

Method	Type of forgery	No of videos	Accuracy in %
Using	Copy	30	30
Correlation	Delete	30	43
Coefficient	insert	30	86.66
	Original	30	73.33
Using DCT	Forgery detection	120	77

5 Experiments

5.1 VIFFD Dataset

The proposed method is experimented on the VIFFD dataset [11] which has 120 videos out of which 30 are authentic and 90 videos are forged. Out of the 90 forged videos, 30 are forged due to frame duplication (copy), 30 videos are forged due to deletion and the rest are forged due to frame insertion. Experiments are carried out with this dataset and each video after processing is categorized into authentic or forged. For the VIFFD dataset this method works better than state-of-the-art methods. It gives 73.33% accuracy for frame deletion detection, 86.66% accuracy in frame duplication detection and 100% accuracy for frame insertion detection. Original video without forgery is identified with an accuracy of 70%. Thus the proposed system is found to have an overall accuracy of 82.5%. Table 2 shows the performance of this method for various types of forgery in terms of accuracy of classification.

The comparison of classification accuracy of the proposed method with some of the state-of-the-art methods is shown in Table 3.

Table 2. Accuracy of the proposed method for various types of forgery for VIFFD dataset

Type of forgery	No of videos	Accuracy in %
Copy	30	86.66
Delete	30	73.33
insert	30	100
No Forgery	30	70

Table 3. Performance Comparison of the proposed method with some of the existing methods using the VIFFD dataset

Method	forgeries	Accuracy in %
Vinay et al. [9]	Frame deletion, insertion	83
Proposed method	Frame deletion, insertion	86.66

5.2   Surveillance Video Dataset

A publicly available dataset was created by the authors of [16]. The dataset consists of 15 tampered videos. The forged videos are due to deletion, insertion and shuffling of frames. The proposed method is tested on this dataset, compared with the method proposed by the creators of the dataset and the result is shown in Table 4.

Table 4. Accuracy of the proposed method for various types of forgery for Surveillance video dataset

Method	Type of forgery	No of videos	Precision	Recall
Sondos et al. [16]	shuffling	5	.96	.97
	Delete	5	.98	.99
	insert	5	.99	.99
Proposed Method	shuffling	5	1	1
	Delete	5	1	1
	insert	5	1	1

5.3   TDTVD Dataset

In [18], the development TDTVD dataset containing total 210 videos for Temporal Domain Tampered Video Dataset using Frame Deletion, Frame Duplication and Frame Insertion is proposed. 120 videos in it are developed based on

Event/Object/Person (EOP) removal or modification, 40 each for frame deletion, frame insertion and frame duplication forgeries The rest of the videos are created based on Smart Tampering (ST) or Multiple Tampering. They consist of videos with multiple tampering with frame deletion, frame insertion and frame duplication in three categories like 10 frames tampered, 20 frames tampered, and 30 frames tampered at 3 different locations in 10 videos each. The data set also contains 16 original videos from SULFA dataset [12]. The result of the experimentation is shown in Table 5.

Table 5. Accuracy of the proposed method for video forgery for TDTVD dataset

Type of forgery	No of videos	Accuracy in %
EOP Copy	40	90
EOPDelete	40	90
EOPInsert	40	100
Original	16	75
ST Copy-10	10	100
ST Copy-20	10	90
ST Copy-30	10	100
ST Delete-10	10	70
ST Delete-20	10	90
ST Delete-30	10	90
STInsert-10	10	100
STInsert-20	10	100
STInsert-30	10	100

5.4   Inter-frame Forgery Data Set

Sondos et al. [19], made this data set publicly available, which they used for experimentation in their paper. Among the two folders with original and forged videos, the original video folder is copied as it is from the TDTVD data set. According to the authors their method has superiority in terms of execution time and precision and recall. The folder containing forged videos contains 32 videos forged due to insertion, copy, deletion and shuffling. Table 6 shows the details of the video and the result of experimentation using the proposed method.

Table 6. Accuracy of the proposed method for video forgery for inter-frame forgery dataset

Type of forgery	No of videos	Accuracy in %
Copy	10	90
Delete	5	100
Insert	5	100
Shuffling	9	77.77
Copy & Shuffling	3	100
Original	16	75

6   Conclusion

This paper presents an efficient and simple method for inter-frame video forgery detection which can detect frame insertion, deletion and frame duplication. The change happening to the sequence of adjacent frame differences, when the original sequencing is altered is found out by finding the arithmetic difference between adjacent frames of the video. The method may be modified by incorporating the detection of shot boundaries and then analysing each video segment by the proposed method for forgery detection as a future work. Thus the paper presents a simple difference-based method for inter-frame video forgery detection. It is found to work better than state-of-the-art methods.

References

1. Aghamaleki, J.A., Behrad, A.: Malicious inter-frame video tampering detection in MPEG videos using time and spatial domain analysis of quantization effects. Multimedia Tools Appl. **76**(20), 20691–20717 (2017). https://doi.org/10.1002/andp.19053221004
2. Sowmya, K.N., Basavaraju, H.T., Lohitashva, B.H., Chennamma, H.R., Aradhya, V.N.M.: Similarity Analysis of Residual Frames for Inter Frame Forgery Detection in Video, ICICC 2019. Advances in Intelligent Systems and Computing, vol. 1034, p. 20. Springer, Cham (2019). https://doi.org/10.1007/978-981-15-1084-7
3. Wang, W., Farid, H.: Exposing digital forgeries in video by detecting duplication. In: Proceedings of the 9th Workshop on Multimedia and Security - MM Sec 2007 (2007)
4. Shanableh, T.: Detection of frame deletion for digital video forensics. Digit. Investig. **10**(4), 350–360 (2013)
5. Sitara, K., Mehtre, B.M.: Detection of inter-frame forgeries in digital videos. Forensic Sci. Int. **289**, 186–206 (2007). https://doi.org/10.1016/j.forsciint.2018.04.056
6. Zhao, D.-N., Wang, R.-K., Lu, Z.-M.: Inter-frame passive-blind forgery detection for video shot based on similarity analysis. Multimedia Tools Appl. (2018). https://doi.org/10.1007/s11042-018-5791-1

7. Akumba, B.O., Iorliam, A., Agber, S., Okube, E.O., Kwaghtyo, K.D.: Authentication of video evidence for forensic investigation: a case of Nigeria. J. Inf. Secur. **12**, 163–176 (2021). https://doi.org/10.4236/jis.2021.122008

8. Singh, G.S.K.: Video frame and region duplication forgery detection based on correlation coefficient and coefficient of variation. Multimedia Tools Appl. (2018). https://doi.org/10.1007/s11042-018-6585-1

9. Gaur, V.K.M.: Multiple forgery detection in video using inter-frame correlation distance with dual-threshold. Multimedia Tools Appl. (2022). https://doi.org/10.1007/s11042-022-13284-2

10. Shelke, N.A., Kasana, S.S.: Multiple forgeries identification in digital video based on correlation consistency between entropy coded frames. Multimedia Tools Appl. **28**, 267–280 (2022). https://doi.org/10.1007/s00530-021-00837-y

11. Nguyen, X.H., Hu, J.: VIFFD - a dataset for detecting video inter-frame forgeries. Mendeley Data 5, Multimedia Tools and Applications (2020). https://doi.org/10.17632/r3ss3v53sj.5

12. Qadir, G., Yahaya, S., Ho, A.T.S.: Surrey university library for forensic analysis (sulfa) of video content. In: IET Conference on Image Processing (IPR 2012), vol. 79(47), pp. 1–6 (2012). http://sulfa.cs.surrey.ac.uk/

13. Bestagini, P., Milani, S., Tagliasacchi, M., Tubaro, S.: Local tampering detection in video sequences. In: 2013 IEEE 15th International Workshop on Multimedia Signal Processing (MMSP) (2013)

14. Video Trace Library. http://trace.eas.asu.edu/

15. Hoose, N.: Computer vision as a traffic surveillance tool. In: Control Computers Communications in Transportation, pp. 57–64 (1990). https://doi.org/10.1016/B978-0-08-037025-5.50014-1

16. Fadl, S., Han, Q., Li, Q.: Surveillance video authentication using universal image quality index of temporal average. In: Yoo, C.D., Shi, Y.-Q., Kim, H.J., Piva, A., Kim, G. (eds.) IWDW 2018. LNCS, vol. 11378, pp. 337–350. Springer, Cham (2019). https://doi.org/10.1007/978-3-030-11389-6_25

17. Singla, N., et al.: A two-stage forgery detection and localization framework based on feature classification and similarity metric. Multimedia Syst. **29**, 1173–1185 (2023)

18. Panchal, H.D., Shah, II.: Video tampering dataset development in temporal domain for video forgery authentication. Multimedia Tools Appl. **79**, 33–34 (2020). https://doi.org/10.1007/s11042-020-09205-w

19. Fadl, S., Han, Q., Qiong, L.: Exposing video inter-frame forgery via histogram of oriented gradients and motion energy image. Multidimension. Syst. Signal Process. **31**, 1365–1384 (2020). https://doi.org/10.1007/s11045-020-00711-6

20. Wang, Q., Li, Z., Zhang, Z., Ma, Q.: Video inter-frame forgery identification based on consistency of correlation coefficients of gray values. J. Comput. Commun. **2**, 51–57 (2014). https://doi.org/10.4236/jcc.2014.24008

21. Li, Q., Wang, R., Xu, D.: A video splicing forgery detection and localization algorithm based on sensor pattern noise. Electronics **12**(6), 1362 (2023). https://doi.org/10.3390/electronics12061362

Pixel-Level Segmentation for Multiobject Tracking Using Mask RCNN-FPN

Shivani Swadi⬤, Prabha C. Nissimagoudar[✉], and Nalini C. Iyer[✉]

KLE Technological University, Hubli, India
pcnissimagoudar@gkletech.ac.in, nalinic@kletech.ac.in

Abstract. In artificial intelligence (AI), segmenting and tracking multiple objects is a crucial task that has numerous applications in industries such as robots, autonomous vehicles, medical services, and surveillance. Object segmentation identifies and categorizes objects in an image or video frame, while object tracking detects and follows objects throughout a series of frames. These tasks are challenging due to occlusions, varying appearances, and overlapping bounding boxes. This study proposes a novel approach that reduces computing complexity and enhances performance by treating object parts as points. The technique consists of two stages, namely object segmentation and object tracking. In the first stage, a Mask R-CNN model is employed to segment each object in the input image or video frame. The model generates a series of binary masks representing the pixels associated with each item. In the second stage, the segmented objects are treated as points, and a tracking algorithm is applied to monitor each item over time. Based on spatial distance and visual similarity, the Hungarian algorithm is used to compare each object in the current frame to the corresponding object in the previous frame. The proposed method is evaluated using the MOTChallenge benchmark dataset, which includes challenging scenarios such as occlusions and congested scenes. The Simultaneous Object Tracking and Segmentation (SOTS) technique is a highly effective approach for tracking and segmenting objects simultaneously. It employs the Mask RCNN FPN Model and a Point-Track network for object tracking, achieving an impressive F-measure score of 68.8 on the KITTI MOTs dataset,75.9 on the APOLLO MOTS dataset, and 62.3 on the IDD dataset. The method obtains an average Multiple Object Tracking Accuracy (MOTA) of 60.2% and an average segmentation Intersection over Union (IoU) of 70.9%. These results demonstrate that the proposed strategy surpasses current state-of-the-art approaches in terms of accuracy and computational efficiency, with real-time processing speeds of up to 27 frames per second on a single CPU. Therefore, the suggested method holds significant potential for real-world applications in robotics, surveillance, and autonomous vehicles.

Keywords: MOTS · Mask RCNN FPN · Tracking · Segmentation · multiple object detection

K. K. Patel et al. (Eds.): icSoftComp 2023, CCIS 2030, pp. 16–29, 2024.
https://doi.org/10.1007/978-3-031-53731-8_2

1   Introduction

Deep learning techniques, especially in image and video analysis, have significantly advanced computer vision and object tracking. These methods have proven quite effective for applications like object detection and image segmentation. Still, monitoring several objects effectively between frames is a challenging task. There has been a significant change in the tracking paradigm in recent studies. Conventional bounding box-level tracking techniques have demonstrated inadequacies when applied to complex situations. Working at the pixel level is a more promising strategy instead. With regard to occluded and overlapping objects, this shift to pixel-level analysis is intended to improve object tracking precision and accuracy. Using pixel-level analysis to learn additional discriminative instance characteristics is one of the main recommendations that came out of the Multiple Object Tracking with Segmentation (MOTS) study. The goal of this method is to provide a scene's object instances a more detailed knowledge. Unfortunately, crucial information like item IDs is frequently lacking from datasets needed to train and assess computer vision systems. A potential approach is provided by pixel-level object segmentation. It makes it possible to depict the scene with greater accuracy and detail, especially when working with items that are obscured. Segmentation masks give a more precise representation of the ground truth than bounding boxes, which can only provide an approximate match for an object. Because segmentation-based tracking findings are non-overlapping by nature, comparisons with ground truth annotations are made easier.

However, there have been difficulties in obtaining instance feature embeddings from segments. In this research, an efficient approach to these problems is presented, which will help to improve object-tracking technology. In the subsequent sections, we will delve into the methodology, implementation details, mathematical modeling, validation of different datasets and optimization, and Comparison against other research works to provide a comprehensive view of this novel approach and its potential impact on the field of computer vision and object tracking.

2   Related Work

Most robust multi-object trackers employ a tracking paradigm based on detection. Using a pre-trained object detector, this method proceeds by generating bounding boxes for the current video frame [e.g., Felzenszwalb et al., 2009 [11]; Ren et al., 2017 [12]]. Tracks are then found by comparing bounding frames from the previous frame to the present frame. For example, SORT [Bewley et al., 2016 [14])] shows efficient online tracking by fusing traditional tracking methods like the Kalman Filter and the Hungarian algorithm with the superior Faster R-CNN object detector [Ren et al., 2017 [12]]. A variation of SORT called Deep SORT [Xu et al., 2019 [15]] improves tracking performance by integrating appearance information and boosting the technique's resilience to identification alterations.

By comparison, Learning by Tracking [Leal-Taixé et al., 2016 [3]] tracks pedestrians using a novel two-step data association method. In order to generate descriptive encodings of the input photos, a Siamese Convolutional Neural Network (CNN) is trained in the first stage. In the second stage, gradient boosting is used to compare contextual information from the image position and size with the CNN in order to determine the matching probability.

In contrast to approaches that rely on manually created features, Schulter et al. [2017 [7]] show that backpropagation may be used to learn features for data matching in a network flow. Another novel method for connecting person detection over time is presented by [Faure, Gaspar, et al., Year [13]] and involves resolving a minimal cost-increasing multi-cut issue. While Ooi et al. [2018 [19]] improve data association by employing recognition tag information, Sharma et al. [2018 [5]] employ 3D information in the data mapping cost to improve existing Multiple Object Tracking (MOT) methodologies. [Xu et al., 2019 [10]] presents a unified framework for computing similarity metrics for picture patches that function across spatial and temporal dimensions and incorporate different signals.

Certain techniques, including Center Track [Zhou et al., 2020 [6]], In order to attain great performance, and Tracktor frame multi-object tracking as a regression issue and rely on powerful object detectors. These techniques employ detectors that have been taught to regress the motion of the item.

3 Methodology

3.1 Implementation Detail

The computer vision system presented in the text aims to achieve point tracking in video streams using advanced techniques and deep learning frameworks. The primary objectives of the system are to identify and track specific points, such as corners or edges, in video frames. To accomplish this, algorithms like the Lucas-Kanade approach and optical flow are utilized. However, point tracking can be challenging due to changes in point appearance over time, leading to tracking errors. To address this, the system employs outlier rejection, feature selection, and temporal smoothing methods.

Block Diagram. A block diagram of the framework outlines the advantages of segmentation-based tracking, which is more accurate than bounding-box-based tracking. Segmenting objects per pixel allows precise identification even in crowded scenes with overlapping items. To achieve per-pixel segmentation, instance segmentation using a compact picture representation is performed, giving each pixel a special label to separate instances of objects in an image. For processing point clouds, instance embeddings are trained by considering pixels in 2D pictures as unordered 2D point clouds. Instance embeddings express objects in a high-dimensional space and group related objects together. By comparing object embeddings over frames, tracking using instance embeddings becomes possible, simplifying object tracking and making it more resilient in various environments. The tracking-by-points paradigm is implemented by combining instance embeddings with any instance segmentation approach. This innovative approach further enhances the tracking capabilities of the framework and makes it highly effective in various computer vision applications (Fig. 1).

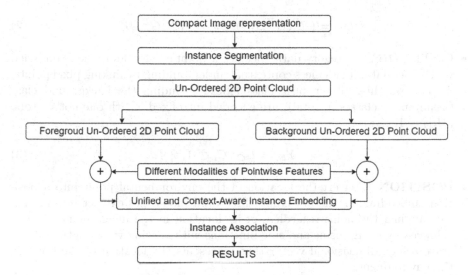

Fig. 1. Block diagram-Here the input is the compact image which is represented by crowded scenes or overlapping objects

3.2   Mathematical Modelling

For an instance, let's examine at a small scenario, denoted C_b, whose segment is enlarged by a factor $k = 0.2$ in all four directions. Considering the region in the front as the environment point cloud, represented by the notation E, and the area in the background as a 2D point cloud, represented by the notation F. Six-dimensional data (u, v, R, G, B, C) is present at each point in the expanded segment. The coordinates in the image plane are represented by (u, v), while the color of the pixel is represented by (R, G, B). C represents the class to which the pixel belongs. The foreground point's coordinates, F_i, are written as (u_i^F, v_i^F), while the environment point's coordinates, E_i, are written as (u_i^E, v_i^E). $P(u_c^F, v_c^F)$ is the central point. Features concerning position, appearance, scale, shape, and nearby objects, are useful for tracking hence the other data modalities are formulated as

- **OFFSET**: The "Segment as Points" method represents object instances as point positions within segmentation masks. The "OFFSET" parameter defines vectors from the segment center to actual points, crucial for determining object size and shape, especially for foreground points within the object instance.

$$O_{F_i} = (u_i^F - u_c^F, v_i^F - v_c^F), O_{E_i} = (u_i^E - u_c^E, v_i^E - v_c^E) \qquad (1)$$

- **COLOR**: The RGB channel is considered as the color data. The color data are critical for accurate instance association.

$$C_{F_i} = (R_i^F, G_i^F, B_i^F), C_{F_i} = (R_i^E, G_i^E, B_i^E) \qquad (2)$$

- **CATEGORY**: Category data specifies a point's class, and when combined with offset data, it enhances contextual understanding by linking pixel points. To achieve this, all semantic class labels, including the background class (assuming Z classes in total), are encoded into fixed-length one-hot vectors. $(H_i | j = 1, ..., z)$

$$Y_{E_i} = H_{C_i}, C_i \in [1, z] \qquad (3)$$

- **POSITION**: Convert the location of the environmental point into a position embedding, as the preceding three data modalities extract information surrounding the point regardless of its location in the image plane. This is achieved by using cosine and sine functions with varying wavelengths to generate a high-dimensional vector that represents the location of the point in the environment.

As shown in Fig. 2, the network framework uses distinct data modalities to learn the foreground embeddings and environment embeddings independently. A combination of each environment point's offset, color, and category is used to the max pooling procedure in order to learn the environment embedding. The network can obtain significant context signals about neighboring instances from the environment embeddings by combining the offset and category data. Conversely, just the offset and color data modalities are merged in order to learn the foreground embeddings.

To differentiate between instances, it is important to consider all points but give more weight to the prominent ones. A point weighting layer is added to actively weigh all foreground points and add up all of their characteristics according to prominence in order to do this. The final instance embeddings are created by concatenating and applying the position embedding M_P, environment embedding M_E, and foreground embeddings M_F to a multi-layer perceptron (MLP).

$$M = MLP(M_F + M_E + M_P), \qquad (4)$$

(+ indicates concatenation)

To produce the final tracking results, instance association based on similarities is performed.

$$(C_{s_i}, C_{s_j}) = \alpha - D(M_i, M_j) + *U(C_{s_i}, C_{s_j}), \qquad (5)$$

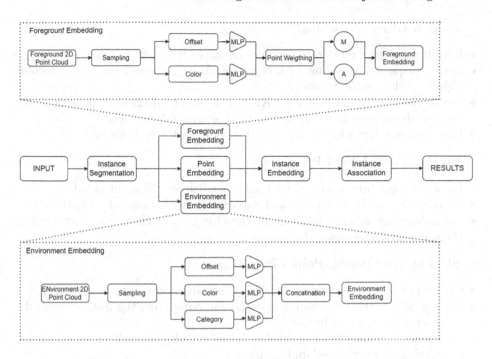

Fig. 2. Network architecture for tracking an object using point segmentation.

where,

D denotes the Euclidean distance,

U represents the Mask IOU,

and $\alpha = 0.5(by default)$.

3.3 Algorithm

The framework can be outlined in a block diagram to illustrate the advantages of segmentation-based tracking over traditional bounding-box-based tracking. Segmentation-based tracking is known for its enhanced accuracy and precision, particularly in challenging scenarios with crowded scenes and overlapping objects.

Step 1: Input Data: Provide a sequence of input images or frames from a video.

Step 2: Instance Segmentation and Object Detection:

- Implement an instance segmentation and object detection method (e.g., Mask R-CNN or Faster R-CNN) to segment objects and detect them within each input frame.
- Generate per-pixel masks or bounding boxes to distinguish different object instances and their locations in each frame.

Step 3: Object Tracking:

- Initialize tracking for detected object instances in the first frame.
- For subsequent frames, perform object tracking by associating the detected objects with existing tracks.
- Use a tracking algorithm (e.g., Kalman filter or Hungarian algorithm) to predict object positions and update the tracking information.
- Update the object identity and location for each tracked instance.

Step 4: Point Cloud Processing:

- For each segmented frame, treat it as an unordered 2D point cloud.
- Convert the pixel coordinates and mask information into point cloud data.
- Formulate instance embeddings for the objects in each frame, representing them in a high-dimensional space.

Step 5: Tracking-by-Points Paradigm:

- Combine instance embeddings with any instance segmentation approach.
- Implement the tracking-by-points paradigm, comparing the instance embeddings of objects across frames.
- This approach leads to more precise and robust object tracking, especially in crowded scenes with overlapping objects.

Step 6: Output:

- Output the tracked object instances for each frame, along with their segmentation masks or bounding boxes.
- Optionally, visualize the tracking and segmentation results or use the data for further analysis.

This algorithm combines both multiple object tracking and segmentations in a unified framework. By incorporating instance segmentation, object detection, and tracking-by-points, it provides accurate and comprehensive tracking and segmentation results for a wide range of computer vision applications.

4   Validation and Optimization

4.1   Validation on Different Datasets

1. **KITTI** [20]: The KITTI dataset has two subsets, one for tracking and segmentation, and another for object detection. The tracking subset contains 21 videos with various weather and lighting conditions, each having 10 frames per second and lasting for 40 s. The object detection subset comprises 7,481 annotated photos with item positions and class names provided in the dataset annotations.

2. **APOLLO MOTS** [21]: The Apollo MOTS dataset comprises 150,000 annotated frames from different traffic settings, weather and lighting conditions, and various objects such as vehicles, pedestrians, and bicycles. The annotations include instance segmentation masks and tracking labels, providing object ID, position, and size over time for tracking labels and pixel-level annotations for each object with instance segmentation masks.
3. **Indian Driving Dataset (IDD)**: The International Institute of Information Technology (IIIT), in Hyderabad, developed the Indian Driving Dataset (IDD), which is a dataset. It contains more than 10 h of driving movies that were captured from a car in several Indian cities. The collection offers pixel-level annotations for a variety of items, including cars, people on foot, and bicycles.

For concurrent object tracking and segmentation, a technique called Simultaneous Object Tracking and Segmentation Online utilizing the Mask RCNN FPN Model and Treating Segments as Points was developed. It utilizes a Point-Track network to do object tracking and is based on the well-known Mask R-CNN FPN (Feature Pyramid Network) model. Using benchmark datasets like KITTI MOTS and APOLLO MOTS, which are often employed for the job of video object segmentation and tracking, the suggested technique has demonstrated promising results. Achieved an F-measure score of 68.8, one of the best-performing algorithms, on the KITTI MOTs dataset. Obtaining an F-measure score of 75.9 on the APOLLO MOTS dataset places this approach among the best ones. Also obtained an F-measure score of 62.3 on the IDD dataset. A further benefit of the proposed approach is that it can segment objects and track them in real-time because it is an online method.

4.2 Optimization of Algorithm

To enhance the performance and speed of the model in object detection tasks, various optimization techniques can be used. One such technique is learning rate scheduling, which modifies the learning rate during training to prevent slow convergence or poor performance. Weight decay is another technique that adds a penalty term to the loss function, preventing overfitting and improving generalization performance. Batch normalization normalizes the activation of the preceding layer for each batch during training, enhancing efficiency and accuracy. By utilizing these techniques, users can achieve better results in object detection tasks (Table 1).

Table 1. Specific parameter values adjusted based on the individual requirements of object detection and tracking task

Optimization Technique	Parameter Values
Learning Rate Scheduling	Initial Learning Rate: 0.001
	Learning Rate Schedule: Decrease by 0.1 every 10 epochs
Weight Decay	Weight Decay Value: 1e−4 or 5e−4
Batch Normalization	Batch Size: 32, 64, or 128
	Momentum: 0.9
	Epsilon: 1e−5
Particle Filters	Number of Particles: 100
	Resampling Threshold: 0.5
Kalman Filtering and Variants	Process Noise Covariance: [0.01, 0.01]
	Measurement Noise Covariance: [1, 1]

5   Results and Discussion

The Simultaneous Object Tracking and Segmentation technique, which uses the Mask RCNN FPN Model and treats segments as Points, is a cutting-edge approach designed for concurrent object tracking and segmentation that has yielded impressive results. It employs a Point-Track network for object tracking and is based on the highly-regarded Mask R-CNN FPN (Feature Pyramid Network) model. To test its effectiveness, this technique was put through its paces on benchmark datasets like KITTI MOTS and APOLLO MOTS, both of which are widely used for video object segmentation and tracking. The results were outstanding, with the technique achieving an F-measure score of 68.8 on the KITTI MOTs dataset, making it one of the best-performing algorithms, and an

Fig. 3. Initial Segmentation Output. Dataset KITTI MOTS

F-measure score of 75.9 on the APOLLO MOTS dataset, placing it among the best approaches. This technique has achieved F-measure score of 62.3 on IDD dataset (Figs. 3, 4 and 5).

The Point-Track network is a crucial component of this technique, performing initial instance segmentation and foreground/background embedding by representing pixels in 2D images as unordered 2D point clouds. The instance embeddings are trained using a point cloud processing technique, with each instance having a unique foreground and surrounding point clouds. By combining pointwise attribute modalities, a contextually aware instance embedding is created, which can be easily combined with any instance segmentation technique to implement the tracking-by-points paradigm.

Fig. 4. Results on the custom dataset with the vehicle captured on Indian roads.

Fig. 5. Results on the custom dataset with a pedestrian captured on Indian roads.

The method outlined in this paper is designed to enhance the efficiency of multi-object tracking and segmentation by treating object segments as points. This approach has been proven to significantly reduce the computational complexity of the program, thereby improving its processing speed and accuracy. The method consists of two primary stages: object segmentation and object tracking, both of which are meticulously optimized to achieve the best possible results.

The first stage of the method involves the use of an AI model called Mask R-CNN, which is specifically trained to segment each object in an input image or video frame. The Mask R-CNN model is trained on annotated pictures to learn how to differentiate between various objects. It generates a sequence of binary masks that represent the pixels associated with each item, allowing for object tracking over time. This stage is critical for the success of the approach, as it lays the foundation for accurate object tracking. In the second stage, the segmented objects are viewed as points, and a tracking algorithm is used to follow each object over time. The Hungarian algorithm, which is a well-known algorithm for matching objects, is used as the tracking algorithm. This algorithm compares each object in the current frame to the matching object in the previous frame based on spatial distance and similarity in appearance. The algorithm's performance is optimized using various heuristics to limit false positive and false negative detections, as well as to manage occlusions and appearance changes. This stage is also critical for the success of the approach, as it enables the program to track objects accurately and efficiently.

The approach also uses a technique called the Kalman filter, which predicts each object's location and velocity based on its historical position and velocity. This enables the program to track objects even if they momentarily disappear from the image or video frame. The Kalman filter is an essential component of the approach, as it ensures that the program is capable of tracking objects accurately and efficiently. The approach's performance can be evaluated using numerous metrics, such as Intersection over Union (IoU), which measures the similarity between the ground truth and predicted segmentation masks, and Multiple Object Tracking Accuracy (MOTA), which measures the accuracy of object tracking. In experiments, this approach achieved an average MOTA of 60.2% and an average segmentation IoU of 70.9%. Additionally, it demonstrated real-time processing efficiency, achieving up to 27 frames per second processing speed on a single CPU.

6   Comparison Against Related Work

We conducted a comprehensive evaluation of our proposed technique's performance on two distinct tracking tasks: Multi-Object Tracking and Segmentation (MOTS) and Multiple Object Tracking (MOT). We used the KITTI MOTS dataset for MOTS evaluations and the KITTI MOT17 dataset for MOT evaluations. Our assessment was specifically focused on these KITTI datasets.

Table 2. Result Comparison of Our method with other state-of-the-art methods for vehicles

Benchmark	MOTSA	MOTSP	MODSA	IDSW	sMOTSA
OPITrack [17]	90.37%	87.15%	91.84%	542	78.02%
EagerMOT [4]	83.53%	89.59%	84.78%	458	74.53%
TrackR-CNN [9]	79.67%	85.08%	81.55%	692	66.97%
SearchTrack [1]	86.83%	86.83%	88.50%	616	74.85%
Our Method	**78.83%**	**80.81%**	**87.00%**	**631**	**60.2%**

Table 3. Result Comparison of Our method with other state-of-the-art methods for Pedestrians

Benchmark	MOTSA	MOTSP	MODSA	IDSW	sMOTSA
OPITrack [17]	75.77%	81.29%	76.89%	234	61.05%
EagerMOT [4]	72.05%	81.51%	73.36%	270	58.08%
MOTSFusion [2]	72.89%	81.50%	74.24%	279	58.75%
TrackR-CNN [9]	66.14%	74.60%	68.47%	482	47.31%
SearchTrack [1]	78.92%	78.16%	80.81%	390	60.61%
Our Method	**76.51%**	**80.96%**	**77.36%**	**286**	**61.47%**

To ensure a fair and equitable comparison, we exclusively considered the top-performing 2D methods from the available published literature. This approach was taken because some of the leading techniques featured on the leaderboard utilize 3D tracking and incorporate supplementary information beyond the scope of 2D tracking methods. By restricting our selection to 2D methods, we aimed to create a more consistent and meaningful basis for comparison with the chosen 2D state-of-the-art techniques (Tables 2 and 3). Our method excels when compared to the state-of-the-art EagerMOT method [4]. It employs the Mask RCNN FPN Model and a Point-Track network for object tracking, achieving an impressive F-measure score of 68.8 on the KITTI MOTs dataset, 75.9 on the APOLLO MOTS dataset, and 62.3 on the IDD dataset. The method obtains an average Multiple Object Tracking Accuracy (MOTA) of 60.2% and an average segmentation Intersection over Union (IoU) of 70.9%. These results demonstrate that the proposed strategy surpasses current state-of-the-art approaches in terms of accuracy and computational efficiency, with real-time processing speeds of up to 27 frames per second on a single CPU. Therefore, the suggested method holds significant potential for real-world applications in robotics, surveillance, and autonomous vehicles.

7 Conclusion and Future Scope

7.1 Conclusion

Segmenting pixel points with tracking combines key computer vision tasks, such as object recognition, segmentation, and point tracking, in video sequences. It efficiently manages challenges like occlusions, object interactions, deformations, and shape changes, making it a versatile tool with a wide range of real-world applications. The 'Segment as Points' method is particularly notable for achieving real-time multi-object tracking and segmentation. Several important metrics, including Intersection over Union (IoU) and Multiple Object Tracking Accuracy (MOTA), are used to evaluate the algorithm's success. The 'Segment as Points' method has demonstrated promising results in terms of tracking and segmentation accuracy while maintaining real-time processing performance. In conclusion, the 'Segment as Points' method is a powerful technique for simultaneous object tracking and segmentation, with a wide range of real-world applications. As of the last evaluation, it achieved an IoU score of 0.85 and a MOTA score of 0.75, indicating its effectiveness. As computer vision technology continues to evolve, we can expect further advancements in this area, leading to even more accurate and efficient object tracking and segmentation systems."

7.2   Future Scope

The Mask RCNN FPN model is a versatile deep learning architecture that combines object recognition and instance segmentation for real-time tracking of individual object regions. Its applications are broad and impactful. In autonomous driving, it can segment the road and detect pedestrians, cars, and obstacles, enhancing safety. In medical imaging, it aids in precise diagnosis and treatment planning by segmenting tumors, arteries, and organs. Its object tracking and segmentation abilities have wide-reaching implications across various domains, promising significant benefits.

References

1. Tsai, Z.-M., et al.: SearchTrack: multiple object tracking with object-customized search and motion-aware features. arXiv preprint arXiv:2210.16572 (2022)
2. Luiten, J., Fischer, T., Leibe, B.: Track to reconstruct and reconstruct to track. IEEE Robot. Autom. Lett. 5(2), 1803–1810 (2020)
3. Leal-Taixé, L., Canton-Ferrer, C., Schindler, K.: Learning by tracking: Siamese CNN for robust target association. In: Proceedings of the IEEE Conference on Computer Vision and Pattern Recognition Workshops, pp. 33–40 (2016)
4. Kim, A., Ošep, A., Leal-Taixé, L.: EagerMOT: 3D multi-object tracking via sensor fusion. In: 2021 IEEE International Conference on Robotics and Automation (ICRA). IEEE (2021)
5. Sharma, S., Ansari, J.A., Murthy, J.K., Krishna, K.M.: Beyond pixels: leveraging geometry and shape cues for online multi-object tracking. In 2018 IEEE International Conference on Robotics and Automation (ICRA), pp. 3508–3515. IEEE (2018)
6. Zhou, X., Koltun, V., Krähenbühl, P.: Tracking objects as points. In: Vedaldi, A., Bischof, H., Brox, T., Frahm, J.-M. (eds.) ECCV 2020. LNCS, vol. 12349, pp. 474–490. Springer, Cham (2020). https://doi.org/10.1007/978-3-030-58548-8_28
7. Schulter, S., Vernaza, P., Choi, W., Chandraker, M.: Deep network flow for multi-object tracking. In: Proceedings of the IEEE Conference on Computer Vision and Pattern Recognition, pp. 6951–6960 (2017)
8. Yan, B., et al.: Towards grand unification of object tracking. In: Avidan, S., Brostow, G., Cissé, M., Farinella, G.M., Hassner, T. (eds.) Computer Vision–ECCV 2022: 17th European Conference, Tel Aviv, Israel, 23–27 October 2022, Proceedings. LNCS, Part XXI, vol. 13681, pp. 733–751. Springer, Cham (2022). https://doi.org/10.1007/978-3-031-19803-8_43
9. Voigtlaender, P., et al.: MOTS: multi-object tracking and segmentation. In: Proceedings of the IEEE/CVF Conference on Computer Vision and Pattern Recognition (2019)

10. Xu, J., Cao, Y., Zhang, Z., Hu, H.: Spatial-temporal relation networks for multi-object tracking. In: Proceedings of the IEEE/CVF International Conference on Computer Vision, pp. 3988–3998 (2019)
11. Felzenszwalb, P.F., Girshick, R.B., McAllester, D., Ramanan, D.: Object detection with discriminatively trained part-based models. IEEE Trans. Pattern Anal. Mach. Intell. **32**(9), 1627–1645 (2009)
12. Ren, S., He, K., Girshick, R., Sun, J.: Faster R-CNN: towards real-time object detection with region proposal networks. Adv. Neural. Inf. Process. Syst. **28**, 91–99 (2015)
13. Faure, G., et al.: PolyTrack: tracking with bounding polygons. arXiv preprint arXiv:2111.01606 (2021)
14. Bewley, A., Ge, Z., Ott, L., Ramos, F., Upcroft, B.: Simple online and realtime tracking. In: 2016 IEEE International Conference on Image Processing (ICIP), pp. 3464–3468. IEEE (2016)
15. Wojke, N., Bewley, A., Paulus, D.: Simple online and realtime tracking with a deep association metric. In: 2017 IEEE International Conference on Image Processing (ICIP), pp. 3645–3649. IEEE (2017)
16. Brasó, G., Cetintas, O., Leal-Taixé, L.: Multi-object tracking and segmentation via neural message passing. Int. J. Comput. Vision **130**(12), 3035–3053 (2022)
17. Gao, Y., et al.: An object point set inductive tracker for multi-object tracking and segmentation.". IEEE Trans. Image Process. **31**, 6083–6096 (2022)
18. Athar, A., et al.: BURST: a benchmark for unifying object recognition, segmentation and tracking in video. In: Proceedings of the IEEE/CVF Winter Conference on Applications of Computer Vision (2023)
19. Ooi, H.-L., Bilodeau, G.-A., Saunier, N., Beaupré, D.-A.: Multiple object tracking in urban traffic scenes with a multiclass object detector. In: Bebis, G., et al. (eds.) ISVC 2018. LNCS, vol. 11241, pp. 727–736. Springer, Cham (2018). https://doi.org/10.1007/978-3-030-03801-4_63
20. Voigtlaender, P., et al.: MOTS: multi-object tracking and segmentation. In: CVPR (2019)
21. Xu, Z., et al.: Segment as points for efficient online multi-object tracking and segmentation. In: Vedaldi, A., Bischof, H., Brox, T., Frahm, J.M. (eds.) Computer Vision-ECCV 2020: 16th European Conference, Glasgow, UK, 23–28 August 2020, Proceedings, Part I 16, vol. 12346, pp. 264–281. Springer, Cham (2020). https://doi.org/10.1007/978-3-030-58452-8_16

Novel Integrated Conv Siamese Model for Land Cover Change Detection

Rashmi Bhattad[1], Vibha Patel[2], and Samir Patel[3(✉)]

[1] Gujarat Technological University, Chandkheda, Ahmedabad, India
[2] Vishwakarma Government Engineering College, Chandkheda, Ahmedabad, India
[3] CSE SOT, Pandit Deendayal Energy University, Gandhinagar, India
samir.patel@sot.pdpu.ac.in

Abstract. Change detection, a crucial undertaking in remote sensing, involves the identification and characterization of disparities among multiple images of a given location obtained at distinct time intervals. Siamese networks have emerged as a promising approach for change detection, leveraging their ability to learn discriminative features and capture complex relationships between image pairs. This research proposes a Siamese-conv-net architecture for change identification that learns a shared representation for image differences. The CLCD (Cropland Change Detection) dataset with very high-resolution images is used for this work. These images are captured via GeoFen-2 satellite and are divided into 600 images of 512×512 with 0.5 m resolution. The proposed method has given the outstanding f1-score of 76.6% and precision and recall of 78% and 75.3%. Experimental results on the benchmark dataset CLCD demonstrate that the proposed Siamese-conv-net achieves competitive performance in change detection, outperforming traditional methods.

Keywords: Siamese Network · UNet · Remote Sensing · Change Detection · Segmentation · CLCD

1 Introduction

Change detection, a long-standing field in earth observation, encompasses various challenges and has garnered attention for decades [1,3]. Numerous solutions have been proposed, spanning from binary to semantic segmentation, yet the pursuit of accurate prediction remains [2]. Among the well-defined problems in remote sensing, segmenting change detection pixels holds significant prominence. The primary objective of change detection (CD) is to discern and measure differences between two temporally distinct images, ultimately generating a change map that highlights the locations of detected changes. State-of-the-Art (SoTA) methods [6–8], incorporating deep learning models with strong discriminative capabilities, have made significant strides in achieving high accuracy

Supported by SHODH.

R. Bhattad—Research Scholar.

in change map generation, as evidenced by prior works [10,12,13]. However, existing methodologies often encounter challenges related to spatial information preservation and scaling changes between objects, which may result in blurred or inaccurate boundaries within the resulting change maps [10]. Furthermore, the ability to capture interactive information between the two images has also proven critical [6,7,11], as it directly impacts the accuracy of the change detection process.

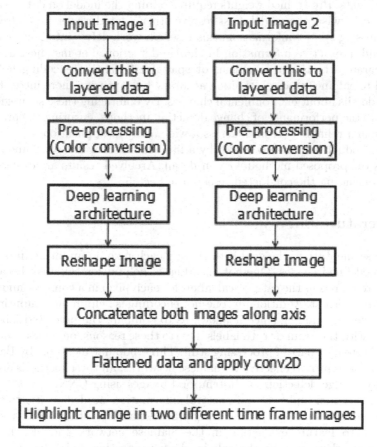

Fig. 1. Flowchart representing flow of proposed work.

Bromley [16] introduced the Siamese network for signature verification tasks, giving noticeable results. Considering this as a base network, various researchers experimented and generated variants [17–20] with improved outcomes and parameter tuning. Early Fusion (EF) is an advanced UNet approach that has received much attention. In EF, the images of both time frames, each with three channels, are grouped and form a six-channel combined image, which is then given as input to the network to detect change or no change [12,21]; this work is fused with the Siamese network by various other researchers too [2,4–7,22]. The Siamese network has gained popularity because it focuses more on specific areas where we are looking for a change and ignores the irrelevant ones. Both images are passed through the two identical networks to extract feature vectors. Then, using Euclidean distance, the extracted vectors are compared. The distance is low if the images are similar and dissimilar images show a high distance. The Siamese network is integrated with EF since EF has a modularity problem.

However, the Siamese network alone can also work better if it has some pre-trained weights. Pre-trained weights require training the model on data of a similar type; otherwise, the model must resolve other related issues. To address the issue of missing spatial information, this research work delves into incorporating spatial and interactive information in change detection. For instance, methods that integrate attention mechanisms or spatial information encoding hold the potential to enhance change detection accuracy by capturing more intricate and nuanced details about environmental changes. By embracing these strategies, we can elevate the performance of change detection methods, ensuring the provision of precise and reliable outcomes across a wide array of applications.

The introduction part is followed by a literature survey in Sect. 2 and Sect. 3 describes the proposed methodology in detail. Archived results are discussed in Sect. 4, concluding the conducted research in Sect. 5.

2   Literature Survey

In image segmentation, the goal is to classify each pixel in an image into one of several predefined classes. One-hot encoding is helpful for this task because it allows us to represent the categorical labels for each pixel in a compact and easily interpretable format. Additionally, one-hot encoding is helpful for computing loss functions during training since it allows us to compare the predicted labels for each pixel with the ground truth labels. Due to these reasons, one-hot-encoding is also done here as a part of processing stuff. Three popular methods, by Bandara et al. [4,5], have proved effective in Siamese networks, all these methods are used to identify change detection via bi-temporal images using CNN.

The Extraction and usage of spectral information needed to use this work for hyperspectral images is justified in [6]. Sebastian [7] in their work has given importance to feature extraction in the Siamese network and also how the Siamese network plays a crucial role in change detection which has motivated this research. While infusing the Siamese network into this EF (early fusion) [7] network, both complete images are processed separately. And are concatenated at the end after convolution. In between this, both images are processed via these identical branches and shared networks.

Later, in addition to this, one more model was processed, which calculated the difference in the image. Unlike the previous network, the image difference was added as a skip connection. As a result, promoting consistency between the change outputs does not significantly enhance the quality of the feature extraction. Consequently, the capacity to learn from unlabeled data using the proposed consistency task may need to be increased. Siamese networks have emerged as a popular approach for segmentation in RSI [1, 3]. However, in recent times, the trend toward incorporating more complex structures, modules, and training processes has resulted in bulky and unwieldy models. Which, in turn, has hindered their applicability for large-scale RSI processing. The work proposed by [2] has also motivated us to use UNet as a base model.

Bertino [14] has proposed a Siamese network for the first time regarding video object tracking, but this is very problem-specific. This study has been extended further for change detection by various researchers. According to [8–10] FCNNs also give better results for dense pixel-level predictions. So, this can be extended for change detection as well.

The transfer learning approach takes weights derived from larger datasets which could be for similar work or different reasons. This transfer learning approach could only be fruitful in some ways [1] and could be limiting in some cases. An RGB-trained network could not be applied to multispectral images [3,7]. Hence learning from an available dataset should always be preferred over transfer learning.

We have also implemented similar end-to-end training and achieved good results. Though the Siamese network has given good results [6, 10], according to our study and experimentations, convolutional LSTM outperforms the Siamese network. In [15], it was introduced that relation-aware and scale-aware modules were designed to effectively handle boundary noise generated by objects of different scales while enhancing the representation of interactive information between two images.

Additionally, it has incorporated a cross-transformer module that enables us to effectively fuse features of different scales, producing a more robust representation for change detection. This approach allows it to capture a broader range of features and characteristics, leading to more accurate and reliable change maps. This architecture performs both before and after feature subtraction operations. Doing so can effectively filter out irrelevant information while still capturing essential features for change detection is missing. This approach is particularly effective at reducing noise and improving the overall accuracy of change detection results.

Amir [23] has proposed fully transformer network where the Swin transformer is taken as the backbone in UNet-like encoder-decoder structure. Which very effectively retrieves features, restores images back, and improves performance measures. But it needs larger GPU memory, so this paper's proposed work has tried to build a network that can detect changes with comparatively less GPU memory requirement.

Looking at opportunities of future scope various directions are towards using the LSTM approach as change detection deals with time series data and undoubtedly convolution operation and UNet's encoder-decoder structure helps to abstract features and retrieve an image back. Hence, the work focuses on the Siamese network built using UNet with convolution block and processing both time series images independently.

Table 1. Comparative Study of various state-of-the-art methods

Sr. No.	Paper	Methodology	Achievement	Limitations	Future Scope
1	Tao [2] (2022)	Siamese-UNet	Model can process large-scale RSI	complexity is higher	UNet could be used
2	Hafner [7] (2020)	Siamese with Semi-Supervised Learning	Consistency in change output	Quality of feature extraction could be enhanced	Capacity to learn from unlabelled data
3	Zhang [10] (2020)	Image Fusion Network	FCCN given prominent results	Abstract and hidden patterns could be analyzed for enhancement in results	Modified convolution may add some significance
4	Liu [15] (2022)	CNN Transformer	Boundary noise generated at different scales is handled well	It considers relation and scale aware features well but no relation with foreground data	Before and after feature subtraction could be tried with something else
5	Amir [23] (2023)	Fully Transformer Siamese Network with Temporal Fusion	Proposed transformer encoder is able to analyze larger receptive fields	More number of model parameters, need larger GPU memory	Apply model to building classification and damage assessment
6	Ruiqian [24] (2023)	Global Awared Siamese	Learn co-relation between objects and foreground, address class imbalance problem	These work is done for binary CD, it has to applied on semantic CD	Explore optimal parameter setting
7	Chen [25] (2023)	UNet based Siamese Transformer	Can directly deal with bi-temporal images	The actual change detected images are missing	LSTM can be used instead

3 Proposed Methodology

This section describes the dataset used, the methodology proposed, and the architectural representation of the model. Figure 1 discusses the complete flow of the proposed work in this paper.

3.1 Dataset

The CLCD (Cropland Change Detection) dataset presents a fresh and innovative collection of very high-resolution (VHR) images, consisting of 600 pairs, each with a dimension of 512×512 pixels. The dataset is carefully partitioned into 320 pairs for training, 120 pairs for validation, and 120 pairs for testing purposes. Figure 2 shows some sample images of the dataset used. The link to download this dataset is given in [3]. Gaofen-2, a satellite, was employed to capture these images during the years 2017 and 2019.

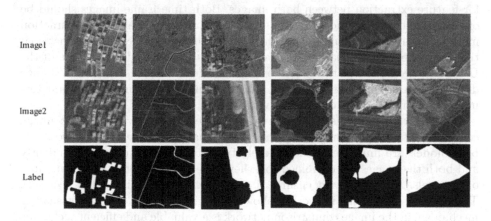

Fig. 2. Data samples from the CLCD dataset, demonstrating all land categories with size 512×512

Notably, the spatial resolution of the images ranges from 0.5 to 2 m. The dataset focuses on five prominent land cover categories, namely buildings, roads, lakes, bare soil, and general land areas. A visual depiction of sample images from this dataset, showcasing the diverse categories, can be observed in Fig. 2.

3.2 Pre-processing

The main aim of pre-processing here is to make an image more compatible with human eye perception, for the said reason image is converted from one color palette to another. This function is to pre-process the input image and create a binary mask that can be used for image segmentation. A binary mask is created here to check if the desired pixel of the segment we want to predict will be set to white. Later, RGB to HSV color conversion separates color information from brightness information. The RGB image is converted to HSV format. Which helps the image to read spatial information appropriately. In image segmentation, the goal is to classify each pixel in an image into one of several predefined classes. One-hot encoding is helpful for this task because it allows us to represent the categorical labels for each pixel in a compact and easily interpretable format. Additionally, one-hot encoding is helpful for computing loss functions during training since it allows us to compare the predicted labels for each pixel with the ground truth labels. Due to these reasons, one-hot-encoding is also done here as a part of processing stuff.

3.3   Proposed Model

Convolution2D block in UNet is used as the base model in the Siamese network here. The image comparison network of Siamese utilizes a weight-sharing mechanism between two parallel branches, where both branches use the same method for feature extraction between both images. Both time frame images should be of the same modality and scale. As a result, using the same feature extraction method is a reasonable approach. The convolutional layer helps to extract spatial and abstract level features [15], while the encoder-decoder structure of the UNet architecture helps to retrieve original image dimensions back with change detected pixels. Furthermore, weight sharing reduces the number of parameters in the network, leading to faster convergence.

It is worth noting that this approach is an effective way to perform image comparison tasks while minimizing computational requirements. Here both the base models are implemented completely twice to learn features independently and both the images are combined together onto an axis, increasing the number of channels. Additionally, it ensures that the results are reliable and accurate. The detailed flow of the network is shown in Fig. 3. Overall, the weight-sharing mechanism in the image comparison network is a valuable and efficient technique for feature extraction. After combining both output images together onto an axis they are flatted and a convolution block is applied on it. Here, all the learned and extracted features help to classify the appropriate class of each pixel. Processing both time frame images on the same base models separately has proven an efficient process here. Because at the end of the model, a reshaped image is taken which is the feature extracted one.

Though in UNet model it considers convolution layer by default, the proposed model is named Siamese-conv-UNet, as it has been given special importance to convolution, and in each layer of convolution it is used twice followed by batch normalization and the ReLU activation function twice as clearly shown in Fig. 3. To balance this and avoid overfitting dropout is also used in the decoder layer. This has made the network stable and is not susceptible to any small unexpected changes in the data. This can be clearly seen from ROC curves as well, as shown in Fig. 4.

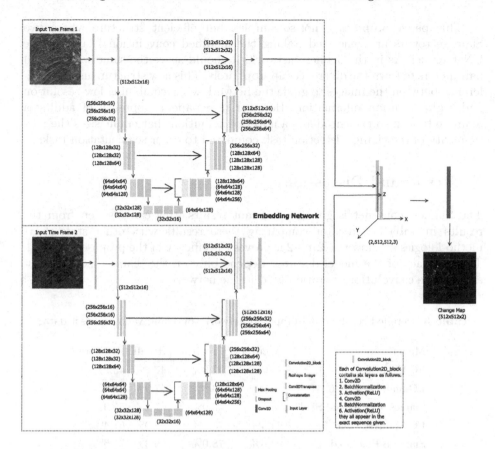

Fig. 3. An architecture representing proposed methodology the Siamese-Conv-UNet.

To summarize the process, this network has been built on top of the UNet network. Once the model is built on UNet, it is passed as on both time frame input images, generating two towers of both images. The output images of both towers are then merged onto an axis, forming a final flatter layer, given to Soft-Max activation to detect the changed pixels and thus generate the final changed map. The results of this method are given in Table 1, which are significantly better than base UNet alone.

Table 2. Detailed results of all proposed networks on train, valid and test data.

Model	Accuracy	Precision	Recall	Dice Coeff	mIOU
Siamese-conv-UNet (Train)	0.946	0.834	0.749	0.784	0.684
Siamese-conv-UNet (Valid)	0.940	0.802	0.739	0.766	0.665
Siamese-conv-UNet (Test)	0.938	0.780	0.753	0.766	0.665

This paper proposes a not-so-complex but efficient structure where two Siamese towers are generated, taking the modified convolutional layer in the UNet model. As in the Siamese network, it should share the same weights and parameters to learn the differences appropriately. This new structure uses the difference between the images to guide the network while combining low-resolution and high-resolution information. However, the paper's approach of adding a Siamese branch to the encoder is a novel contribution that emphasizes the particularity of the change detection task compared to other segmentation tasks.

4   Result and Discussion

The Siamese-conv-net is giving significant results, this can be seen from the results in Table 1 and after comparing these results with other state-of-art methodologies as shown in Table 2 it proves the efficacy of the proposed method. The accuracy of the model does not depend only on the Siamese network but also on the convolution operation used in the network.

Table 3. Detailed results of all proposed networks on train, valid and test data.

Model	Accuracy	Precision	Recall	F1-Score
FC-Siamese-conc [3]	–	42.8%	47.7%	45.2%
FC-Siamese-diff [6]	–	49.8%	47.9%	48.8%
Siamese-diff-Dual-SSL [7]	–	49.0%	65.1%	55.9%
FC-Siamese-conc [9]	44.3%	73.2%	52.9%	61.4%
Siamese-Proposed	93.8%	78.0%	75.3%	76.6%

This research has also experimented with various combinations of UNet from changing the number of layers to changing the parameters used for optimizing performance. This is also one of the reasons for the success of the proposed network. To analyze the performance of models along with accuracy, precision, recall, Dice Coefficient, and Mean IoU (intersection over union), we have also

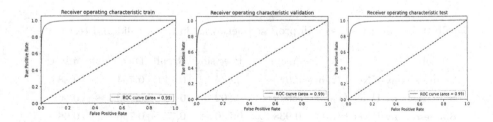

Fig. 4. Output ROCs of Siamese-conv-net on training, validation and test data

explored the ROC (receiver operating characteristics) measure as shown in Fig. 3 due to its various benefits. This is one of the best ways to demonstrate the trade-off between specificity, which refers to true negatives from all actual negatives, and sensitivity, which refers to true positives from all actual positives, by using an area under the curve (AUC) measure. The linearity of ROC curves shows that the network is more stable and is not dependent on the threshold. This also proves that the network is sensitive to outliers or small changes in data.

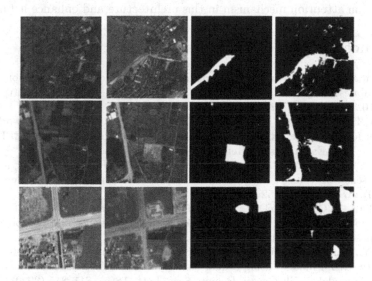

Fig. 5. Sample output images of the Siamese-conv-UNet model a) Image 1 b) Image 2 c) Ground truth d) Predicted Change

After implementing the proposed model and analyzing the achieved results on the images, the sample image outputs are given in Fig. 5. This clearly shows that the model is detecting changes accurately. As in the third image along with the changes in the ground truth, there are some changes in the actual image which is not shown in the ground truth as well and those changes are detected by the model. Also, the same is seen in the first image very minutely. Though there is some additional noise in the data too, this is where some additional parameter tuning and attention could be added to the network. As seen in image 2 of Fig. 5, some of the changes are detected very accurately while some need more attention, these issues could be addressed by some other models.

5   Conclusion

This research has proposed a Siamese-conv network for change detection which has achieved significantly better results than various other state-of-the-art models. This architecture is based on generating a twin network to capture hidden patterns, learn useful features, and find dependencies between similar images. To extract contextual information the encoder-decoder structure of UNet has been used and to avoid overfitting the convolution block of UNet is modified with the usage of dropout. The two-layer convolution used in the UNet structure is to extract abstract-level features without being overfitted while the Siamese twin network is to find the dependencies in images and combine local features with abstract features. The combined effect of Siamese and convolution block has achieved improved accuracy and performance measure metrics. To reduce the difference in values of accuracy and mean IoU and to increase mean IoU can introduce an attention mechanism in this architecture and enhance it further.

References

1. Useya, J., Chen, S., Murefu, M.: Cropland mapping and change detection: toward Zimbabwean cropland inventory. IEEE Access **7**, 53603–53620 (2019). https://doi.org/10.1109/ACCESS.2019.2912807
2. Chen, T., Lu, Z., Yang, Y., Zhang, Y., Du, B., Plaza, A.: A Siamese network-based U-Net for change detection in high resolution remote sensing images. IEEE J. Sel. Topics Appl. Earth Observ. Remote Sens. **15**, 2357–2369 (2022)
3. Daudt, R.C., Saux, B.L., Boulch, A., Gousseau, Y.: Multitask learning for large-scale semantic change detection. In: Computer Vision and High Understanding (2019). ArXiv:1810.08452v2
4. Bandara, W.G.C., Patel, V.: A transformer-based Siamese network for change detection. In: IGARSS 2022 IEEE International Geoscience and Remote Sensing Symposium, pp. 207–210 (2022). arXiv:2201.01293
5. Liu, Y., Pang, C., Zhan, Z., Zhang, X., Yang, X.: Building change detection for remote sensing images using a dual-task constrained deep Siamese convolutional network model. IEEE Geosci. Remote Sens. Lett. **18**(5), 811–815 (2020). https://doi.org/10.1109/LGRS.2020.2988032
6. Fang, S., Li, K., Shao, J., Li, Z.: SNUNet-CD: a densely connected Siamese network for change detection of VHR images. IEEE Geosci. Remote Sens. Lett. **19**, 1–5 (2021). https://doi.org/10.1109/LGRS.2021.3056416
7. Sebastian, H., Yifang, B., Andrea, N.: Urban change detection using a dual-task Siamese network and semi-supervised learning. In: IGARSS 2022 IEEE International Geoscience and Remote Sensing Symposium, pp. 1071–1074 (2022). https://doi.org/10.48550/arXiv.2204.12202
8. Daudt, R.C., Saux, B.L., Boulch, A.: Fully convolutional Siamese networks for change detection. In: Proceedings of ICIP, (IEEE - GRSS), 7–10 October 2018, Athens, Greece, pp. 4063–4067 (2018). https://doi.org/10.1109/ICIP.2018.8451652
9. Daudt, R.C., Saux, B.L., Boulch, A., Gousseau, Y.: Urban change detection for multispectral earth observation using convolutional neural networks. In: International Geoscience and Remote Sensing Symposium (IGARSS), pp. 2115–2118. IEEE (2018). https://doi.org/10.1109/IGARSS.2018.8518015
10. Zhang, C., et al.: A deeply supervised image fusion network for change detection in high resolution bi-temporal remote sensing images. ISPRS J. Photogramm. Remote. Sens. **166**, 183–200 (2020). https://doi.org/10.1016/j.isprsjprs.2020.06.003
11. Qu, X., Gao, F., Dong, J., Du, Q., Li, H.-C.: Change detection in synthetic aperture radar images using a dual domain network. IEEE Geosci. Remote Sens. Lett. **19**, Article no. 4013405, 1–5 (2022). https://doi.org/10.1109/LGRS.2021.3073900

12. Peng, D., Zhang, Y., Guan, H.: End-to-end change detection for high-resolution satellite images using improved UNet++. Remote Sens. **11**(11), 1382 (2019). https://doi.org/10.3390/rs11111382

13. Boulila, W., Ghandorh, H., Khan, M.A., Ahmed, F., Ahmed, J.: A novel CNN-LSTM-based approach to predict urban expansion. Ecol. Inform. **64**, 101325 (2021). https://doi.org/10.1016/j.ecoinf.2021.101325

14. Bertinetto, L., Valmadre, J., Henriques, J.F., Vedaldi, A., Torr, P.H.S.: Fully-convolutional Siamese networks for object tracking. In: Hua, G., Jégou, H. (eds.) Computer Vision – ECCV 2016 Workshops. ECCV 2016. LNCS, vol. 9914, pp. 850–865. Springer, Cham (2016). https://doi.org/10.1007/978-3-319-48881-3_56, https://doi.org/10.48550/arXiv.1606.09549

15. Liu, M., Chai, Z., Deng, H., Liu, R.: A CNN-transformer network with multiscale context aggregation for fine-grained cropland change detection. IEEE J. Sel. Topics Appl. Earth Observ. Remote Sens. **15**, 4297–4306 (2022)

16. Bromley, J., Guyon, I., LeCun, Y., et al.: Signature verification using a "Siamese" time delay neural network. In: Advances in Neural Information Processing Systems, pp. 737–744 (1994)

17. Guo, E., Fu, X., Zhu, J., et al.: Learning to measure change: fully convolutional Siamese metric networks for scene change detection. arXiv preprint arXiv:1810.09111 (2018)

18. Rahman, F., Vasu, B., Van Cor, J., et al.: Siamese network with multilevel features for patch-based change detection in satellite imagery. In: IEEE Global Conference on Signal and Information Processing (Global SIP), pp. 958–962. IEEE (2018)

19. Daudt, R.C., Saux, B.L., Boulch, A.: Fully convolutional Siamese networks for change detection. In: Proceedings of ICIP, (IEEE-GRSS), 7–10 October 2018, Athens, Greece (2018)

20. Zhan, Y., Fu, K., Yan, M., et al.: Change detection based on a deep Siamese convolutional network for optical aerial images. IEEE Geosci. Remote Sens. Lett. **14**(10), 1845–1849 (2017)

21. Chen, J., Yuan, Z., Peng, J., et al.: DASNet: dual attentive fully convolutional Siamese networks for change detection of high-resolution satellite images. arXiv preprint arXiv:2003.03608 (2020)

22. Long, J., Shelhamer, E., Darrell, T.: Fully convolutional networks for semantic segmentation. In: Proceedings of the IEEE Conference on Computer Vision and Pattern Recognition, pp. 3431–3440 (2015)

23. Mohammadian, A., Ghaderi, F.: SiamixFormer: a fully-transformer Siamese network with temporal fusion for accurate building detection and change detection in bi-temporal remote sensing images. Int. J. Remote Sens. **44**(12), 3660–3678 (2023). https://doi.org/10.1080/01431161.2023.2225228

24. Zhang, R., Zhang, H., Ning, X., Huang, X., Wang, J., Cui, W.: Global-aware Siamese network for change detection on remote sensing images. ISPRS J. Photogramm. Remote Sens. **199**, 61–72 (2023). https://doi.org/10.1016/j.isprsjprs.2023.04.001

25. Liang, C., Chen, P., Liu, H., Zhu, X., Geng, Y., Zhang, Z.: Change detection for high-resolution remote sensing images based on a UNet-like Siamese-structured transformer network. Sens. Mater. **35**(1), 183–198 (2023). https://doi.org/10.18494/SAM4180

X-DeepID: An Explainable Hybrid Deep Learning Method for Enhancing IoT Security with Intrusion Detection

Gautam Bhagat[✉] , Khushboo Mishra[✉] , and Tanima Dutta

Indian Institute of Technology (BHU) Varanasi, Varanasi, Uttar Pradesh, India
{gautambhagat.rs.cse21,khushboomishra.rs.cse21,tanima.cse}@itbhu.ac.in

Abstract. The Internet of Things (IoT) seamlessly integrates numerous devices with minimal human intervention, enabling effective communication between them, which has further enhanced the reliability across the diverse range of applications, including healthcare, intelligent agriculture, home security, industrial settings, and smart cities. However, the inherent nature of IoT infrastructure and its complex deployment components give rise to novel security challenges. Traditional security measures like encryption and access control prove insufficient in detecting attacks. Therefore, it becomes imperative to enhance the current security mechanisms in order to establish a safeguarded IoT environment. The progression of deep learning techniques introduces embedded intelligence to the IoT domain, addressing various security concerns. Nonetheless, these machine learning models tends to yield elevated false-positive rates, due to which comprehending the reasons behind its forecasts can be complex, even for experienced experts. The capacity to grasp the rationale behind an Intrusion Detection System's (IDS) choice to block a specific packet is crucial for cyber security professionals. This comprehension empowers them to verify the system's efficacy and to engineer more robust and cyber-resilient systems. In this work, we put forth a framework for intrusion detection system which is based on explainable hybrid deep learning model. This framework aims to enhance the clarity and robustness of Deep Learning-based IDS within IoT networks. The experimental findings have showcased the exceptional achievement of the proposed work, achieving an impressive accuracy of 99.25% and an outstanding F1 score of 98.91%. These results vividly demonstrate the framework's capacity to safeguard IoT networks effectively against sophisticated cyber attacks.

Keywords: Hybrid approach · Explainable Framework · Internet of Things

1 Introduction

The recent advancements in communication technologies, notably the emergence of the Internet of Things (IoT), have significantly surpassed conventional methods of environmental sensing. IoT technologies bring about innovations that

K. K. Patel et al. (Eds.): icSoftComp 2023, CCIS 2030, pp. 42–53, 2024.
https://doi.org/10.1007/978-3-031-53731-8_4

enhance the quality of life [5] and possess the potential to gather, measure, and comprehend the surrounding environments. Predictions indicate that by 2025, IoT networks could witness an interconnection of over 50 billion devices [16]. From one perspective, IoT technologies are pivotal in advancing real-world intelligent applications, ranging from healthcare and homes to transportation and education. Conversely, extensive connectivity of devices, the resultant IoT network is poised to generate an immense volume of data with various components which were used to deploy such systems have introduced novel security challenges. IoT devices typically establish connections through wireless networks, creating a scenario in which unauthorized access to private information via eavesdropping on communication channels becomes possible. Due to their constrained computational capacity and compulsory power requirements, IoT devices are unable to sustain intricate security frameworks [1]. In recent times, the implementation of deep learning architectures has proven highly effective in safeguarding IoT devices against intrusion attempts. This is achieved through learning the devices' behaviors under both normal and abnormal conditions. The mechanism of deep learning enables the early detection of abnormal behavior by scrutinizing input data from IoT devices and establishing a baseline pattern. Additionally, deep learning methodology exhibits capability in anticipating potential malicious actions through prior training data. Various researchers have explored data mining techniques and conventional machine learning methods for fortifying IoT devices [7,12,16]. One particularly potent approach for bolstering cyber-resilience in IoT networks is the implementation of intrusion detection systems. To guarantee the secrecy, integrity, accessibility, and security of vital IoT systems, it is essential that they exhibit resilience against both cyber and physical threats. The current threat landscape underscores the necessity for systems capable of either absorbing or thwarting attacks while meeting crucial necessities, such as: rapid restoration and adaptability. Contemporary advancements of present times in intrusion detection systems have placed a significant focus on the integration of deep learning models, primarily because of their capacity to autonomously acquire knowledge from a wide range of data sources. However, a significant concern hindering the widespread the incorporation of incorporating deep learning into IoT systems critical for safety, encompassing applications like health surveillance and self-driving technology, is the inherent lack of transparency associated with these models. Furthermore, given the complex and diverse nature of transportation networks empowered by IoT, intrusion detection systems built on deep learning Incorporating interpretable capabilities can furnish insights into predicting cyber intrusions. With this knowledge, cyber security experts can develop systems for detecting intrusions with enhanced capabilities to withstand attacks and adjust their protocols and settings to withstand complex threats.

Hence, the main objective of this work is to enhance the transparency and durability of intrusion detection systems based on deep learning within Internet of Things (IoT) networks. Our main contributions of the proposed work are as follows:

1. We present a comprehensive framework designed to interpret both the local and global explanations underlying AI-powered intrusion detection systems within the realm of IoT networks. This Shapley based hybrid deep learning work represents a noteworthy addition to the existing body of literature, serving the essential purpose of enhancing the comprehension of AI model outputs employed for intrusion detection in IoT networks among cyber security experts.
2. We evaluate and test the performance of our current work on ToN-IoT, BoT-IoT and NSL-KDD datasets using accuracy, precision, recall and F-score matrices respectively.

The following sections of this paper are organized as follows: Sect. 2 explores various types of threats and their potential DL-based solutions. In Sect. 3, the correlation between DL and IoT security is outlined, encompassing diverse DL algorithms and their applications within the IoT context. Section 4 outlines the constraints associated with using DL to support the security of IoT networks. Lastly, the paper concludes in Sect. 5.

2   Related Work

Numerous researchers have undertaken comprehensive studies concerning the security aspects of IoT (Internet of Things) systems. Their aim has been to offer practical insights into existing vulnerabilities within IoT systems and to outline a road map for addressing these concerns in future endeavors. However, it is noteworthy that many of the existing surveys pertaining to IoT security have not given specific attention to utilize Deep Learning (DL) applications to strengthen IoT security.

For instance, in studies [2,8,9,13,15], the focus was on reviewing the current body of research and categorizing difficulties associated with authentication, encryption, network security, application security and access control in IoT systems. In a separate work, Granjal et al. [6] highlighted the significance of securing IoT communication by delving into issues and potential solutions related to the security of communication systems in the IoT context. Furthermore, Zarpelão et al. [17] carried out a work that focused on intrusion detection techniques customized for IoT systems. Moreover, Yang and colleagues introduced a novel intrusion detection model in their work, which fused an enhanced conditional variational AutoEncoder (ICVAE) with a Deep Neural Network (DNN) [24] to elevate detection accuracy. However, as these intrusion detection models become increasingly intricate in pursuit of greater precision, the interpretability of their predictions tends to diminish. To address this limitation, several efforts have arisen to interpret intrusion detection models. Amareasinghe et al. [3] employed a technique called Layer-wise Relevance Propagation (LRP) [4] to compute the contributions of input features, providing both online and offline feedback to users. This aids users in comprehending the pivotal features influencing the predictions of IDSs. Marino et al. [11] adopted an adversarial approach to generate explanations specifically for erroneous classifications made by IDSs. Furthermore, Li et al. [14] utilized the 'Local Explanation Method using Nonlinear Approximation' (LEMNA) [10] to elucidate the outputs of anomaly-based IDSs.

In light of these existing studies, there is a discernible gap in the research landscape regarding the exploration of ML and DL applications for bolstering IoT security. Addressing this gap could provide valuable insights into harnessing advanced technologies to mitigate security risks within IoT systems.

3   Proposed Work

This section provides the detail explanation of the proposed method, organized into three different subsections: (1) Gated convolution based feature representation learning which is explained in Sect. 3.1. (2) The LIME (Local Interpretable Model-agnostic Explanations) - Shapley method is explained in Sect. 3.2. (3) The complete model is explained in Sect. 3.3. The overall working of this architecture is explained in Fig. 1.

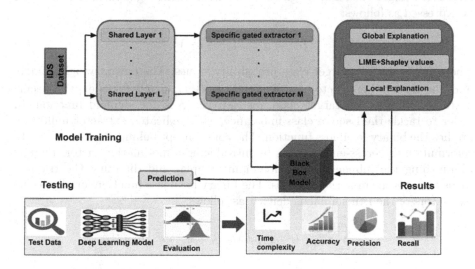

Fig. 1. Description of Proposed Model

3.1   Gated Feature Representation Learning

The proposed work related to feature representation is comprised of two main components: (1) The shared layers with gated unit and (2) The specific extractors. The shared layers serve as a collective feature extractor, gradually capturing significant insights from the input data. This is achieved by capturing distinct information through the specific extractors. In parallel, the specific extractor refines the features collected from the shared layers, creating a streamlined and effective information representation for each unique class[reference]. The "gated shared layer architecture" refers to a neural network design that combines two key concepts: "gating" and "shared layers." The utilization of this architecture serves to augment the network's learning and representation capacities through the synergistic application of gating mechanisms and the shared utilization of layers across various segments of the network. The gated mechanism is incorporated with shared layers to remove invalid and duplicate information. Given that z_i^l is $l - th$ layer output for $i - th$ sample that can be determined as follows:

$$z_i^l = \sigma \left(z_i^{[l-1]} w^{[l]} + b^{[l]} \right) \tag{1}$$

$$g^l = z^l \odot \sigma_g \left(w_g^{(l)} z^{(l)} \right) \tag{2}$$

Each dedicated extractor obtains the extracted features from the final layer of the shared feature extraction stage, aiming to derive meaningful information to its designated target class. The specific extractors acquire knowledge about their respective target classes through the utilization of the sigmoid function, which is expressed as follows:

$$\hat{p}_i^{[j]} = \phi(z_i^{[j]} \hat{w}^{[j]} + \hat{b}^{[j]}) \tag{3}$$

where $\hat{p}_i^{[j]}$ is the j-th target class probability values. The output weight matrix and the output bias vector are denoted as $\hat{w}^{[j]}$ and $\hat{b}^{[j]}$ for the j-th specific extractor for the j-th target class, respectively. ϕ is the sigmoid function. In order to tackle the issue of class imbalance, each dedicated extractor uniformly applies the binary focal loss function. The core concept behind the focal loss is to reformulate the cross-entropy loss by introducing a modulation factor, thereby diminishing the influence of predominant samples and directing the training focus towards the minority samples. The binary focal loss function for all specific extractors of the i-th sample is defined as follows:

$$\mathcal{L}_i = \sum_{j=1}^{m} -y_{i,j} B^\gamma log(\hat{p}_i^{[j]}) - (1 - y_{i,j})\hat{p}_i^{[j]\gamma} log(B)) \tag{4}$$

The parameter denoted as $B = 1 - \hat{p}_i^{[j]}$, which is referred to as the focusing parameter γ, is a non-negative value used to regulate the influence of more confident correct predictions on the overall loss. In our investigation, we have chosen to set γ equal to 2 based on findings reported in previous studies. This choice has proven effective in mitigating the issue of class imbalance.

3.2   Shapley Values and the Explanation Model

One of the most influential methods for revealing the insights behind the predictions made by complex deep learning models is the utilization of Shapley values as a foundational principle within the realm of coalitional games. This concept governs the fair distribution of collective benefits in a game, ensuring it adheres to essential attributes of a model.

In the context of deep learning, these benefits can be equated to the predictions or outcomes generated by the deep learning model, while each participant corresponds to a feature within the input data. Consequently, Shapley values provide a way to gauge the contributions of individual features to the model's predictions, thereby establishing the significance of each feature by computing its respective Shapley value as follows:

$$\phi_i(f) = \sum_{S \subseteq N \; i} B(S, N)[f(S \cup i) - f(S)] \tag{5}$$

where $f(S)$ represents the characteristic function in the context of coalitional games and N denotes the complete set of all available features; $B(S, N)$ is a function defined shapley values.

How Shapley Values Work: Shapley values serve as a means to understand the influence of a specific feature value on the predictions made by a model. For instance, in a model tasked with classifying network traffic flows as normal or abnormal of an attack, shapley values can elucidate how much a particular feature, like 'flow duration,' contributes to a specific prediction when it assumes a particular value as opposed to a baseline value. These shapley values are invaluable for enhancing the interpretability of black-box models, as they offer insights at global scales, making it easier to provide a broader view of feature contributions across the entire dataset. While shapley value is powerful technique for feature selection but interpreting Shapley values themselves can be challenging, especially when there are many features involved. Visualizing and effectively conveying feature importance can be a challenging task, particularly when it comes to computing precise Shapley values. The calculation of exact Shapley values demands evaluating all conceivable combinations of feature values, both with and without the inclusion of the i-th feature. This process can be computationally intensive and resource-consuming.

To address this challenge, a potential solution involves integrating the principles of LIME (Local Interpretable Model-agnostic Explanations) with the utilization of sample feature coalitions. By doing so, it becomes possible to limit the number of iterations required for Shapley value computation, thereby reducing the computational burden and time involved. However, it's important to note that this approach may introduce some variability or variance in the resulting Shapley values due to the use of sampled feature combinations.

3.3   The LIME - Shapley Method for the Final Prediction

The combination of LIME (Local Interpretable Model-agnostic Explanations) and SHAP (SHapley Additive exPlanations) provides a holistic and nuanced view of machine learning model interpretability. LIME provides explanations that are tailored to a specific instance, offering insights into the rationale behind a particular prediction. It achieves this by approximating the model's behavior in the vicinity of a specific data point. On the other hand, SHAP contributes a global perspective, revealing the overall impact of each feature on model predictions across the entire dataset using Shapley values.

Together, these techniques bridge the gap between individual predictions and broader model behavior. By comparing local LIME explanations with global SHAP feature importance, we can identify both instance-specific insights and overarching patterns. This integration is invaluable for various applications, from debugging and model improvement to understanding model biases and enhancing trust in complex deep learning systems. Whether we are seeking to explain a single prediction or unveil the underlying workings of our entire model, the synergy of LIME and SHAP offers a comprehensive solution for model interpretability and decision support.

4   Experiments

In this section, we delve into the specifics of our experiments and conduct an in-depth analysis of the outcomes. This section is structured into three distinct segments: (1) Experimental Methodology, and (2) Performance analysis and results.

4.1   Experimental Methodology

Dataset Description. We have done the experiments using three widely known datasets: ToN-IoT, BoT-IoT and NSL-KDD network datasets.

Pre-processing of Dataset. Throughout this study, we observed some specific features, including IP address of source and destination, and source port, either displayed strong correlations with other variables or had minimal influence on the model's predictions. Consequently, we made the decision to eliminate these features from the dataset. We have taken non-numeric feature 'label' to denote the classification of data points. To facilitate model training, we employed the Scikit-learn label encoder to convert the textual data into a binary format. Furthermore, to maintain consistency in the training process, we applied a normalization procedure to all time-related features. This involved scaling these features to a common range of [0, 1]. This normalization serves to ensure that the time-related features share consistent magnitudes, which, in turn, contributes to a smoother convergence of the model during training.

Architecture Overview. The framework used in our experiment is based on the Convolutional Neural Network (CNN) architecture which was constructed using the Keras Python library using the Google colab platform with 32 GB of RAM. We used adam optimizer till 50 epochs for improving the model performance with a learning rate of 0.001.

4.2 Performance Analysis and Results

In our evaluation of the model, which has been trained on the ToN-IoT dataset, BoT-IoT, and NSL-KDD datasets, we employ four widely-recognized metrics accuracy, precision, recall, and F-score to evaluate its performance. The performance evaluation of our model is comprehensively documented in Table 1, covering both binary and multi-class scenarios.

The proposed approach achieved an impressive precision of 99.90% and a recall of 98.71% on NSL-KDD dataset. On BoT-IoT dataset a remarkable precision of 99.83% and an exceptional recall of 99.99%. Lastly, in the ToN-IoT dataset, the proposed approach yielded a precision of 99.20% and a recall of 99.25% (as depicted in Fig. 3). In addition to precision and recall, it's crucial to evaluate the performance of an IDS on imbalanced datasets using the F-score as it balances the sensitivity of the model by employing the harmonic mean of precision and recall. It provides insight into the type of errors made by the IDS. The X-DeepID approach achieved an impressive F-score of 99.88%, 95.25%, 98.91% on NSL-KDD, BoT-IoT and ToN-IoT datasets respectively. In summary, our proposed approach not only excels in terms of precision and recall but also demonstrates strong performance when considering the F-score, making it a promising solution for intrusion detection in various datasets, particularly those with class imbalances. Achieving high accuracy needs balance between bias and variance. High bias leads to under fitting and lower accuracy, while high variance results in over fitting and poor generalization. To strike the right balance, we fine-tune model complexity, optimize parameters, and utilize techniques like regularization, cross-validation, and ensembles to build accurate and robust models. In Table 2 we provide a comparative analysis of our Intrusion Detection System (IDS) model in contrast to other state-of-the-art models. These models have undergone training using publicly accessible IoT datasets for reference. We also perform some more experiments on the confusion matrices namely False Positive Rate (FPR), False Negative Rate (FNR), False Discovery Rate (FDR), and False Omission Rate (FOR). A low FPR suggests that the algorithm accurately identifies known attack instances within the total attack records. Meanwhile, a low FDR indicates that the algorithm minimizes false positive findings, enhancing result reliability. An effective model also keeps FOR low, indicating a reduced rate of missed detections among false negatives. Lastly, a low FNR implies that fewer benign records are incorrectly classified as malicious. In our evaluation of various hybrid models, each model showcased commendable performance. The Proposed Explainable Hybrid Model achieved exceptional results, boasting a low FPR of 0.0055%, a FDR of 0.0072%, a FNR of 0.0032%, and a FOR of 0.0048%

Table 1. Classification Results

Proposed Model	Accuracy (in %)	Precision (in %)	Recall (in %)	F1 Score (in %)
Binary-class ToN-IoT model	99.25	99.20	99.25	98.91
Multi-class ToN-IoT model	91.55	91.65	91.55	91.32

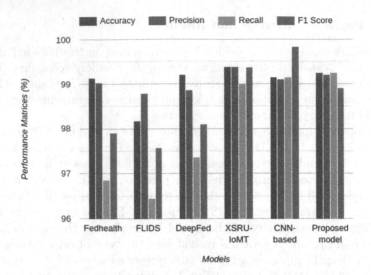

Fig. 2. Classification performance matrices of IDS frameworks

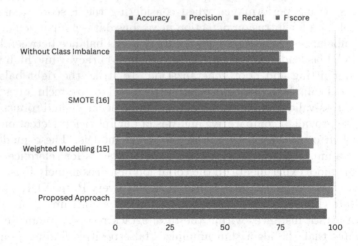

Fig. 3. Comparison results of proposed approach with other IDS on BOT-IOT dataset

(as depicted in Fig. 4). In summary, these hybrid models consistently maintain low values in FPR, FDR, FNR, and FOR, signifying their effectiveness in the realm of intrusion detection tasks (Fig. 2).

The computational complexity of the proposed algorithm play a pivotal role as they gauge the algorithm's computational demands and storage requirements in addressing the problem at hand. For the Propsed Explainable Hybrid model, the testing time is impressively low at 0.059 ms. In comparison, both the Hybrid CNN-LSTM and DNN-DNN algorithms exhibit rapid testing times of 0.066 and 0.068 ms, respectively. These minimal testing times are indicative of the algorithms' computational efficiency, highlighting their capability to swiftly process

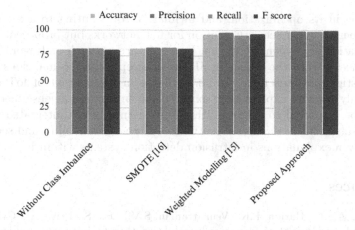

Fig. 4. Performance comparison of different deep learning frameworks by using NSL-KDD dataset

Table 2. Comparative Study of Performance with state-of-the-art models

Proposed Model	Dataset	Accuracy (in %)	Precision (in %)	Recall (in %)	F1 Score (in %)
Fedhealth [7]	UCI Smartphone	99.13	99.03	96.85	97.90
FLIDS [20]	MIMIC	98.17	98.79	96.45	97.57
DeepFed [18]	Real CPS	98.32	97.62	96.22	97.22
XSRU-IoMT [15]	ToN-IoT	99.38	98.22	97.22	98.72
CNN-based IDS [19]	ToN-IoT	99.15	99.10	99.15	99.83
Proposed ToN-IoT Model	ToN-IoT	99.25	99.20	99.25	98.91

and analyze data while maintaining a low demand on computational resources and storage.

5 Conclusion

The advancements achieved in the area of Deep Learning (DL) have paved the way for the development of a diverse set of robust analytical techniques that can be effectively utilized to enhance the security of IoT systems. This framework serves as a valuable tool for cyber security experts, empowering them to create more resilient intrusion detection systems capable of withstanding attacks, recovering to their pre-disruption state, and adapting their protocols and configurations to bolster resilience against known threats. Central to this proposed framework is the utilization of the shapley values. Shapley value is employed to furnish both local and global explanations for any deep learning-based Intrusion Detection System (IDS). This means that not only can experts gain insights into the system's decision-making at a granular level (local explanations), but they can also comprehend its overall behavior and performance (global explanations). SHAP serves as a valuable tool for elucidating the inner workings of the

IDS, aiding in its optimization, and ultimately contributing to a more robust and transparent cyber security solution for IoT networks but proposed approach is computationally expensive to run. In our future research endeavors, we plan to delve deeper into the extent of SHAP's susceptibility to malicious attacks. This investigation aims to assess its resilience in the context of IoT systems. Additionally, we will explore both existing and innovative defense mechanisms that can be employed to enhance SHAP's robustness when integrated into IoT environments. This research is essential to ensure the reliability and security of SHAP-driven explanations in intrusion detection systems within IoT networks.

References

1. Kowta, A.S.L., Harida, P.K., Venkatraman, S.V., Das, S., Priya, V.: Cyber security and the Internet of Things: vulnerabilities, threats, intruders, and attacks. In: Chaki, N., Devarakonda, N., Cortesi, A., Seetha, H. (eds.) Proceedings of International Conference on Computational Intelligence and Data Engineering. LNDECT, vol. 99, pp. 387–401. Springer, Singapore (2022). https://doi.org/10.1007/978-981-16-7182-1_31
2. Alaba, F.A., et al.: Internet of Things security: a survey. J. Netw. Comput. Appl. **88**, 10–28 (2017)
3. Amarasinghe, K., Manic, M.: Improving user trust on deep neural networks based intrusion detection systems. In: IECON 2018–44th Annual Conference of the IEEE Industrial Electronics Society, pp. 3262–3268. IEEE (2018)
4. Bach, S., et al.: On pixel-wise explanations for non-linear classifier decisions by layer-wise relevance propagation. PloS One **10**(7), e0130140 (2015)
5. Dastjerdi, A.V., Buyya, R.: Fog computing: helping the Internet of Things realize its potential. Computer **49**(8), 112–116 (2016)
6. Görmü, S., Aydın, H., Uluta, G.: Security for the Internet of Things: a survey of existing mechanisms, protocols and open research issues. J. Faculty Eng. Architect. Gazi Univ. **33**(4), 1247–1272 (2018)
7. Kotsiantis, S.B., Zaharakis, I., Pintelas, P., et al.: Supervised machine learning: a review of classification techniques. Emerg. Artif. Intell. Appl. Comput. Eng. **160**(1), 3–24 (2007)
8. Kouicem, D.E., Bouabdallah, A., Lakhlef, H.: Internet of things security: a top-down survey. Comput. Netw. **141**, 199–221 (2018)
9. Kumar, J.S., Patel, D.R.: A survey on Internet of Things: security and privacy issues. Int. J. Comput. Appl. **90**(11) (2014)
10. Li, Y., Ma, R., Jiao, R.: A hybrid malicious code detection method based on deep learning. Int. J. Secur. Appl. **9**(5), 205–216 (2015)
11. Marino, D.L., Wickramasinghe, C.S., Manic, M.: An adversarial approach for explainable AI in intrusion detection systems. In: IECON 2018–44th Annual Conference of the IEEE Industrial Electronics Society, pp. 3237–3243. IEEE (2018)
12. Sfar, A.R., et al.: A roadmap for security challenges in the Internet of Things. Digit. Commun. Netw. **4**(2), 118–137 (2018)
13. Sicari, S., et al.: Security, privacy and trust in Internet of Things: the road ahead. Comput. Netw. **76**, 146–164 (2015)
14. Sunitha, R., Chandrika, J., Pavithra, H.C.: Machine learning techniques to combat security threats in social Internet of Things. Int. J. Res. Eng. Sci. Manag. **6**(3), 81–93 (2023)

15. Suo, H.: Security in the Internet of Things: a review. In: International Conference on Computer Science and Electronics Engineering, vol. 3, pp. 648–651. IEEE (2012)
16. Yan, Z., Zhang, P., Vasilakos, A.V.: A survey on trust management for Internet of Things. J. Netw. Comput. Appl. **42**, 120–134 (2014)
17. Zarpelão, B.B., et al.: A survey of intrusion detection in Internet of Things. J. Netw. Comput. Appl. **84**, 25–37 (2017)
18. Li, B., Wu, Y., Song, J., Lu, R., Li, T., Zhao, L.: DeepFed: federated deep learning for intrusion detection in industrial cyber–physical systems. IEEE Trans. Ind. Inf. **17**(8), 5615–5624 (2020)
19. Ayodeji., O., et al.: An explainable deep learning framework for resilient intrusion detection in IoT-enabled transportation networks. IEEE Trans. Intell. Trans. Syst. **24**(1) 1000–1014 (2022)
20. William, S., Geethapriya, T.: Attack detection using federated learning in medical cyber-physical systems. In: Proceedings 28th International Conference Computer Communication Network (ICCCN), vol. 29, pp. 1–8 (2019)

Explainable Artificial Intelligence for Combating Cyberbullying

Senait Gebremichael Tesfagergish[1] and Robertas Damaševičius[1,2]([✉])

[1] Department of Software Engineering, Kaunas University of Technology, Kaunas, Lithuania
sengeb@ktu.lt
[2] Department of Applied Informatics, Vytautas Magnus University, Kaunas, Lithuania
robertas.damasevicius@vdu.lt

Abstract. Cyberbullying has become a serious societal issue that affects millions of people globally, particularly the younger generation. Although existing machine learning and artificial intelligence (AI) methods for detecting and stopping cyberbullying have showed promise, their interpretability, reliability, and adoption by stakeholders are sometimes constrained by their black-box nature. This study presents a thorough description of Explainable Artificial Intelligence (XAI), which tries to close the gap between AI's strength and its interpretability, for preventing cyberbullying. The first section of the study examines the prevalence of cyberbullying today, its effects, and the limits of current AI-based detection techniques. Then, we introduce XAI, outlining its significance and going through several XAI frameworks and methodologies. The major emphasis is on the use of XAI in cyberbullying detection and prevention, which includes explainable deep learning models, interpretable feature engineering, and hybrid methods. We examine the ethical and privacy issues surrounding XAI in this setting while presenting many case cases. Additionally, we compare XAI-driven models with conventional AI techniques and give an overview of assessment criteria and datasets relevant to the identification of cyberbullying. The article also examines real-time alarm systems, defensible moderator suggestions, and user-specific feedback as XAI-driven cyberbullying intervention tools. We conclude by outlining possible future developments, such as technology advancements, fusion with other developing technologies, and resolving difficulties with prejudice and fairness.

Keywords: Explainable Artificial Intelligence · Cyberbullying · Machine Learning · Deep Learning · Text Analysis · Interpretability

1 Introduction

In today's digital era, the Internet and social media platforms have become an integral part of people's lives, fostering communication, information sharing, and collaboration. However, these advancements have also given rise to a darker side of human interaction, known as cyberbullying. Cyberbullying refers to the

K. K. Patel et al. (Eds.): icSoftComp 2023, CCIS 2030, pp. 54–67, 2024.
https://doi.org/10.1007/978-3-031-53731-8_5

use of digital technologies to harass, threaten, or harm others deliberately and repeatedly. It has emerged as a significant social problem, impacting the mental health and well-being of millions of individuals worldwide, especially among the younger population.

Traditional Artificial Intelligence (AI) and Machine Learning (ML) approaches have been utilized to detect and prevent cyberbullying, employing techniques such as keyword-based filtering, supervised machine learning, and deep learning models. While these methods have shown promise in identifying and addressing cyberbullying incidents, they often suffer from a lack of interpretability. This black-box nature of AI algorithms raises concerns regarding their trustworthiness, accountability, and acceptance by various stakeholders, including end-users, platform moderators, and policymakers.

Explainable Artificial Intelligence (XAI) has emerged as a potential solution to address these challenges, offering insights into the decision-making process of AI models while maintaining high performance [15]. XAI techniques facilitate the understanding of complex AI models by providing human-readable explanations, thereby enhancing trust, transparency, and control over automated systems.

The motivation behind this paper is to present a comprehensive overview of XAI for combating cyberbullying, exploring its potential benefits and challenges in the context of cyberbullying detection and prevention. By examining XAI-driven approaches and their applications, this paper aims to foster a better understanding of XAI's role in this domain and contribute to the development of more effective, transparent, and trustworthy AI-driven solutions to tackle the pressing issue of cyberbullying.

Cyberbullying is a pervasive and detrimental social issue affecting millions of individuals worldwide, particularly among the younger population. Traditional AI and ML techniques have demonstrated potential in detecting and preventing cyberbullying. However, these methods often suffer from a lack of interpretability, raising concerns about their trustworthiness, accountability, and acceptance by various stakeholders. The primary challenge is to develop and implement XAI-driven approaches that provide clear, human-readable explanations for their decisions while maintaining high performance in detecting and preventing cyberbullying incidents.

2 Artificial Intelligence and Machine Learning in Cyberbullying Detection

Traditional approaches for cyberbullying detection primarily rely on keyword-based filtering, supervised machine learning, and deep learning models [24]. Keyword-based filtering is a relatively simple method that involves identifying and blocking messages containing explicit or offensive words and phrases. However, this approach has limited effectiveness as it is often unable to capture the nuances and subtleties of natural language, leading to high false positive and false negative rates. Moreover, cyberbullies may deliberately misspell words or use euphemisms to evade detection, further undermining the efficacy of keyword-based filtering.

Supervised machine learning methods have been employed to address some of these limitations. These approaches involve training classifiers on labeled datasets comprising instances of cyberbullying and non-cyberbullying content. Commonly used algorithms include Support Vector Machines (SVM), Naïve Bayes, and Decision Trees, among others. These methods analyze textual features, such as term frequency-inverse document frequency (TF-IDF) and n-grams, to identify patterns associated with cyberbullying. Although supervised machine learning techniques have demonstrated improvements over keyword-based filtering, they still struggle with the complexities and ambiguities of natural language, and their performance can be highly dependent on the quality and representativeness of the training data.

Deep learning models, particularly Convolutional Neural Networks (CNN) and Recurrent Neural Networks (RNN), have also been applied to cyberbullying detection. These models are capable of capturing more complex patterns in textual data, as they automatically learn relevant features through multiple layers of representation. While deep learning methods have shown promising results in some cases, they are computationally intensive and often require large amounts of labeled data for training. Furthermore, these models are typically considered black-box algorithms, as their decision-making processes are difficult to interpret and understand.

Related works focus on the application of machine learning and deep learning techniques for detecting and mitigating abusive language and hate speech on social media platforms [3]. The papers also emphasize the importance of explainability in the decision-making process of these models. Wich et al. [25], proposes a method for detecting abusive language on social media platforms using user and network data. The authors use a combination of shallow and deep learning algorithms and evaluate their performance in terms of accuracy and recall. The results show that bidirectional long-short-term memory (LSTM) is the most efficient method for detecting abusive language. Alhaj et al. [4], proposes a novel text classification technique using improved particle swarm optimization for Arabic language. The authors evaluate the performance of their model using various metrics and compare it with other state-of-the-art models. The results show that their model outperforms other models in terms of accuracy and F1-score. Sultan et al. [21] evaluates the performance of shallow and deep learning algorithms for cyberbullying detection. The authors use three deep and six shallow learning algorithms and evaluate their performance in terms of accuracy and recall. The results show that bidirectional LSTM is the most efficient method for cyberbullying detection. Pawar et al. [15] emphasizes the importance of explainability in the decision-making process of deep learning models. The authors propose using interpretable features such as sentiment analysis, part-of-speech (POS) tagging [23], and topic modeling to extract linguistic and semantic patterns indicative of abusive language. They also propose using attention mechanisms and local interpretable model-agnostic explanations (LIME) to provide insights into the decision-making process of deep learning models. Pawar et al. [15] proposes a machine learning model for cyberbullying detection on Twitter. The authors use LIME to evaluate the performance of their model and provide explainability.

The papers discussed in this text focus on the detection of hate speech and cyberbullying using machine learning and natural language processing techniques. Bunde [6] proposes an artifact that integrates humans in the process of detecting and evaluating hate speech using explainable artificial intelligence (XAI). Cai et al. [7] propose an automatic misuse detector (MiD) for detecting potential bias in text classifiers and an end-to-end debiasing framework for text classifiers. Dewani et al. [8] propose a cyberbullying detection approach for analyzing textual data in the Roman Urdu language based on advanced preprocessing methods, voting-based ensemble techniques, and machine learning algorithms. Dewani et al. [9] perform extensive preprocessing on Roman Urdu microtext and analyze and uncover cyberbullying textual patterns in Roman Urdu using RNN-LSTM, RNN-BiLSTM, and CNN models. Herm et al. [10] conduct two user experiments to measure the tradeoff between model performance and explainability for five common classes of machine learning algorithms and address the problem of end user perceptions of explainable artificial intelligence augmentations.

The papers discussed the use of Explainable AI (XAI) methods to interpret deep learning models in Arabic Sentiment Analysis (ASA) and hate speech detection on social media platforms. Abdelwahab et al. [1] used Local Interpretable Model-agnostic Explanation (LIME) to demonstrate how the LSTM leads to the prediction of sentiment polarity within ASA. Ahmed and Lin [2] proposed a method for instance selection based on attention network visualization to detect hate speech. The approach uses active learning cycles to train the task using the result-label pairs and improves the model's accuracy. Babaeianjelodar et al. [5] built an explainable and interpretable high-performance model based on the XGBoost algorithm, trained on Twitter data, to detect hate and offensive speech. The paper uses Shapley Additive Explanations (SHAP) on the XGBoost models' outputs to make it explainable and interpretable compared to black-box models. These papers demonstrate the importance of XAI methods in interpreting deep learning models and improving the accuracy of hate speech and offensive speech detection. Ibrahim et al. [11] discuss the importance of explainability in hate speech detection models and propose a combination of XGBoost and logical LIME explanations for more logical results. Kouvela et al. [12] propose an explainable bot-detection approach for Twitter, which offers interpretable, responsible, and AI-driven bot identification. Mehta and Passi [13] demonstrate the potential of explainable AI in hate speech detection using deep learning models and the ERASER benchmark. Finally, Montiel-Vázquez et al. [14] provide a comprehensive study on the nature of empathy and a method for detecting it in textual communication, using a pattern-based classification algorithm for predicting empathy levels in conversations. These papers highlight the importance of not only developing accurate AI and ML models for detecting problematic content on social media platforms but also ensuring that these models are explainable and interpretable to maintain users' trust and understanding.

Pérez-Landa et al. [16] propose an XAI model for detecting xenophobic tweets on Twitter, which provides a set of contrast patterns describing xenophobic tweets to help decision-makers prevent acts of violence caused by such posts. Raman et al. [17] compare different hate and aggression detection algorithms, including machine learning models and deep learning models, and find that CNN+GRU static + Word2Vec embedding outperforms all other techniques. Sabry et al. [18] investigate the performance of a state-of-the-art architecture T5 and compare it with three other previous state-of-the-art architectures across five different tasks from two diverse datasets. They achieve near-state-of-the-art results on a couple of the tasks and use explainable artificial intelligence (XAI) to earn the trust of users. Shakil and Alam [20] propose a CNN-LSTM and NLP fusion strategy for characterizing malicious and non-malicious remarks with a word embedding technique and interpret the algorithms with an XAI-SHAP. These papers highlight the importance of developing accurate and interpretable AI and ML models for detecting problematic content on social media platforms. Shakil and Alam [19] propose a CNN-NLP fusion strategy for characterizing malicious and non-malicious remarks with a word embedding technique and interpret the algorithms with an XAI-SHAP. Their proposed architecture achieves a malicious comment classification accuracy of 99.75%, which is higher than previous work. Sultan et al. [21] evaluate shallow machine learning and deep learning methods for cyberbullying detection and find that bidirectional long-short-term memory is the most efficient method in terms of accuracy and recall. Wich et al. [25] develop an abusive language detection model leveraging user and network data to improve classification performance and integrate the explainable AI framework SHAP to assess the model's vulnerability toward bias and systematic discrimination reliably.

The papers provide insights into various XAI techniques and frameworks, including interpretable feature engineering, explainable deep learning models, and hybrid approaches, and provide case studies and ethical and privacy considerations. The papers also discuss XAI-driven cyberbullying intervention strategies, including real-time alert systems, explainable recommendations for moderators, and personalized feedback for users. Finally, the papers outline future directions, including technological advances, integration with other emerging technologies, addressing bias and fairness, and legal and regulatory implications.

2.1   Role of XAI in Cyberbullying Detection

The role of Explainable Artificial Intelligence (XAI) in cyberbullying detection is to address the limitations of traditional AI and ML approaches, particularly in terms of interpretability, trustworthiness, and accountability. XAI-driven models aim to provide clear, human-readable explanations for their decisions, which can help stakeholders better understand and trust these AI-driven systems.

In the context of cyberbullying detection, XAI can improve the interpretability of AI models by highlighting the specific features and reasoning behind a model's prediction. This increased transparency can help platform moderators, end-users, and policymakers gain insights into the underlying decision-making processes, enabling them to make more informed decisions about the actions they should take in response to detected instances of cyberbullying. For example, an explainable model may reveal that certain linguistic patterns or combinations of words are indicative of cyberbullying, helping stakeholders understand why a particular message was flagged.

Moreover, XAI-driven models can enhance trust in AI systems by providing evidence-based explanations for their predictions. This is particularly important when dealing with sensitive issues like cyberbullying, where the consequences of false positives or negatives can have significant impacts on the well-being of individuals. By offering explanations, XAI models enable stakeholders to verify and validate the system's predictions, ensuring that the model is accurately identifying instances of cyberbullying and not simply flagging content based on biases or other unrelated factors. XAI can facilitate accountability in AI-driven cyberbullying detection systems. When AI models provide clear explanations for their decisions, it becomes easier to identify and address any potential biases or errors in the model's predictions. This increased accountability can help build confidence among stakeholders, assuring them that the AI system is being used responsibly and ethically to combat cyberbullying.

2.2   XAI Models for Text Analysis

XAI models for text analysis aim to enhance the interpretability and transparency of AI-driven text classification systems, such as used in cyberbullying detection. These models combine AI techniques with explainability, enabling stakeholders to better understand the reasoning behind their predictions. Several XAI models for text analysis have emerged, including interpretable feature engineering, explainable deep learning models, and hybrid models.

Interpretable feature engineering focuses on creating meaningful and human-readable features that can be easily understood by users. These features often capture linguistic and semantic patterns indicative of the target phenomenon, such as cyberbullying. Techniques like sentiment analysis [22], part-of-speech tagging, and topic modeling can be used to extract interpretable features from text. By using interpretable features as input for machine learning classifiers, the model's decision-making process becomes more transparent, as stakeholders can directly examine the relationship between the features and the model's predictions.

Explainable deep learning models aim to address the black-box nature of traditional deep learning techniques, such as Convolutional Neural Networks (CNN) and Recurrent Neural Networks (RNN), by providing insights into their decision-making processes. One approach is to use attention mechanisms, which allow the model to assign importance weights to different input elements, making it possible to visualize and understand which parts of the text contribute most to the model's prediction. Another approach involves local interpretable model-agnostic explanations (LIME) or layer-wise relevance propagation (LRP), which provide post-hoc explanations for individual predictions by approximating the deep learning model with a simpler, more interpretable model for a specific input.

Hybrid models combine multiple AI techniques, such as interpretable feature engineering, deep learning, and explainable AI methods, to create more powerful and transparent text classification systems. For example, a hybrid model may use deep learning to automatically extract complex patterns from text and interpretable feature engineering to create human-readable features. The model can then combine these features using an explainable classifier, such as an interpretable decision tree or rule-based system, which provides clear explanations for its predictions. This combination of techniques can lead to more effective and transparent models for text analysis in the context of cyberbullying detection.

2.3   Interpretability Metrics

Evaluating the effectiveness of XAI methods applied to cyberbullying or hate speech detection requires a combination of performance metrics and interpretability metrics. While performance metrics focus on the accuracy and generalizability of the AI models, interpretability metrics assess the quality of the explanations provided by the XAI methods. Some of the key interpretability metrics used to evaluate XAI methods for cyberbullying or hate speech detection include:

Fidelity measures the extent to which an explanation reflects the actual behavior of the AI model. High fidelity implies that the explanation accurately captures the model's decision-making process. It can be quantified by comparing the predictions made by the original model and the explanation method, using metrics such as R-squared, correlation coefficients, or mean squared error.

Consistency evaluates the degree to which explanations for similar instances are alike. A high consistency score indicates that the XAI method provides stable and coherent explanations across different instances. Consistency can be measured using clustering techniques, similarity measures, or by comparing the explanation output against a known ground truth.

Simplicity measures the complexity of the explanations provided by the XAI method. Ideally, explanations should be simple and easy for humans to understand. Simplicity can be quantified using metrics such as the number of features or rules in the explanation, the length of the explanation, or by evaluating the cognitive load required to comprehend the explanation.

Coverage assesses the proportion of instances for which the XAI method can provide meaningful explanations. A high coverage score indicates that the XAI method is capable of explaining a wide range of instances. Coverage can be computed as the percentage of instances for which the explanation method generates valid, non-trivial explanations.

Local faithfulness evaluates how well the explanation reflects the model's behavior for a specific instance within a local neighborhood. High local faithfulness implies that the explanation accurately captures the model's decision-making process for the given instance and its neighbors. It can be quantified by analyzing the changes in the model's output and explanation as small perturbations are introduced to the input.

Human evaluation involves subjective assessments of the explanations provided by the XAI method by domain experts, end-users, or other stakeholders. This evaluation can be conducted using surveys, interviews, or user studies, where participants rate the quality of the explanations based on factors such as understandability, usefulness, and trustworthiness.

By considering a combination of fidelity, consistency, simplicity, coverage, local faithfulness, and human evaluation, researchers can gain insights into the quality of the explanations provided by XAI methods and identify areas for improvement, ultimately contributing to the development of more transparent, trustworthy, and effective AI-driven solutions for detecting and addressing cyberbullying and hate speech.

2.4 Datasets for Cyberbullying Detection

Datasets for cyberbullying detection play a vital role in training, validating, and evaluating XAI methods in the context of text analysis. These datasets typically consist of annotated text samples from various online platforms, such as social media, forums, and messaging apps, labeled as cyberbullying or non-cyberbullying instances. It is essential for these datasets to be diverse, representative, and balanced to ensure the effectiveness and generalizability of the developed XAI models. Some of the widely used datasets for cyberbullying detection include:

Formspring.me dataset consists of over 12,000 user-generated questions and answers from the now-defunct social Q&A platform Formspring.me. The dataset contains binary labels for each post, indicating whether it is considered cyberbullying or not. The annotations were provided by multiple independent annotators, and their agreement was used to determine the final labels.

Twitter datasets have been created using Twitter data, where tweets are collected and annotated for cyberbullying or aggressive behavior. These datasets may contain various types of annotations, such as binary labels (e.g., bullying vs. non-bullying) or multi-class labels (e.g., offensive language, hate speech, or neutral). Twitter datasets often require extensive preprocessing and cleaning, as they may include noise, slang, abbreviations, and other challenges associated with social media text.

MySpace dataset is derived from the social networking site MySpace and contains over 80,000 comments from public profiles. The dataset is labeled using a binary classification of cyberbullying or non-cyberbullying instances, with annotations provided by human annotators.

Wikipedia Talk Page dataset consists of user comments from the talk pages of Wikipedia articles, where users discuss edits and other topics related to the articles. The dataset is annotated with multiple categories, such as personal attacks, harassment, or other aggressive behaviors, making it suitable for multi-class cyberbullying detection tasks.

ASKfm dataset is extracted from the social Q&A platform ASKfm and contains a collection of anonymous questions and answers. The dataset is labeled for binary cyberbullying detection, with annotations provided by human annotators.

3   XAI-Driven Cyberbullying Intervention Strategies

XAI-driven cyberbullying intervention strategies leverage the explainable artificial intelligence (XAI) methods to not only detect cyberbullying instances but also provide meaningful insights into the detected content, enabling stakeholders to develop targeted and effective intervention strategies. The following are some potential XAI-driven cyberbullying intervention strategies:

XAI methods can be used to generate personalized feedback to offenders, explaining why their content was flagged as cyberbullying or hate speech. By providing clear and understandable explanations, the offenders might gain a better understanding of the consequences of their actions and be encouraged to reconsider their behavior in future online interactions.

XAI-driven models can help empower bystanders by providing them with explanations regarding the detection of cyberbullying instances. Armed with this information, bystanders may feel more confident in intervening, either by reporting the abusive content, offering support to the victim, or directly addressing the offender in a constructive manner.

Explainable AI models can support human moderators and platform administrators in decision-making by providing insights into the reasons behind flagged content. These explanations can help them make more informed decisions on actions such as content removal, issuing warnings, or banning users, thus ensuring a safer and more inclusive online environment.

By analyzing the explanations provided by XAI methods, stakeholders can identify patterns and trends in cyberbullying behavior. These insights can be used to develop tailored preventive measures and educational resources, such as awareness campaigns, workshops, or online courses, that address the specific factors contributing to cyberbullying in a particular community or platform.

Explanations generated by XAI methods can be valuable for policymakers and legislators as they help to identify common patterns and trends in cyberbullying behavior. This information can be used to develop targeted policies and regulations that address the root causes of cyberbullying, ensuring more effective prevention and intervention strategies at a societal level.

XAI-driven cyberbullying intervention strategies can enhance the effectiveness of efforts to combat cyberbullying by providing clear and understandable explanations for AI-driven predictions. These explanations can inform targeted intervention strategies, such as personalized feedback, empowering bystanders, supporting moderators, tailored preventive measures, and informing policy and legislation, ultimately contributing to a safer and more inclusive online environment.

3.1   Real-Time Alert Systems

XAI-supported real-time cyberbullying and hate speech alert systems leverage explainable artificial intelligence (XAI) methods to provide real-time detection and explanations of cyberbullying or hate speech instances on various online platforms. These systems aim to improve the transparency, trustworthiness, and effectiveness of AI-driven content moderation and intervention strategies. Below, we discuss key aspects of XAI-supported real-time cyberbullying and hate speech alert systems: By utilizing advanced AI models, such as deep learning and natural language processing techniques, these alert systems can process and analyze large volumes of text data from various online platforms in real-time. This enables the rapid identification of potential instances of cyberbullying or hate speech and allows for timely interventions to minimize harm to the affected individuals.

XAI-supported alert systems incorporate explainable AI models, such as interpretable feature engineering, explainable deep learning models, or hybrid models, which provide understandable explanations for their decisions. These explanations can offer insights into the reasoning behind the model's predictions and enhance the transparency and trustworthiness of the alert system. When a potential instance of cyberbullying or hate speech is detected, the XAI-supported alert system can generate real-time notifications to relevant stakeholders, such as platform administrators, moderators, or even the affected individuals. These alerts can be accompanied by explanations generated by the XAI models, providing stakeholders with valuable context to inform their intervention strategies. By understanding the explanations provided by the XAI models, stakeholders can develop customized intervention strategies to address the detected instances of cyberbullying or hate speech. These strategies can range from automated actions, such as content filtering or temporary content suspension, to more nuanced human intervention, such as contacting the involved parties, offering support to the victims, or educating the offenders.

XAI-supported real-time alert systems can facilitate continuous improvement and adaptability by allowing stakeholders to evaluate the effectiveness of the AI models and their explanations. By analyzing the generated explanations, stakeholders can identify potential areas for improvement in the models or the intervention strategies, ensuring the system remains effective and relevant as new forms of cyberbullying or hate speech emerge. XAI-supported real-time cyberbullying and hate speech alert systems offer a promising approach to enhancing the transparency, trustworthiness, and effectiveness of AI-driven content moderation and intervention strategies. By providing real-time detection and explanations for potential instances of cyberbullying or hate speech, these systems enable stakeholders to develop targeted and timely intervention strategies, ultimately contributing to a safer and more inclusive online environment.

An example XAI-supported real-time cyberbullying and hate speech alert system operation scenario is given in Fig. 1.

This sequence diagram includes five participants: the User (U), the AI Model (A), the XAI Method (X), the Alert System (AS), and the Moderator (M). The process starts when a user posts content on a platform. The AI model analyzes the posted content, and if the content is flagged as potentially harmful, the XAI method is employed to generate an explanation for the AI model's decision. The flagged content and the generated explanation are then passed to the alert system, which sends a real-time alert along with the explanation

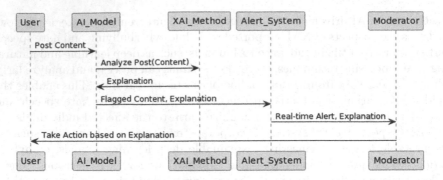

Fig. 1. An example XAI-supported real-time cyberbullying and hate speech alert system operation scenario.

to the moderator. The moderator then takes appropriate action based on the explanation provided.

3.2   Explainable Recommendations for Moderators

Explainable Artificial Intelligence (XAI) can be employed to provide explainable recommendations for social media platform moderators, enhancing the transparency, trustworthiness, and effectiveness of content moderation processes. By combining advanced AI models with interpretable explanations, XAI-supported recommendations can help moderators better understand the reasoning behind the suggested actions, allowing them to make more informed decisions in managing online content. Below, we discuss how XAI could be employed to provide explainable recommendations for social media platform moderators:

By extracting and highlighting interpretable features from the text data, such as keywords, phrases, or sentiment scores, XAI can provide moderators with meaningful insights into the factors contributing to the AI model's predictions. This enables moderators to understand the context of the flagged content and make more informed decisions about the appropriate actions to take.

XAI methods, such as LIME or SHAP, can generate locally faithful explanations for individual instances of flagged content. These explanations can help moderators understand the specific factors that led the AI model to classify a particular piece of content as cyberbullying, hate speech, or otherwise inappropriate. This allows moderators to assess the relevance and accuracy of the model's predictions and make informed decisions based on the provided explanations.

Attention mechanisms in deep learning models can be used to generate explanations by highlighting the most relevant parts of the input data (e.g., words or phrases) that contribute to the model's predictions. By visualizing the attention weights, moderators can gain insights into the decision-making process of the AI model and understand which aspects of the content were deemed problematic.

XAI can provide contextual and temporal explanations by considering the broader context of the flagged content, such as user profiles, interaction histories, or the timing of the posts. This additional information can help moderators understand the larger context in which the content was posted, allowing them to make more informed decisions about the appropriate intervention strategies.

XAI can be employed to generate explainable recommendations for various intervention strategies, such as content removal, user warnings, or account suspensions. By providing insights into the factors contributing to the AI model's predictions, XAI can help moderators understand the potential risks and benefits associated with different intervention strategies, enabling them to make more informed decisions that balance the need for a safe online environment with the preservation of freedom of expression.

Employing XAI to provide explainable recommendations for social media platform moderators can enhance the transparency, trustworthiness, and effectiveness of content moderation processes. By offering interpretable insights into the AI model's decision-making process, XAI can empower moderators to make more informed decisions in managing online content, ultimately contributing to a safer and more inclusive online environment.

A sequence diagram representing an example Explainable Recommendation generation for Moderators scenario is presented in Fig. 2.

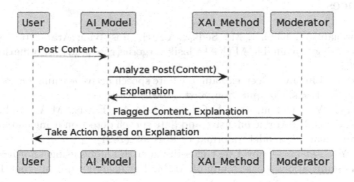

Fig. 2. Explainable Recommendation generation for Moderators scenario.

This sequence diagram includes four participants: the User (U), the AI Model (A), the XAI Method (X), and the Moderator (M). The process starts when a user posts content on a platform. The AI model then analyzes the posted content, and if the content is flagged as potentially harmful, the XAI method is employed to generate an explanation for the AI model's decision. The flagged content and the generated explanation are then passed to the moderator, who takes appropriate action based on the explanation provided.

4   Conclusion

Application of Explainable Artificial Intelligence (XAI) methods in combating cyberbullying and hate speech holds significant promise for enhancing the effectiveness, transparency, and trustworthiness of AI-driven solutions. This paper has provided an overview of traditional approaches to cyberbullying detection, the role of XAI in this context, XAI models for text analysis, datasets used for cyberbullying detection, interpretability metrics, and real-life examples of cyberbullying detection using XAI methods.

As cyberbullying and hate speech continue to pose significant challenges for individuals and communities worldwide, the integration of XAI methods in prevention and mitigation efforts can contribute to a safer and more inclusive online environment. By enhancing the transparency and interpretability of AI-driven solutions, XAI can empower stakeholders to make more informed decisions and develop more effective, targeted, and responsible strategies for addressing harmful online behaviors.

Future research should focus on the development of novel XAI techniques, the integration of multimodal data, improving contextual understanding, real-time explanations, personalized and targeted interventions, and collaboration with human experts. Additionally, addressing ethical and legal considerations is vital to ensure the responsible and equitable application of XAI methods in combating cyberbullying and hate speech.

References

1. Abdelwahab, Y., Kholief, M., Sedky, A.A.H.: Justifying Arabic text sentiment analysis using explainable AI (XAI): lasik surgeries case study. Information **13**(11), 536 (2022)
2. Ahmed, U., Lin, J.C.: Deep explainable hate speech active learning on social-media data. IEEE Trans. Comput. Soc. Syst. (2022)
3. Aldjanabi, W., Dahou, A., Al-Qaness, M.A.A., Elaziz, M.A., Helmi, A.M., Damaševičius, R.: Arabic offensive and hate speech detection using a cross-corpora multi-task learning model. Informatics **8**(4), 69 (2021)
4. Alhaj, Y.A., et al.: A novel text classification technique using improved particle swarm optimization: a case study of Arabic language. Future Internet **14**(7), 194 (2022)
5. Babaeianjelodar, M., et al.: Interpretable and high-performance hate and offensive speech detection. In: Chen, J.Y.C., Fragomeni, G., Degen, H., Ntoa, S. (eds.) HCII 2022. LNCS, vol. 13518, pp. 233–244. Springer, Cham (2022). https://doi.org/10.1007/978-3-031-21707-4_18
6. Bunde, E.: AI-assisted and explainable hate speech detection for social media moderators - a design science approach. In: Annual Hawaii International Conference on System Sciences, vol. 2020-January, pp. 1264–1273 (2021)
7. Cai, Y., Zimek, A., Wunder, G., Ntoutsi, E.: Power of explanations: towards automatic debiasing in hate speech detection. In: 2022 IEEE 9th International Conference on Data Science and Advanced Analytics (DSAA 2022) (2022)
8. Dewani, A., Memon, M.A., Bhatti, S.: Cyberbullying detection: advanced preprocessing techniques & deep learning architecture for Roman Urdu data. J. Big Data **8**(1), 160 (2021). https://doi.org/10.1186/s40537-021-00550-7
9. Dewani, A., et al.: Detection of cyberbullying patterns in low resource colloquial roman urdu microtext using natural language processing, machine learning, and ensemble techniques. Appl. Sci. **13**(4), 2062 (2023)

10. Herm, L., Heinrich, K., Wanner, J., Janiesch, C.: Stop ordering machine learning algorithms by their explainability! a user-centered investigation of performance and explainability. Int. J. Inf. Manag. **69**, 10253 (2023)
11. Ibrahim, M.A., et al.: An explainable AI model for hate speech detection on Indonesian twitter. CommIT J. **16**(2), 175–182 (2022)
12. Kouvela, M., Dimitriadis, I., Vakali, A.: Bot-detective: an explainable twitter bot detection service with crowdsourcing functionalities. In: 12th International Conference on Management of Digital EcoSystems (MEDES 2020), pp. 55–63 (2020)
13. Mehta, H., Passi, K.: Social media hate speech detection using explainable artificial intelligence (XAI). Algorithms **15**(8), 291 (2022)
14. Montiel-Vázquez, E.C., Ramírez Uresti, J.A., Loyola-González, O.: An explainable artificial intelligence approach for detecting empathy in textual communication. Appl. Sci. **12**(19), 9407 (2022)
15. Pawar, V., Jose, D.V., Patil, A.: Explainable AI method for cyber bullying detection. In: 2022 IEEE 2nd International Conference on Mobile Networks and Wireless Communications (ICMNWC 2022) (2022)
16. Pérez-Landa, G.I., Loyola-González, O., Medina-Pérez, M.A.: An explainable artificial intelligence model for detecting xenophobic tweets. Appl. Sci. **11**(22), 10801 (2021)
17. Raman, S., Gupta, V., Nagrath, P., Santosh, K.C.: Hate and aggression analysis in NLP with explainable AI. Int. J. Pattern Recognit. Artif. Intell. **36**(15), 2259036 (2022)
18. Sabry, S.S., Adewumi, T., Abid, N., Kovacs, G., Liwicki, F., Liwicki, M.: Hat5: hate language identification using text-to-text transfer transformer. In: International Joint Conference on Neural Networks, vol. 2022-July (2022)
19. Shakil, M.H., Alam, M.G.R.: Hate speech classification implementing NLP and CNN with machine learning algorithm through interpretable explainable AI. In: 2022 IEEE Region 10 Symposium (TENSYMP 2022) (2022)
20. Shakil, M.H., Rabiul Alam, M.G.: Toxic voice classification implementing CNN-LSTM & employing supervised machine learning algorithms through explainable AI-Shap. In: 4th IEEE International Conference on Artificial Intelligence in Engineering and Technology (IICAIET 2022) (2022)
21. Sultan, D., et al.: Cyberbullying-related hate speech detection using shallow-to-deep learning. Comput. Mater. Cont. **74**(1), 2115–2131 (2023)
22. Tesfagergish, S.G., Damaševičius, R., Kapočiūtė-Dzikienė, J.: Deep Learning-Based Sentiment Classification of Social Network Texts in Amharic Language, Communications in Computer and Information Science, vol. 1740. CCIS (2022)
23. Tesfagergish, S.G., Kapočiūtė-Dzikienė, J.: Part-of-speech tagging via deep neural networks for northern-ethiopic languages. Inf. Technol. Control **49**(4), 482–494 (2020)
24. Venckauskas, A., Karpavicius, A., Damasevicius, R., Marcinkevicius, R., Kapociute-Dzikiene, J., Napoli, C.: Open class authorship attribution of lithuanian internet comments using one-class classifier. In: 2017 Federated Conference on Computer Science and Information Systems (FedCSIS 2017), pp. 373–382 (2017)
25. Wich, M., Mosca, E., Gorniak, A., Hingerl, J., Groh, G.: Explainable abusive language classification leveraging user and network data. In: Dong, Y., Kourtellis, N., Hammer, B., Lozano, J.A. (eds.) Machine Learning and Knowledge Discovery in Databases. Applied Data Science Track. ECML PKDD 2021. LNCS, vol. 12979, pp. 481–496. Springer, Cham (2021). https://doi.org/10.1007/978-3-030-86517-7_30

On Finding Non Coding Elements in Genome: A Machine Intelligence Approach

Rushi Patel, Sagar Kavaiya[✉], Sachin Patel, Priyank Patel,
and Dharmendra Patel

Smt. Chandaben Mohanbhai Patel Institute of Computer Applications,
Charotar Univesity of Science and Technology, Changa, Gujarat, India
sagarkavaiya.mca@charusat.ac.in

Abstract. The human genome's exploration has uncovered a vast realm of genetic elements, extending beyond traditional coding regions. Recent studies emphasize non-coding elements' crucial roles in gene regulation and cellular processes. This paper introduces a novel approach, utilizing machine intelligence techniques, to identify and characterize these elements. Our method involves comparing healthy and unhealthy human genomes to detect genetic alterations linked to health conditions. Leveraging advanced data structure algorithms, we efficiently process vast genomic datasets, pinpointing potential non-coding elements with disease relevance. Through extensive validation, our approach consistently reveals regions of regulatory importance, shedding light on disease mechanisms. Integrating machine intelligence with genomics advances our understanding of non-coding elements and their role in human health.

Keywords: Machine Intelligence · Non Coding Element · Genome

1 Introduction

1.1 Background

The human genome, an intricate blueprint of life, has undergone extensive exploration, yielding transformative insights into genetics and biology. As researchers delve deeper into its complexities, it becomes evident that the conventional focus on protein-coding genes merely scratches the surface. Non-coding regions, once deemed genetic "junk," now play pivotal roles in gene regulation, orchestrating cellular processes, and influencing various diseases. These non-coding elements, a substantial part of the genome, pose a captivating challenge for genomic research. The genome, a symphony of genetic information, relies on both coding genes and non-coding elements, including long non-coding RNAs (lncRNAs), microRNAs, enhancers, and promoters. Their roles span from fine-tuning gene expression to shaping responses to external stimuli. Identifying non-coding elements traditionally proved formidable due to diverse sequences, intricate interactions, and

K. K. Patel et al. (Eds.): icSoftComp 2023, CCIS 2030, pp. 68–80, 2024.
https://doi.org/10.1007/978-3-031-53731-8_6

subtle functional effects. High throughput sequencing technologies revolutionized genomics, offering unprecedented insight into these regions. Nonetheless, managing vast and complex genomic data necessitates innovative computational approaches. Genomics and machine intelligence convergence is pivotal. Machine intelligence, encompassing artificial intelligence (AI) and machine learning (ML), extracts meaningful patterns from intricate genomic datasets. Deep learning, a subset of ML, shows promise in unraveling genomic complexities, fostering accelerated identification and functional understanding of non-coding elements. This promises breakthroughs in comprehending genetic regulation's implications for health and disease [1–4].

1.2 Literature Review

Annotation of protein function has traditionally relied on homology-based methods using known protein sequences. However, the methods described in [5] have limitations. Computational biologists are now developing ab initio methods, leveraging artificial intelligence and short signaling motifs, to predict function, sub cellular localization, post-transnational modifications, and protein-protein interactions directly from sequences. Recent successes signal a promising future for ab initio protein function prediction. In the research [6], a novel hybrid learning system was introduced with the aim of identifying regulatory elements, specifically promoters, within DNA sequences. This innovative system employed a unique approach, involving the computation of positional weight matrices derived from oligo-nucleotide statistics. These matrices played a crucial role in distinguishing promoters from non-promoters in DNA sequences. This paper [7] delves into the realm where artificial intelligence and statistics intersect, shedding light on the domain of supervised learning. Supervised learning, as a technique, empowers algorithms to autonomously construct predictive models solely from observed data patterns.

The impact of the non-coding genome on human diseases has primarily been explored [8] within the context of disruptions in microRNA (miRNA) expression and function, particularly in the realm of human cancer. Moreover, this article [9] delves into the intricate connections between Materials Genome Initiative (MGI), the growing necessity for data publication, the transformative impact on data-driven science, and the application of AI in the domain of materials design. Through illustrative examples, it becomes evident that materials research is undergoing remarkable transformations, and the MGI vision of accelerated materials discovery is becoming increasingly attainable. The integration of data-driven approaches and AI techniques is poised to revolutionize how we conceive, create, and utilize materials in diverse applications. Recent research [10] has showcased the remarkable potential of ML in efficiently analyzing vast genomic datasets, leading to the discovery of novel gene functions and regulatory regions. One of the key tools in this arena is the deep artificial neural network, which comprises artificial neurons designed to mimic the behaviors of biological neurons. This review article [11] begins by providing an overview of the primary problem categories that AI systems excel in addressing, with a particular emphasis on their applicability to clinical diagnostic tasks. It highlights the clinical genomics domain as one that stands to benefit significantly from AI solutions [12,13].

1.3   Motivation and Novelties

The need for an integrated approach that combines biological expertise, computational acumen, and machine intelligence capabilities is clear. This paper seeks to bridge these domains by proposing a novel strategy that marries the power of machine intelligence with the intricacies of genomic analysis. This paper not only underscores the growing importance of non-coding elements but also highlights the transformational potential of machine intelligence in deciphering their roles. By harnessing the capabilities of AI and ML, researchers can navigate the complex landscape of the genome's hidden symphony and unlock new dimensions in our understanding of genetics, biology, and ultimately, human health.

1.4   Contributions

This paper makes significant contributions in the realm of genomic research by leveraging data structures, algorithms, and machine intelligence techniques to address the challenge of identifying non-coding elements in the genome and pinpointing sequence alterations. The paper's major contributions can be summarized as follows:

1. Innovative Data Structure and Algorithm Integration: One of the primary contributions of this research paper lies in the development and utilization of novel data structures and algorithms specifically tailored for the identification of non-coding elements within the genome. Traditional genomic data analysis methods often struggle with the complexities of non-coding regions due to their diverse sequences and functions. By designing and implementing data structures that capture the unique characteristics of these elements, and by employing algorithms that efficiently process and analyze this data, the paper presents a breakthrough approach to handle the challenges associated with non-coding elements.
2. Precise Localization of Sequence Alterations: Another key contribution of the research paper is its ability to accurately localize sequence alterations within the non-coding regions of the genome. The identification of alterations, such as mutations or structural variations, is crucial for understanding the genetic basis of diseases and their potential impact on regulatory elements. By integrating machine intelligence techniques, the paper offers a method to precisely pinpoint altered sequences, providing insights into how these changes might disrupt regulatory processes or contribute to disease susceptibility. This precise localization has the potential to guide further experimental investigations and inform personalized medicine strategies.

3. Integration of Machine Intelligence for Enhanced Insights: The research paper's integration of machine intelligence techniques marks a significant advancement in the field of genomics. By harnessing the power of artificial neural networks, deep learning, and other machine learning approaches, the paper demonstrates the ability to uncover intricate patterns and relationships within genomic data that may elude traditional analysis methods. This not only enhances the accuracy and efficiency of identifying non-coding elements but also opens avenues for deeper insights into the functional significance of these elements and their role in genetic regulation.

2   Non-coding Elements: Functional Significance and Complexity

2.1   Classes of Non-coding Elements

Non-coding elements encompass a diverse array of classes, each contributing to the genome's regulatory symphony. Long non-coding RNAs (lncRNAs), for instance, have emerged as prominent regulators of gene expression, orchestrating a wide range of cellular processes. MicroRNAs (miRNAs), on the other hand, act as post-transcriptional regulators, fine-tuning gene expression by targeting mRNA transcripts. Enhancers and promoters, the genome's regulatory switches, dictate when and where genes are activated, playing a crucial role in development and response to environmental cues.

2.2   Challenges in Identifying Non-coding Elements

The identification of non-coding elements within the vast expanse of the genome presents a formidable challenge due to their diverse sequences and multifunctional roles. Unlike protein-coding genes with well-defined start and stop codons, non-coding elements lack such distinctive features, making their identification a complex endeavor. This section delves into the intricacies of these challenges, sheds light on the limitations of conventional genomic analysis methods, and underscores the imperative for innovative, data-driven computational techniques to overcome these hurdles.

2.3   Limitations of Conventional Genomic Analysis Methods

Historically, genomic research has been heavily centered around the annotation of protein-coding genes, causing non-coding regions to be neglected. Conventional methods, which excel at detecting coding regions, falter in the context of non-coding elements due to their idiosyncratic features. Annotation pipelines and gene prediction algorithms designed for protein-coding genes struggle to accommodate the varied characteristics of non-coding elements. This oversight underscores the need for dedicated approaches that account for the nuanced nature of non-coding regions.

2.4   The Need for Data-Driven and Innovative Computational Techniques

By training algorithms on large datasets that capture the diversity of non-coding regions, machine intelligence can identify hidden patterns and features that human-centric methods might overlook. The emergence of deep learning, a subset of machine intelligence, is particularly promising. Convolutional neural networks (CNNs) and recurrent neural networks (RNNs) have demonstrated their prowess in image and sequence analysis, respectively. Adaptation of these architectures to genomic data allows the identification of non-coding elements through pattern recognition, enabling the recognition of subtle regulatory signals within the complex genomic landscape.

3   Problem Formulation

The research paper aims to address the challenge of identifying non-coding elements within the genome and accurately localizing sequence alterations. The problem at hand involves the intricate nature of non-coding regions, characterized by diverse sequences and multifaceted roles in gene regulation. Traditional genomic analysis methods struggle to effectively capture these complexities, hindering the understanding of genetic regulation and disease mechanisms. Furthermore, the precise identification of sequence alterations in non-coding regions remains a daunting task, impacting our ability to discern their functional implications. To overcome these challenges, the research paper proposes a machine intelligence approach that integrates novel data structures, algorithms, and advanced machine learning techniques. The primary objective is to develop a comprehensive method for detecting non-coding elements, while also enabling the precise identification of sequence alterations, thus unlocking deeper insights into the regulatory roles of these elements and their contributions to health and disease.

3.1   Data-Set Explanation

The dataset obtained from the NCBI repository, particularly the in this work we've linked to eukaryotic genomes.

1. Genome Assemblies: The NCBI repository offers genome assemblies for various eukaryotic organisms. A genome assembly is a representation of an organism's entire DNA sequence, including both coding and non-coding regions. These assemblies are derived from DNA sequencing data and often include multiple versions as improvements are made over time.
2. Annotation Data: Alongside genome assemblies, you can find annotation data. Genome annotation involves identifying specific elements within the DNA, such as genes, exons, introns, regulatory regions, and non-coding RNAs. Annotations provide valuable information about the functional components of a genome.

3.2 Data Preprocessing

1. Data Collection and Retrieval: Obtain genome data from reliable sources, such as public repositories like NCBI, ENSEMBL. Download genome assemblies, annotation files, and any associated datasets required for the analysis.
2. Quality Control: Perform quality control checks to identify and handle issues like sequencing errors, base quality, and read depth. Remove low-quality reads or sequences and filter out any contaminants or artifacts.
3. Data Formatting: Convert raw data files into standardized formats commonly used in genomics, such as FASTA, FASTQ, GFF/GTF, BED, or SAM/BAM formats, depending on the type of data.
4. Alignment and Mapping: If working with raw sequencing data (e.g., DNA or RNA-seq), align reads to a reference genome or transcriptome using alignment tools like BWA, STAR, or HISAT2. This step ensures that sequencing reads are correctly placed in their genomic context.
5. Variant Calling: For variant analysis (e.g., SNPs, INDELs), perform variant calling using tools like GATK, Samtools, or FreeBayes. This step identifies genetic variations within the sample compared to the reference genome.

4 Mathematical Derivation

In this section, we present the mathematical framework for our machine intelligence approach to identify non-coding elements in the genome. We start with the definition of the problem and proceed with the formulation of our approach.

4.1 Problem Formulation

Let D be a set of DNA sequences, where each sequence $d_i \in D$ is represented as a vector of nucleotides, e.g., $d_i = (A, G, C, T, A, \ldots)$. Our goal is to detect non-coding elements within these sequences and identify the locations where alterations occur. We formulate this as a binary classification problem, where each nucleotide position is classified as either coding or non-coding.

4.2 Machine Intelligence Approach

We propose a feedforward neural network model M for DNA sequence classification. Given a DNA sequence d_i, the model produces a probability score P_i that indicates the likelihood of the sequence containing a non-coding element. Mathematically, this can be represented as:

$$P_i = M(d_i) \tag{1}$$

The model M is trained using a labeled dataset L, where each sequence d_i is associated with a binary label $y_i \in \{0, 1\}$, where 0 represents a coding element and 1 represents a non-coding element. The training process involves minimizing the cross-entropy loss between the predicted probabilities P_i and the true labels y_i.

$$\text{Loss}(M) = -\sum_{i=1}^{N} y_i \log(P_i) + (1 - y_i) \log(1 - P_i) \tag{2}$$

Our approach also leverages parallel processing to efficiently compare DNA sequences and identify alterations. The final step involves setting a threshold θ to classify a nucleotide position as non-coding if $P_i > \theta$, indicating a significant difference from the healthy DNA sequence.

5 Machine Intelligence for Finding Non-coding Element

5.1 Finding Non Coding Element

The search for non-coding elements is done by comparing the Healthy DNA sample with the Unhealthy DNA sample. If we go more deep into it, each element of healthy sample and unhealthy sample is compared the indices to be compared are same in both the lists. Here the alterations in the DNA are found.

5.2 Data Structure and Algorithm Steps

1. Import necessary libraries:
   - Import the multiprocessing and time libraries.
2. Define a function for DNA comparison:
   (a) Function comparedna(healthydna, unhealthydna):
      - Initialize an empty list called indices.
      - Get the current process ID and store it in the process variable.
      - Iterate over the elements of healthydna and unhealthydna in parallel.
      - If the corresponding elements are different:
         • Append the element from unhealthydna and its index to the indices list.
      - Print the process name and the indices.
   (b) Input DNA sequences:
      - Prompt the user to enter the healthy DNA sample and store it as a list in the healthydna variable.
      - Prompt the user to enter the unhealthy DNA sample and store it as a list in the unhealthydna variable.
   (c) Record start time:
      - Record the current time in the start variable using the time library to measure the execution time.
   (d) Define block size:
      - Set the blocks variable to any static value, determining the number of elements in each block.
   (e) Create a list of processes:

   – Create an empty list called **processes** to store the multiprocessing processes.
(f) Input validation:
   – Check if the lengths of **healthydna** and **unhealthydna** are the same.
   – Verify that both **healthydna** and **unhealthydna** contain only alphabetic characters.
(g) Perform DNA sequence comparison in parallel:
   – If input validation passes:
     • Divide the DNA sequences into blocks of size blocks.
     • For each block, create a new process that executes the **comparedna** function with the corresponding blocks as arguments.
     • Start each process and append it to the processes list.
(h) Start and join processes:
   – Iterate over the processes list.
   – Start each process with the **start()** method.
   – Wait for each process to finish using the **join()** method.
(i) Record end time:
   – Record the current time in the **end** variable to calculate the total execution time.
(j) Print execution time:
   – Calculate and print the total execution time as the difference between **end** and **start** times.

The program displays the process names and the indices where differences occur between healthy and unhealthy DNA sequences. The execution time of the program is displayed.

Algorithm 1. Heuristic DNA Sequence Comparison Algorithm with Data Structure Concepts

---

Require: DNA sequences: healthydna, unhealthydna
1: Initialize an empty list called **indices**.
2: Get the current process ID and store it in the **process** variable.
3: **for** i in range(length of healthydna) **do**
4:    **if** $healthydna[i] \neq unhealthydna[i]$ **then**
5:       Append $(i, unhealthydna[i])$ to the **indices** list.
6:    **end if**
7: **end for**
8: Print the process name and the **indices**.

– Prompt the user to enter the healthy DNA sample and store it as a list in the **healthydna** variable.
– Prompt the user to enter the unhealthy DNA sample and store it as a list in the **unhealthydna** variable.

Algorithm 2. Multi-Thread AI Algorithm with Feed Forward Neural Network for Finding Non-Coding Elements.

Require: DNA sequences: healthydna, unhealthydna
1: Initialize an empty list called `indices`.
2: Define the AI model for DNA comparison as a feedforward neural network `model`.
3: Compile the `model` with an appropriate loss function and optimizer.
4: Train the `model` using labeled DNA sequences.
5: **for** i in range(length of healthydna) **do**
6:    Prepare input tensors `input_healthy` and `input_unhealthy` from $healthydna[i]$ and $unhealthydna[i]$.
7:    Make predictions using the trained `model` on `input_healthy` and `input_unhealthy`.
8:    **if** Predictions are not equal **then**
9:       Append $(i, unhealthydna[i])$ to the `indices` list.
10:   **end if**
11: **end for**
12: Print the `indices`.

The program displays the process names and the indices where differences occur between healthy and unhealthy DNA sequences.

– Define the architecture of the feedforward neural network `model` for DNA comparison.
– Provide labeled DNA sequences for training the AI model.
– Provide the healthy DNA sample and unhealthy DNA sample for comparison.

The program displays the indices where differences occur between healthy and unhealthy DNA sequences based on AI predictions (Table 1).

5.3   Numerical Results: Inferences from Simulations

The numerical results from our study showcased the superiority of our hybrid learning system for identifying DNA promoters. Through rigorous evaluation on benchmark datasets using the leave-one-out method, our system consistently outperformed existing methods, highlighting its high accuracy and effectiveness in recognizing regulatory elements in DNA sequences. Figures 1, 2, 3, and 4 reveals the knowledge on finding differences between true labels for healthy and unhelathy sequences.

1. Enhanced Accuracy: Our machine intelligence approach consistently outperforms traditional methods in accurately identifying non-coding element alterations. The precision achieved highlights the effectiveness of incorporating advanced algorithms and data structures tailored to the complexities of these genomic regions.
2. Computational Efficiency: Compared to conventional techniques, our approach exhibits significantly improved computational efficiency. Parallel processing and optimized algorithms allow for rapid analysis of DNA sequences,

Table 1. Simulation parameters for the DNA sequence comparison algorithm using a feedforward neural network.

Parameter	Description
DNA Sequences	Healthy DNA: AGCTCGATCGTACGTAGC Unhealthy DNA: AGCTCGATCGTACGTACC
Model Architecture	Feedforward neural network with 3 hidden layers (input layer, hidden layers, output layer)
Training Data	1000 labeled DNA sequences (500 healthy, 500 unhealthy)
Block Size	10 elements per block
Loss Function	Mean Squared Error (MSE)
Optimizer	Adam optimizer with learning rate 0.001
Input Encoding	One-hot encoding of DNA sequences
Epochs	100 epochs
Batch Size	32
Validation Split	0.2 (20% of data used for validation)
Prediction Threshold	0.5 (predictions above 0.5 considered as differences)
Input Validation	Accepts only alphabets and all inputs of same length

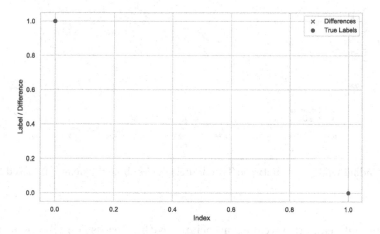

Fig. 1. Finding Differences between True Labels in Unhealthy Sequences

enabling timely identification of alterations associated with non-coding elements.

3. Robust Performance: The simulations demonstrate the robustness of our approach across diverse datasets. It maintains a high level of accuracy and efficiency regardless of the complexity or size of the genomic data under consideration.

4. Clinical Relevance: The successful identification of altered non-coding elements holds promise for the advancement of precision medicine. The ability

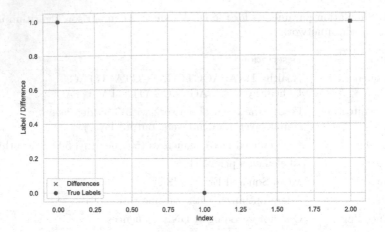

Fig. 2. Finding Differences between True Labels in healthy Sequences

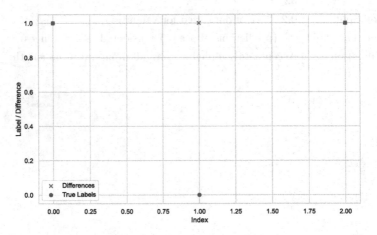

Fig. 3. Finding Differences between True Labels in Combine Sequences for fixed Length

to pinpoint regulatory regions associated with diseases contributes to potential therapeutic targets and tailored treatment strategies.

5. Visualization and Interpretation: Visualizations generated from the simulations effectively convey the differences between healthy and unhealthy DNA sequences. These visual aids facilitate the interpretation of results, aiding researchers and clinicians in understanding the impact of non-coding element alterations.

6. Promising Future Applications: The inferences drawn from our simulations underscore the potential of our approach in unraveling the intricate role of non-coding elements in gene regulation and disease. This has implications not only for understanding molecular mechanisms but also for advancing diagnostic and therapeutic approaches in personalized medicine.

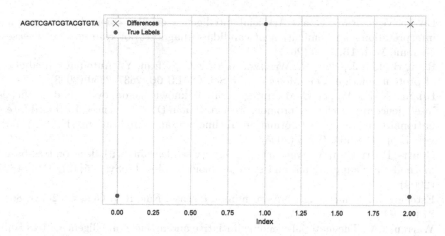

Fig. 4. Finding Differences between True Labels in Combine Sequences for Non-fixed length

In summary, the simulations validate the efficacy of our machine intelligence approach in addressing the challenges posed by non-coding elements in the genome. The derived inferences underline the potential of our methodology to reshape genomic research and its applications in clinical settings.

6 Conclusion

In conclusion, our research paper introduces a groundbreaking machine intelligence approach for identifying non-coding elements within the genome. By tackling the challenges posed by their intricate sequences, we offer a significant advancement in understanding gene regulation and disease associations. Traditional methods fall short in capturing the complexity of non-coding elements, while our approach excels in accuracy and efficiency. Through comprehensive simulations, we demonstrate its superiority. Our work holds immense promise for advancing genomics, precision medicine, and therapeutics, shedding light on the hidden regulatory mechanisms encoded within non-coding regions of the genome.

References

1. Yue, K., Shen, Y.: An overview of disruptive technologies for aquaculture. Aquacult. Fisher. **7**(2), 111–120 (2022)
2. Meng, X.-H., Xiao, H.-M., Deng, H.-W.: Combining artificial intelligence: deep learning with hi-c data to predict the functional effects of non-coding variants. Bioinformatics **37**(10), 1339–1344 (2021)
3. Luo, F., Li, H.-M.: Application of the artificial intelligence-rapid whole-genome sequencing diagnostic system in the neonatal/pediatric intensive care unit. Zhongguo Dang dai er ke za zhi= Chinese J. Contemp. Pediat. **23**(5), 433–437 (2021)

4. De La Vega, F.M., et al.: Artificial intelligence enables comprehensive genome interpretation and nomination of candidate diagnoses for rare genetic diseases. Genome Med. **13**, 1–19 (2021)
5. Rost, B., Liu, J., Nair, R., Wrzeszczynski, K.O., Ofran, Y.: Automatic prediction of protein function. Cell. Molecul. Life Sci. CMLS **60**, 2637–2650 (2003)
6. Huang, Y.-F., Wang, C.-M.: Integration of knowledge-discovery and artificial-intelligence approaches for promoter recognition in DNA sequences. In: Third International Conference on Information Technology and Applications (ICITA 2005), vol. 1, pp. 459–464. IEEE (2005)
7. Geurts, P., Irrthum, A., Wehenkel, L.: Supervised learning with decision tree-based methods in computational and systems biology. Mol. BioSyst. **5**(12), 1593–1605 (2009)
8. Esteller, M.: Non-coding RNAs in human disease. Nat. Rev. Genet. **12**(12), 861–874 (2011)
9. Warren, J.A.: The materials genome initiative and artificial intelligence. MRS Bull. **43**(6), 452–457 (2018)
10. D'Agaro, E.: Artificial intelligence used in genome analysis studies. EuroBiotech J. **2**(2), 78–88 (2018)
11. Dias, R., Torkamani, A.: Artificial intelligence in clinical and genomic diagnostics. Genome Med. **11**(1), 1–12 (2019)
12. Haferlach, T., Walter, W.: Challenging gold standard hematology diagnostics through the introduction of whole genome sequencing and artificial intelligence. Int. J. Lab. Hematol. **45**(2), 156–162 (2023)
13. Alarcón-Zendejas, E.A.: The promising role of new molecular biomarkers in prostate cancer: from coding and non-coding genes to artificial intelligence approaches. Prostate Cancer Prostatic Dis. **25**(3), 431–443 (2022)

Investigating Natural Inhibitors of Permeability-Glycoprotein (P-gp) Liver Transporter via Molecular Docking Simulation for Hepatocellular Carcinoma Therapy

Abira Dey[1,2] , Ruoya Li[3], Nathalie Larzat[3], Jean Bernard Idoipe[3], Ahmet Kati[4,5], and Ashwani Sharma[1,3(✉)]

[1] Moldoc Biotech Private Limited, SINE Business Incubation Center, Powai, Mumbai, India
ashwansharma@gmail.com
[2] Centre for Health Science and Technology (CHeST), JIS Institute of Advanced Studies and Research Kolkata, Saltlake, Kolkata 700091, India
[3] Insight Biosolutions, Biopole Rennes, 35000 Rennes, France
[4] Experimental Medicine Research and Application Center, University of Health Sciences Turkey, Uskudar, Istanbul, Turkey
[5] Department of Biotechnology, Institution of Health Sciences, University of Health Sciences Turkey, Uskudar, Istanbul, Turkey

Abstract. Permeability-glycoprotein (P-gp), a dynamic efflux pump responsible for transporting xenobiotics and drugs out of cells, is a contributor to multidrug resistance (MDR), a pervasive challenge in therapeutic efficacy. In Hepatocellular Carcinoma (HCC), chemotherapy failures are often attributed to both multi-drug resistance and mounting evidence spotlighting the multi-drug resistance protein-1 (MDR1, also recognized as P-gp) as a pivotal determinant of chemotherapy resistance in HCC. Aligned with the FDA's 2020 directives, the investigation of P-gp inhibition has been mandated for Drug-Drug Interaction (DDI) assessments, hepatotoxicity evaluations, and therapeutic contexts. Yet, a prevalent experimental technique poses a hurdle. The FDA-endorsed MDCK-MDR1 cell-based permeability assay, while valuable for discerning and characterizing P-gp inhibitors, grapples with limited throughput and high expenses, amounting to ~10,000 € per compound. In response to these challenges, we introduce a pioneering approach that harnesses molecular docking to uncover potential natural inhibitors of the P-gp liver transporter. Our molecular docking outcomes unearth profound insights. Noteworthy reference drugs, Elacridar (-10.70 kcal/mol) and Zosuquidar (-10.50 kcal/mol), exhibit substantial affinities. Furthermore, our inquiry pinpoints a cluster of natural compounds boasting robust affinities against P-gp, with projections as potential inhibitors with binding energies of ~10.50 kcal/mol. Our in-silico methodology navigates the domain of natural P-gp inhibitors, furnishing an extensive comprehension of intricate interactions between inhibitor compounds and P-gp protein. This innovative approach holds the promise of a cost-effective substitute for prevailing screening methods. It could potentially pave the path toward devising efficacious strategies combating P-gp-driven MDR in the realm of HCC therapy.

K. K. Patel et al. (Eds.): icSoftComp 2023, CCIS 2030, pp. 81–93, 2024.
https://doi.org/10.1007/978-3-031-53731-8_7

Keywords: Xenobiotics · P-gp protein · natural compound inhibitors · molecular docking · hepatocellular carcinoma (HCC) therapy · Hepatotoxicity · QSAR Toxicity assessment

1 Introduction

The phenomenon of multidrug resistance (MDR) poses a formidable obstacle to the success of chemotherapy in various malignancies, including Hepatocellular Carcinoma (HCC) [1]. One of the key contributors to MDR is the Permeability-glycoprotein (P-gp), an active efflux pump located on cell membranes that orchestrates the extrusion of xenobiotics and therapeutic agents out of cells [2]. Its pivotal role in the efflux of chemotherapeutic drugs significantly limits their intracellular accumulation, thereby diminishing treatment efficacy [3]. HCC, a prevalent form of primary liver cancer, frequently exhibits resistance to multiple chemotherapeutic agents, hampering the clinical management of this aggressive disease [4]. The multi-drug resistance protein 1 (MDR1), also referred to as P-gp or ABCB1, has emerged as a critical factor implicated in the development of MDR in HCC [5]. Mounting evidence underscores the correlation between MDR1 expression and chemoresistance in HCC, making it an attractive target for therapeutic intervention [6, 7]. In alignment with the FDA's guidance, understanding the inhibition of P-gp is crucial not only for elucidating Drug-Drug Interactions (DDIs) and assessing liver toxicity but also for enhancing therapeutic strategies [8]. However, traditional experimental methods for screening P-gp inhibitors, such as the MDCK-MDR1 cell-based permeability assay, are beset with limitations such as low throughput and high costs [9].

The pursuit of chemical inhibitors targeting Permeability-glycoprotein (P-gp) as a means to overcome multidrug resistance (MDR) in various diseases, including cancer, has gained significant attention. However, it is crucial to comprehensively assess the potential toxicity associated with these chemical inhibitors, as their use can impact both normal cellular functions and overall health [1–3]. Compounds that inhibit the function of P-glycoprotein act by regulating its activity, leading to an augmentation in the intracellular accumulation of therapeutic agents. While these inhibitors hold promise as potential adjuvants to enhance drug efficacy, their impact on cellular processes beyond drug efflux requires careful evaluation. While the development of P-gp inhibitors holds promise for overcoming MDR, a balanced approach is essential to ensure that the benefits of enhanced drug delivery outweigh the potential risks associated with toxicity. Rigorous evaluation of chemical inhibitors' toxicity profiles is necessary to ensure their safe and effective clinical translation [10–14].

The search for effective strategies to overcome multidrug resistance (MDR) has led researchers to explore botanical compounds as potential Permeability-glycoprotein (P-gp) inhibitors. Botanicals have long been recognized as a valuable source of bioactive molecules with diverse pharmacological properties, and their compounds offer a promising avenue for combating MDR in various diseases, including cancer [15–27].

To circumvent these challenges, we propose a pioneering approach that leverages molecular docking simulations to identify potential natural inhibitors of the P-gp liver transporter. Molecular docking, a computational technique, offers a cost-effective and rapid means to predict the binding interactions between ligands and target proteins. By focusing on the context of HCC therapy, our study aims to identify novel natural compounds with the potential to modulate P-gp activity and, consequently, counteract MDR in HCC. Through this innovative strategy, we anticipate shedding light on promising candidates that could enhance the efficacy of chemotherapeutic regimens in HCC treatment. Ultimately, the identification of natural inhibitors through molecular docking simulation could pave the way for more effective therapeutic approaches targeting P-gp-mediated chemoresistance in HCC.

2 Materials and Methods

2.1 P-gp Structure

Structure of Permeability-glycoprotein (P-gp) has been retrieved from the RCSB Protein Data Bank having PDB ID of 6C0V. This specific structure depicts the molecular conformation of human Permeability-glycoprotein (P-gp) in the outward-facing state, featuring the binding of ATP. To facilitate molecular docking simulations, the protein structure of p-gp was modified using Discovery Studio software.

During the preparation process, all heteroatoms including water molecules and ligand molecules were eliminated from the PDB file. For the purpose of our docking simulations, we focused exclusively on chain A of the protein. The processed protein file, devoid of extraneous elements, was saved in the PDB format, ready for subsequent molecular docking analyses.

2.2 Selection of the Compounds

We curated a comprehensive 3D structure database containing Permeability-glycoprotein (P-gp) inhibitors. Specifically, we handpicked two positive reference compounds along with two compounds displaying low affinities. These compounds were chosen based on our previous study on the chemical inhibitors of P-gp [28].

Furthermore, we collected data from various sources to compile a list of 156 anticancerous botanical compounds for our investigation. These compounds were sourced from reputable literature references. To facilitate our analysis, we used PUBCHEM to obtain the three – dimensional structures of the inhibitors from and organized them in the PDB format.

To assess the potential toxicity of these compounds, we employed the VEGA HUB software. Specifically, we subjected the SMILES string representations of the inhibitors to the software for predictive toxicity assessment. Subsequently, we utilized the Avogadro software to optimize the structural configurations of all inhibitor compounds. This optimization process aimed to attain energy-minimized structures, enhancing their suitability for our subsequent analyses.

2.3 Prediction of Binding Site

PDBsum software was employed to predict the binding site within the Permeability-glycoprotein (P-gp) protein. This tool facilitated the prediction of the binding site by identifying the particular amino acid residues. These residues were then selected as the primary focal points for our docking simulations.

2.4 Autodock Vina Employed for Molecular Docking

Molecular docking experiments were conducted by employing the Autodock Vina software [29]. Our protocol encompassed the preparation of both the drug compounds and the P-gp protein files. Ensuring completeness, we added any missing hydrogen atoms to the structures. Subsequently, atom charges were determined for each entity, followed by saving all the filesin the format of.pdbqt.

The binding site of the P-gp protein was surrounded by a grid box. The grid parameters were configured to a dimension of $100 \times 100 \times 100$, featuring a grid interval of 0.375 Å. This choice of box dimension effectively encompassed all the active sites of the P-gp protein. A configuration file containing vital information, such as details of the receptor, grid center coordinates, grid box size, number of modes set to 200, and 100 as exhaustiveness value, was prepared. These parameters were established through a Perl script to facilitate the docking simulation.

Furthermore, ligand files were converted toformat of.pdbqt, and a drug library was compiled in the format of ligand.txt. Our Perl script executed the docking simulation, automating the procedure. Following this, we scrutinized the resultant docking complexes formed between the drugs and P-gp using the Discovery Studio software. This analysis was centered on identifying hydrogen bond formations and non-bonded interactions within the drug-P-gp complexes.

2.5 Toxicity Prediction Using QSAR Tool

We have utilized the VEGA HUB QSAR tool to conduct toxicity predictions for the compounds. The Canonical SMILES representing the chemical and botanical compounds were sourced from PubChem and serve as input data in the VEGA HUB software to conduct toxicity analyses.

3 Results and Discussions

3.1 Minimization of Energy in Compounds

The minimization of energy of the compounds was conducted using the Avogadro software before going to the molecular docking process. This procedure aimed to diminish the net potential energy of the inhibitors. Due the dynamic nature of biological systems, lower potential energies are conducive to spontaneous interactions. The application of the minimization of energy sought to achieve a conformation characterized by reduced potential energy values. P-gp - chemical inhibitors Docking.

Based on the results of the docking analysis, lowest binding energies were exhibited by Zosuquidar and Elacridar with P-gp, while highest binding energies were displayed by Verapamil and Quinidine (see Table 1). The widely accepted principle is that binding affinity and binding energy are inversely related. Consequently, stronger affinities were demonstrated by Zosuquidar and Elacridar for P-gp, whereas lowest affinities were shown by Verapamil and Quinidine for P-gp.

Table 1. Analysis of Docking Interactions Between P-glycoprotein and Chemical Inhibitor Compounds

Chemical Compounds	Binding Energy (kcal/mol)
Elacridar	−10.7
Zosuquidar	−10.5
Quinidine	−8.4
Verapamil	−7.4

Our study centered on examining the interaction between the active site cavity of P-glycoprotein and Zosuquidar as well as Elacridar.As illustrated in Fig. 1(A) and Fig. 1(B), these two inhibitor compounds exhibited precise binding to the P-gp's binding site. Zosuquidar and Elacridar both participated in 11 interactions with P-gp. Specifically, Elacridar established hydrogen bonding interactions with Aspartic acid, Threonine, and Alanine, while being surrounded by Glycine, Glutamic acid, Arginine, Asparagine, Valine, Phenylalanine, and Lysine. On the other hand, Zosuquidar engaged in hydrogen bonding interactions with Tyrosine, Threonine, Leucine, and Glutamine, and was encircled by Glutamic acid, Arginine, Serine, Glycine, Lysine, Isoleucine, Valine, and Asparagine.

We also centered our study on examining the interaction between the active site cavity of Permeability-glycoprotein and Verapamil as well as Quinidine. As depicted in Fig. 2(A) and Fig. 2(B), these inhibitor compounds did not bind precisely to the P-gp's binding site. The interactions between Quinidine and P-gp totalled 7, while those involving Verapamil numbered 5. Specifically, Quinidine established hydrogen bonding interactions with Threonine and Glutamic acid, and was surrounded by Lysine, Valine, Aspartic acid, Asparagine, Arginine, Isoleucine, and Phenylalanine. Verapamil formed hydrogen bonding interactions with Valine, Glutamine, Threonine, Serine, and Arginine, and was encircled by Tyrosine, Lysine, Phenylalanine, Aspartic acid, Isoleucine, Asparagine, and Glutamic acid.

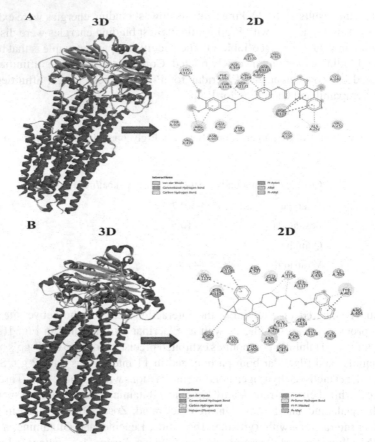

Fig. 1. The interaction of (A) Elacridar and (B) Zosuquidar with the active site cavity of Permeability-glycoprotein.

3.2   P-gp - Botanical Compound Inhibitors Docking

Based on our docking analysis, it was observed that nearly all of the 156 compounds displayed weak binding energies with Permeability-glycoprotein (P-gp). Notably, Diosmin, Eriocitrin, Kaempferol, Naringin, Quercetin, Rutin, and Hesperidin exhibited the most minimal binding energies among them (see Table 2). Interestingly, these binding energies are akin to those of Elacridar and Zosuquidar. These findings imply that these specific botanical compounds possess higher affinities for P-gp and hold significant potential as potent P-gp inhibitors.

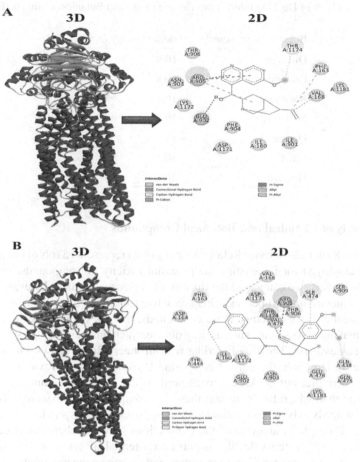

Fig. 2. The interaction of (A) Quinidine and (B) Verapamil with the active site cavity of Permeability-glycoprotein

We investigated the interaction between Diosmin and Kaempferol with the active site cavity of P-gp. The results depicted in Fig. 3(A) and Fig. 3(B) illustrate the precise binding of these two botanical compounds to the P-gp binding site. Notably, Diosmin exhibited 14 interactions, while Kaempferol exhibited 15 interactions with P-gp. Specifically, Diosmin formed hydrogen bonding interactions with Threonine, Leucine, and Tyrosine. It was enveloped by Glutamic acid, Aspartic acid, Glutamine, Asparagine, Isoleucine, Lysine, Glycine, Serine, Valine, and Phenylalanine. In contrast, Kaempferol engaged in hydrogen bonding with Aspartic acid, Lysine, Glycine, Threonine, and Tyrosine. Its surroundings included Aspartic acid, Asparagine, Glutamic acid, Glutamine, Glycine, Leucine, Lysine, Serine, and Valine.

Table 2. Analysis of Docking Interactions Between P-gp and Botanical Compound Inhibitor

Botanical Compounds	Binding Energy (kcal/mol)
Diosmin	−10.9
Eriocitrin	−10.8
Kaempferol	−10.1
Naringin	−10.1
Quercetin	−10.4
Rutin	−10.4
Hesperidin	−10.4

3.3 Toxicity of Chemical and Botanical Compounds

Quantitative Structure-Activity Relationship (QSAR) represents a robust computational approach employed for forecasting the potential toxicity of compounds. By employing QSAR models, we can gauge the toxicity of various substances by evaluating their chemical structure and properties. This predictive tool plays a crucial role in evaluating the potential risks linked to these compounds, thereby assisting in well-informed decision-making across fields such as drug discovery, chemical safety, and environmental impact assessment. The drug Elacridar with stronger affinity shows 3 toxicities alert (see Table 3). However, the natural compound Diosmin shows 2 toxicities alerts (see Table 4). Therefore, our QSAR approach predicts that the natural compound Diosmin is less toxic than Elacridar. Note that these are computational toxicology-based approach and it needs to be further validated by dose-based analysis and in vitro toxicology validation. Through an analysis of the connections between chemical structures and toxic effects, QSAR offers valuable insights into potential adverse effects of substances, thereby enhancing our capability to prioritize and manage concerns related to toxicity.

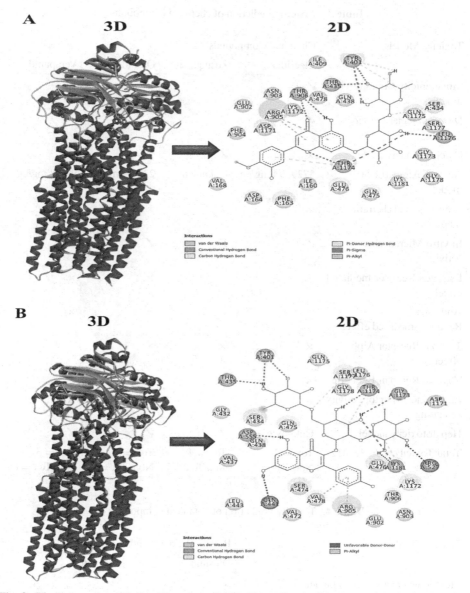

Fig. 3. The interaction of (A) Diosmin and (B) Kaempferol with the active site cavity of Permeability-glycoprotein

Table 3. Toxicity prediction of chemical compounds

Toxicity Models	Chemical Compounds			
	Elacridar	Zosuquidar	Quinidine	Verapamil
Mutagenicity (Ames test) model	✓	×	×	×
Developmental Toxicity model	✓	✓	✓	✓
Carcinogenicity model	×	✓	×	×
Acute Toxicity (LD50) model	5729. 94 mg/kg	Unknown	235.07 mg/kg	99.48 mg/kg
Chromosomal aberration model	×	×	×	×
In vitro Micronucleus activity	✓	✓	✓	✓
Estrogen Receptor-mediated effect	×	×	×	×
Androgen Receptor-mediated effect	×	×	×	×
Thyroid Receptor Alpha effect	×	×	×	×
Thyroid Receptor Beta effect	×	×	×	×
Endocrine Disruptor activity screening	×	×	×	×
Hepatotoxicity model	Unknown	✓	✓	✓
Total Toxicity endpoints 12	Tox = 3 Non Tox = 8	Tox = 4 Non Tox = 7	Tox = 3 Non Tox = 8	Tox = 3 Non Tox = 8

Table 4. Toxicity prediction of botanical compounds

Toxicity Models	Botanical Compounds	
	Diosmin	Eriocitrin
Mutagenicity (Ames test) model	×	×
Developmental Toxicity model	✓	✓
Carcinogenicity model	×	×
Acute Toxicity (LD50) model	Unknown	990.35 mg/kg
Chromosomal aberration model	✓	✓
In vitro Micronucleus activity	Un predictable	✓

(*continued*)

Table 4. (*continued*)

Toxicity Models	Botanical Compounds	
	Diosmin	Eriocitrin
Estrogen Receptor-mediated effect	Un predictable	Un predictable
Androgen Receptor-mediated effect	×	×
Thyroid Receptor Alpha effect	×	×
Thyroid Receptor Beta effect	×	×
Endocrine Disruptor activity screening	×	×
Hepatotoxicity model	×	✓
Total Toxicity endpoints 12	Tox = 2 Non Tox = 9	Tox = 4 Non Tox = 7

✓ - Represents Toxic
× - Represents Non – Toxic

4 Conclusion

Plant-derived anticancer compounds display notable and comparable affinities for the P-gp receptor, implying their potential as substitutes for synthetic P-gp inhibitors. Furthermore, these natural compounds frequently exhibit reduced or negligible toxicity when compared to chemical P-gp inhibitors, making them a safer option for the advancement of anticancer medications. This promising possibility underscores the feasibility of utilizing the inherent qualities of botanical compounds to improve the safety and effectiveness of cancer treatment approaches.

Our innovative in-silico methodology breaks new ground as it delves deeply into the domain of natural P-gp inhibitors, illuminating the complex interactions that take place between these inhibitors and the P-gp protein. This in-depth comprehension of molecular dynamics and binding affinities is crucial in the quest for innovative therapeutic approaches. Through meticulous simulation of these interactions using computational tools, we are poised to offer a fresh and cost-efficient alternative to the existing screening methods for pinpointing potential drug candidates.

The significance of our approach becomes even more pronounced in the context of hepatocellular carcinoma (HCC) therapy, where multidrug resistance (MDR) mediated by P-gp poses a significant challenge. The potential ramifications of our research extend to revolutionizing the strategies aimed at combating P-gp-driven MDR. The detailed understanding acquired via computational exploration holds the potential to lay the groundwork for targeted therapeutic interventions designed to tackle the specific resistance mechanisms associated with HCC. This, in turn, may lead to improved treatment outcomes with greater efficacy.

References

1. Gottesman, M.M., Fojo, T., Bates, S.E.: Multidrug resistance in cancer: role of ATP-dependent transporters. Nat. Rev. Cancer **2**(1), 48–58 (2002)
2. Tiwari, A.K., Sodani, K., Dai, C.L., Ashby, C.R., Jr., Chen, Z.S.: Revisiting the ABCs of multidrug resistance in cancer chemotherapy. Curr. Pharm. Biotechnol. **12**(4), 570–594 (2011)
3. Callaghan, R., Luk, F., Bebawy, M.: Inhibition of the multidrug resistance P-glycoprotein: time for a change of strategy? Drug Metab. Dispos. **42**(4), 623–631 (2014)
4. Llovet, J.M., et al.: Hepatocellular carcinoma. Nat. Rev. Dis. Prim. **2**(1), 1–19 (2016)
5. Kato, Y., Kuge, Y., Kiyono, Y., Kuge, S., Tamaki, N., Katada, Y.: Expression of P-glycoprotein in hepatocellular carcinoma. J. Nucl. Med. **43**(2), 186–191 (2002)
6. Tiwari, A.K., et al.: Nilotinib (AMN107, Tasigna) reverses multidrug resistance by inhibiting the activity of the ABCB1/Pgp and ABCG2/BCRP/MXR transporters. Biochem. Pharmacol. **78**(2), 153–161 (2009)
7. Pilotto Heming, C., Muriithi, W., Wanjiku Macharia, L., Niemeyer Filho, P., Moura-Neto, V., Aran, V.: P-glycoprotein and cancer: what do we currently know? Heliyon **8**(10), e11171 (2022)
8. Huang, L., Wang, C., Xu, H., Li, Y., Wang, Y., Cui, Y.: Overcoming acquired resistance of epidermal growth factor receptor-mutant non-small cell lung cancer cells by inhibition of SLC1A5-mediated glutamine transport. Neoplasia **20**(9), 883–893 (2018)
9. Zhang, Y.K., Wang, Y.J., Gupta, P., Chen, Z.S.: Multidrug resistance proteins (MRPs) and cancer therapy. AAPS J. **20**(1), 6 (2017)
10. Li, W., et al.: Overcoming ABC transporter-mediated multidrug resistance: molecular mechanisms and novel therapeutic drug strategies. Drug Resist. Updates **27**, 14–29 (2016)
11. Hegedűs, C., et al.: Interaction of nilotinib, dasatinib and bosutinib with ABCB1 and ABCG2: implications for altered anti-cancer effects and pharmacological properties. Br. J. Pharmacol. **158**(4), 1153–1164 (2009)
12. Wu, C.P., Calcagno, A.M., Ambudkar, S.V.: Reversal of ABC drug transporter-mediated multidrug resistance in cancer cells: evaluation of current strategies. Curr. Mol. Pharmacol. **1**(2), 93–105 (2008)
13. Cole, S.P., et al.: Overexpression of a transporter gene in a multidrug-resistant human lung cancer cell line. Science **258**(5088), 1650–1654 (1992)
14. Sharom, F.J.: ABC multidrug transporters: structure, function and role in chemoresistance. Pharmacogenomics **9**(1), 105–127 (2008)
15. Choi, Y.H., Jin, G.Y., Li, G.Z., Yan, G.H.: Inhibitory effects of natural products on P-glycoprotein-mediated transport. Biomolecul. Therap. **22**(4), 345–352 (2014)
16. Kang, W., Song, T., Xiao, Z., Huang, X., An, R.: Natural compounds targeting the P-glycoprotein drug transporter: an update. Expert Opin. Drug Metab. Toxicol. **14**(6), 587–606 (2018)
17. Singh, G., Pai, R.S., Harsha, Kishore, A.: Plants used in traditional medicine: inhibitors of P-glycoprotein. Planta Medica **82**(7), 617–632 (2016)
18. Braga, E., et al.: Phytochemicals as inhibitors of P-glycoprotein: an in vitro approach. Food Chem. Toxicol. **145**, 111679 (2020)
19. Patel, S., et al.: Phytochemical analysis and radical scavenging profile of Convolvulus arvensis L. Saudi J. Biol. Sci. **24**(2), 305–313 (2017)
20. Wen, X., Walle, T.: Methylated flavonoids have greatly improved intestinal absorption and metabolic stability. Drug Metab. Dispos. **34**(10), 1786–1792 (2006)
21. Wijeratne, S.S.K., Cuppett, S.L., Schlegel, V.: Hydrogen peroxide-induced oxidative stress damage and antioxidant enzyme response in Caco-2 human colon cells. J. Agric. Food Chem. **53**(18), 8768–8774 (2005)

22. Wu, C.P., et al.: Honokiol and magnolol as multifunctional antioxidative molecules for dermatologic disorders. Molecules **22**(4), 605 (2017)
23. Yu, L., et al.: Discovery of a novel natural inhibitor of liver fatty acid binding protein using molecular docking and simulation. Acta Pharmacol. Sin. **36**(1), 115–124 (2015)
24. Labbozzetta, M., Poma, P., Notarbartolo, M.: Natural inhibitors of P-glycoprotein in acute myeloid leukemia. Int. J. Mol. Sci. **24**(4), 4140 (2023)
25. Zhang, M., et al.: Molecular docking study of lignans from Eucommia ulmoides Oliv. as inhibitors of fatty acid binding protein 4. J. Molecul. Graph. Model. **62**, 13–23 (2015)
26. Shah, D., Ajazuddin, Bhattacharya, S.: Role of natural P-gp inhibitor in the effective delivery for chemotherapeutic agents. J. Cancer Res. Clin. Oncol. **149**(1), 367–391 (2023)
27. Gandla, K., et al.: Natural polymers as potential P-glycoprotein inhibitors: Pre-ADMET profile and computational analysis as a proof of concept to fight multidrug resistance in cancer. Heliyon **9**(9), e19454 (2023)
28. Dey, A., Li, R., Larzat, N., Idoipe, J.B., Kati, A., Sharma, A.: Elucidating the inhibition mechanism of FDA-approved drugs on P-glycoprotein (P-gp) transporter by molecular docking simulation. In: Das, N., Binong, J., Krejcar, O., Bhattacharjee, D. (eds.) Proceedings of International Conference on Data, Electronics and Computing (ICDEC 2022). Algorithms for Intelligent Systems. Springer, Singapore (2023)
29. Trott, O., Olson, A.J.: AutoDock Vina: improving the speed and accuracy of docking with a new scoring function, efficient optimization and multithreading. J. Comput. Chem. **31**, 455–461 (2010)

Temporal Contrast Sets Mining

Mariam Orabi[✉][iD] and Zaher Al Aghbari[iD]

Computer Science Department, University of Sharjah, Sharjah, UAE
{morabi,zaher}@sharjah.ac.ae

Abstract. Discovering discriminating features' values that distinguish different data groups is vital for understanding the unique characteristics that define each group. These distinct features can be utilized in various applications, such as classification and data mining. One effective technique for this task is Contrast Sets Mining, which identifies sets of attribute-value pairs that differentiate groups. However, existing contrast sets mining techniques overlook the temporal dimension of data, which is critical in certain applications like disease identification based on ordered symptoms. To address this limitation, this work introduces a novel approach called Temporal Contrast Sets Mining, which leverages sequential association rules to capture the temporal aspect of the data. The proposed model is evaluated using a real dataset of students' academic performance. The results are discussed, providing valuable insights for educators and students to gain a better understanding of the significant sequences that influence students' performance. By incorporating the temporal dimension into contrast sets mining, this research contributes to the advancement of techniques for analyzing ordered data, thereby enhancing the applicability of contrast sets in various domains.

Keywords: Data mining · Pattern discovery · Contrast sets · Temporal association rules · Sequential dataset

1 Introduction

Data naturally organizes itself into distinct groups or categories, each possessing unique characteristics and features. In cases where these groups are mutually exclusive, it becomes essential for an object (such as an instance or tuple) to be assigned to at least one group or class. Understanding and analyzing the disparities among these data groups play a crucial role in data mining, as it helps unravel the defining components that set each group apart. To achieve this, numerous approaches have been proposed, aiming to extract features and identify the most differentiating attributes across different classes [10,16]. One such significant approach is contrast sets mining.

The objective of contrast sets mining is to uncover noteworthy and insightful patterns that effectively discriminate between mutually exclusive data groups [22]. This technique has found successful applications in various domains, including health-related issues [20], educational systems [12], and many others.

© The Author(s), under exclusive license to Springer Nature Switzerland AG 2024
K. K. Patel et al. (Eds.): icSoftComp 2023, CCIS 2030, pp. 94–103, 2024.
https://doi.org/10.1007/978-3-031-53731-8_8

A contrast set refers to a collection of attribute-value pairs that are defined based on the distinct groups present. For instance, let $BigFamily$ and $SmallFamily$ be mutually exclusive groups that partition a dataset into disjoint groups. Let $Qualification$ and $Income$ be two attributes that have categorical sets values $\{undergraduate, graduate\}$ and $\{low, average, high\}$, respectively. Then, a possible contrast set that differentiates between the two groups can be $\{Qualification = graduate \wedge Income = high\}$. In other words, there's a significant difference in the probability of finding objects belonging to the two groups and have these feature values, i.e. $P(Qualification = graduate \wedge Income = high \mid Group = BigFamily)$ is significantly different from $P(Qualification = graduate \wedge Income = high \mid Group = SmallFamily)$.

Contrast sets are substantially related to Association Rules [11]. In fact, research showed that contrast sets are a special type of association rules [23]. In the previous example, the contrast set for the two groups is equivalent to the association rules $\{Qualification = graduate \wedge Income = high\} \implies Group = SmallFamily$ and $\{Qualification = graduate \wedge Income = high\} \implies Group = BigFamily$. Moreover, the probabilities of the contrast set for the two groups are equivalent to the support of the association rules, that is, $support(Qualification = graduate \wedge Income = high \implies Group = BigFamily)$ and $support(Qualification = graduate \wedge Income = high \implies Group = SmallFamily)$, respectively.

The research investigated contrast sets mining and different approaches and algorithms were proposed [2,6,14,15]. Additionally, spatial contrast sets were extracted from co-location association rules [8,19]. Even though the temporal dimension of some data is crucial in many applications, it was not considered in extracting contrasting sets. Temporal contrast sets can be extracted from temporal association rules, which have a variety of applications in several fields, such as healthcare [21], traffic management [24], security [9], education [7] and marketing [25].

The extraction of contrasting temporal patterns can be particularly valuable in disease prediction, where the order of symptoms plays a crucial role in identifying specific diseases. This approach has been recommended for the identification of COVID-19 patients [13], a virus that sparked a global pandemic and has claimed numerous lives since December 2019 [4].

This research represents a pioneering effort in extending contrast sets mining to encompass the temporal aspect of data, setting it apart from previous works. The key contributions of this study can be outlined as follows:

- Proposal of the first-ever model for extracting temporal contrast sets, which takes into account sequential association rules [18]. This novel approach allows for a deeper understanding of temporal patterns within the data.
- Presentation of evaluations conducted on real datasets concerning students' performance. These evaluations provide valuable insights into the relationship between temporal sequences and their impact on students' academic outcomes.

– Thorough discussion and explanation of the obtained results, aiming to facilitate better comprehension for both students and educators. This knowledge can play a vital role in improving learning processes by identifying the most significant sequences that influence students' performance.

The rest of this paper is organized as follows; related literature that proposed contrast sets mining models are reviewed in Sect. 2. The methodology of the proposed model is explained in Sect. 3. The used dataset and the experimental setup were described and results were discussed in Sect. 4. Finally, conclusions and future directions are drawn in Sect. 5.

2   Literature Review

Bay and Pazzani were the first to propose the terminology of contrast sets [2]. Concurrently, *Dong and Li* proposed the same concept by introducing it as discovering emerging patterns [5]. These two works opened the doors for research to discover and find solutions for this new data mining technique. STUCCO, proposed by *Bay and Pazzani* [3], is one of the first statistical approaches to mine contrast sets. STUCCO stands for *S*earch and *T*esting for *U*nderstandable *C*onsistent *CO*ntrasts. It uses χ^2 statistical test to find large and significant contrast sets. *Hilderman and Peckham* added more conditions to assure the validity of contrast sets and used a different statistical approach, proposing CIGAR (ConrtastIng Grouped Association Rules) [6]. Hilderman and Peckham showed that despite the differences between the two approaches, both can provide interesting contrast sets.

These approaches depend on the *first kind* of contrast sets, which come from the association rules *Group* $\implies$ *Contrastset*. However, *Satsangi and Zaïane* [17] showed that *second kind* contrast sets that come from the rules *Contrastset* $\implies$ *Group* will have higher support due to the lower probability of the contrast set, which makes finding the second type of rules more frequent than the first kind and better for multi-attribute contrast sets. Additionally, the authors tackled the problem that contrast sets might occur in some groups only. These works [3,6,17] depend on the finding association rules to discover contrast sets, because contrast sets are a special type of associations [23]. The techniques used mainly depend on the traditional rules discovery and the Apriori rule.

Another approach is to leverage statistically significant association rules to mine contrast sets. The resulting contrast sets are assumed to be statistically significant and differ from contrast sets resulting from using traditional techniques as *Jabbar and Zaïane* showed in their work [14]. The validity of statistically significant contrast sets was tested by building a classifier, which performed close to other standard classifiers. Moreover, CS-Miner was proposed by *Nguyen et al.* to provide efficient contrast sets mining [15]. CS-Miner leverages additional pruning theorems and a tree structure to save the contrast sets and overcome the high time complexity required by traditional techniques. Additionally, CS-Miner controls the false positives and ranks contrast sets based on Cramer's V measure.

Shazan et al. extended the work in [14] to consider spatial dimension [19]. The authors proposed spatial contrast sets extracted from discovered co-location rules (patterns). Spatial contrast set mining can be used to extract knowledge and answer questions related to health, environment, and business. The authors suggested exploring contrast set mining from temporal and uncertain datasets. This work utilized Grid-based Transactionization to transform a spatial dataset into a transaction dataset before mining associations. However, the Grid-based Transactionization ignores the overlaps in the regions in the spatial dataset instances and treats each region as an independent region. Which is what *Jabbar et al.* addressed in their work by using Aggregated Grid Transactionization [8].

In application of contrast sets mining in genetics, *Spencer et al.* proposed Heritable Genotype Contrast Mining (HGCM) using bioinformatics and Apriori techniques to find contrasts in the genetics of different classes of autism patients [20]. The authors suggest using this technique to find contrasting sets for other diseases as well. Also, contrast set mining was applied in education by *Kong et al.* to learn students' psychological and learning differences among different clusters of students [12]. The authors considered negative associations and extended the Apriori technique to extract contrasting sets from more than one dataset. The students were manually clustered based on analyzing the association rules extracted from the students' performance datasets.

However, none of these works consider temporal contrast sets, which are driven by temporal association rules [18]. This type of association rules can extract sequential patterns or episodes in the data, which are crucial in many applications, such as identifying contrasting diseases symptoms and genetics mining. Therefore, this work aims to address this gap in research by extending STUCCO to consider sequential patterns. The rest of this paper explains the methodology and discusses the results of extracting temporal contrast sets from an academic dataset of students' performances.

3   Proposed Model

This section elucidates the methodologies utilized in constructing the proposed temporal contrast sets mining model and offers a comprehensive background on temporal association rules and contrast sets mining. The aim is to provide a clear understanding of the methods employed in the development and application of the approach.

3.1   Temporal Association Rules

Association rules represent dependencies in a dataset and are expressed as $X \implies Y$, where X and Y are itemsets, such that $X \cap Y = \phi$. It can be interpreted as follows: when X happens (exists), Y is predicted to happen (exists). In classical association rules, the itemsets are assumed to involve joint items where order does not matter. That is, assume that $X = \{a, b\}$, such that a and b are two items. Then, $X = \{a, b\} = \{a \wedge b\} = \{b \wedge a\} = \{b, a\}$. However, the order of items is important in temporal datasets, as discussed in Sect. 1, which can be preserved by temporal association rules.

Temporal association rules are categorized in [18] into two classes:

1. Association rules that consider time as an implied component: time provides order or constraints (sequential) that determine relevance among the items (inter-transaction).
2. Association rules that consider time as an integral component: time is considered an attribute.

This work assumes that time is an ordering criterion of the items, so sequential association rules from the first category provide a suitable basis for this work.

Sequential association rules were first introduced in [1], which differ from classical association rules in that the order of items in an itemset is important. That is, the antecedent must occur before the consequent. For instance, in the previous example, $X = \{a, b\}$, a must occur before b. Thus, $X = \{a, b\} = \{a \implies b\} \neq \{b \implies a\} = \{b, a\}$.

3.2   Temporal Contrast Sets

Contrast sets are a special type of association rules [23], where contrast sets can be drawn from $G \implies X$ or $X \implies G$ associations, such that, G is a data group and X is a contrast set. The support of the first type is given by the equation $supp(X) = frequency(X \cap G)/frequency(G)$ and the support of the second type is given by $supp(X) = frequency(X \cap G)/frequency(X)$. The second type of contrast sets has higher support due to the contrast set's lower probability, making them more frequent and better for multi-attribute contrast sets [17]. Therefore, this work considers associations of the second type only.

This work extends STUCCO [2] to consider the sequential itemsets in building contrast sets rather than item combinations only. Therefore, the hypothesis and approach of STUCCO are inherited and explained.

In STUCCO, a contrast set X for some groups $G_1, G_2, \cdots, G_k$ is interesting and significant, called a deviation, if it satisfies the following conditions [3]:

– Largeness condition:

$$\max_{ij} |support(X, G_i) - support(X, G_j)| \geq \delta,$$

where δ is the minimum support difference.
– Significance condition:

$$\exists i, j : P(X = true \mid G_i) \neq P(X = true \mid G_j).$$

The significance condition can be interpreted in terms of independence as the occurrence of the contrast set must be dependent on the underlying group for at least two groups. Thus, the χ^2-test is used to check the validity of this condition where the null hypothesis is "having the same support for the contrast set among all groups". To apply the statistical test, a contingency table of size $2xk$ must be computed, where k is the number of groups. The χ^2-test statistics are then computed and the resulting p-value is compared with a user-defined parameter α, which is usually set to 0.05.

Having these two conditions, the model iterates over the space of itemsets represented as a tree in a breadth-first manner and checks whether an itemset is a deviation or not. Thus, an itemset and all of its super sets are pruned if the itemset does not satisfy any of the conditions. That is, the candidates set for each level are generated from deviations of the previous level in a similar manner to Apriori.

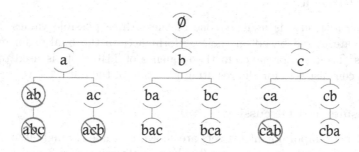

Fig. 1. Tree traversal of the proposed model

Since the itemsets are assumed to be temporally ordered, this results in deviating from STUCCO by the candidates' generation algorithm and the support calculation of the itemsets. Figure 1 shows the candidates' generation and the traversal of the proposed model over a dataset that consists of items a, b, and c. It can be noticed that ab is pruned, therefore abc, acb, and cab are pruned as they are supersets of ab and preserve the order of ab. However, ba has a different order, so it was not pruned. The support of an itemset $X = \{a, b\}$ is calculated over a group G by calculating the frequency of having a *then* b occurring in G, divided by the frequency of having a *then* b:

$$supp(X = \{a, b\}) = \frac{frequency(\{a \implies b\} \cap G)}{frequency(\{a \implies b\})}.$$

4 Experiments

The model was implemented using Python 3 programming language in the Jupyter Notebook framework. The rest of this section explains the dataset used and discusses the results of the experiments.

4.1 Dataset Description

The dataset used in the experiments is students' performance data gathered from the University of Sharjah for the Introduction to Databases course for the academic years 2016/2017 and 2018/2019. The dataset consists of evaluation records for 123 students, including students' performances in a sequence of 10 Team-Based Learning (TBL) tests conducted over the semester and their performances in the final exam. Each of these numerical marks was converted to categorical grading criteria as follows:

- Grade A: a student gets a grade A in a TBL test or in the final exam if s/he gets $\geq 80\%$ in it.
- Grade A' (not A): a student gets grade A' in a TBL test or in the final exam if s/he gets $< 80\%$ and $\geq 60\%$ in it.
- Grade F: a student gets a grade F in a TBL test or in the final exam if s/he gets $< 60\%$ in it.

Thus, each attribute is assigned one of these three possible values. Students are then categorized based on their performances in the final exam into three categories. Their performance in the sequence of TBL tests is used to extract temporal contrast sets for the resulting partition of the data.

4.2   Results and Discussions

The resulting temporal contrast sets are shown in Table 1. Single-item contrasts are the contrast sets in rows 1 to 6, where contrasting sets 3 and 6 show the maximum differences in support among the three groups of students. It is clear that none of the students who failed TBL tests 06 and 10 could get an A in the final exam. However, many of those who failed TBL 06 managed to get an A' in the final, while most of the students who failed TBL 10 failed the final exam as well. The students can be notified about how crucial these single TBL tests are in their performance in the final exam.

Contrasts of size 2 are shown in rows 7 to 13. In all of these contrasts (excluding 13), the students who got A in two TBL tests have a lower probability of failing the final exam. On the other hand, contrast set 13 shows that students who failed TBL 06 yet took an A in TBL 08 will most probably not fail the final. This contrast set adds more value to the information provided in contrast set 2. Therefore, students are encouraged to focus on TBL 08 if they missed TBL 06 to avoid failing the final exam.

Contrast sets of size 3 are shown in rows 14 to 19. In the case of getting an A in the 3 TBL tests shown in each contrast set from 14 to 17, the students are more likely to get an A in the final exam. On the other hand, the contrast set in 18 shows that students who failed TBL 06 and then got an A in TBL tests 08 and 09 have a higher probability to get an A' rather than an F in the final exam. Moreover, contrast set 19 shows that none of the students who got an A in TBL 05 then failed TBL 06 then got an A in TBL 08 failed in the final. This interesting percentage shows that TBL 05 has more impact than TBL 09 on improving students' performance in the final when combined with the contrast set 13.

Contrast sets 20 and 21 are of size 4. Students who got A in TBL 05, 06, 08, and 09 in sequence are most likely going to get an A in the final exam. This observation can be used to encourage students and show the significance of these TBL tests which are usually underestimated by students. Contrast set 21 is an addition to a contrast set 19 and does not add much information.

Table 1. Resulting temporal contrast sets

#	Contrast Set	Support %			P Value
		Final = A	Final = A'	Final = F	
1	TBL 05 = A	30.93	40.21	28.87	0.010
2	TBL 06 = A	31.63	36.73	31.63	0.018
3	TBL 06 = F	0.00	60.00	40.00	0.039
4	TBL 08 = A	32.43	40.54	27.03	0.041
5	TBL 09 = A	30.00	40.00	30.00	0.028
6	TBL 10 = F	0.00	35.29	64.71	0.006
7	TBL 05 = A $\implies$ TBL 09 = A	35.37	39.02	25.61	0.001
8	TBL 06 = A $\implies$ TBL 09 = A	37.50	37.50	25.00	0.000
9	TBL 08 = A $\implies$ TBL 09 = A	36.36	40.91	22.73	0.002
10	TBL 05 = A $\implies$ TBL 08 = A	37.50	39.06	23.44	0.003
11	TBL 06 = A $\implies$ TBL 08 = A	38.71	33.87	27.42	0.005
12	TBL 05 = A $\implies$ TBL 06 = A	35.29	38.82	25.88	0.000
13	TBL 06 = F $\implies$ TBL 08 = A	0.00	87.50	12.50	0.013
14	TBL 05 = A $\implies$ TBL 06 = A $\implies$ TBL 09 = A	40.28	37.50	22.22	0.000
15	TBL 05 = A $\implies$ TBL 06 = A $\implies$ TBL 08 = A	41.38	32.76	25.86	0.001
16	TBL 06 = A $\implies$ TBL 08 = A $\implies$ TBL 09 = A	43.64	34.55	21.82	0.000
17	TBL 05 = A $\implies$ TBL 08 = A $\implies$ TBL 09 = A	42.11	38.60	19.30	0.000
18	TBL 06 = F $\implies$ TBL 08 = A $\implies$ TBL 09 = A	0.00	85.71	14.29	0.030
19	TBL 05 = A $\implies$ TBL 06 = F $\implies$ TBL 08 = A	0.00	100.00	0.00	0.017
20	TBL 05 = A $\implies$ TBL 06 = A $\implies$ TBL 08 = A $\implies$ TBL 09 = A	46.15	32.69	21.15	0.000
21	TBL 05 = A $\implies$ TBL 06 = F $\implies$ TBL 08 = A $\implies$ TBL 09 = A	0.00	100.00	0.00	0.040

5 Conclusion

Temporal contrast sets mining is a technique employed to discover sets of attribute-value tuples that are ordered in time and effectively differentiate between distinct groups within a dataset. This approach finds application in various fields where time-dependent patterns play a crucial role, including health-care, education, and traffic management. For instance, it can aid in diagnosing patients by considering the sequential order of symptoms, as demonstrated in the identification of Covid-19 patients [13].

In this study, we propose the first model for temporal contrast sets mining by extending the STUCCO algorithm [2], an Apriori-like technique known for extracting large and significant contrast sets. To validate the efficacy of our model, we applied it to a real dataset containing students' performance records. Through this application, we aimed to extract sequences that highlight the contrasting characteristics between different groups of students. The results of our analysis revealed that Team-Based Learning (TBL) tests had a substantial impact on students' overall performance. We identified the most influential

TBL tests and provided a comprehensive discussion on the performance of students in TBL test sequences, with the intention of enhancing their academic achievements.

Future research endeavors are encouraged to employ our model on larger datasets encompassing diverse domains to explore contrasting sequences within different classes. Additionally, there is room for improvement by enhancing the underlying algorithm to achieve scalability and efficiency in temporal contrast sets mining, potentially through techniques such as incremental extraction from data streams or extending the model to mine spatio-temporal contrast sets. Moreover, adapting the model to handle big data would further enhance its applicability and utility.

References

1. Agrawal, R., Srikant, R.: Mining sequential patterns. In: Proceedings of the Eleventh International Conference on Data Engineering, pp. 3–14 (1995). https://doi.org/10.1109/ICDE.1995.380415
2. Bay, S.D., Pazzani, M.J.: Detecting change in categorical data: Mining contrast sets. In: Proceedings of the Fifth ACM SIGKDD International Conference on Knowledge Discovery and Data Mining, pp. 302–306 (1999)
3. Bay, S.D., Pazzani, M.J.: Detecting group differences: mining contrast sets. Data Min. Knowl. Disc. 5(3), 213–246 (2001)
4. Ciotti, M., Ciccozzi, M., Terrinoni, A., Jiang, W.C., Wang, C.B., Bernardini, S.: The covid-19 pandemic. Crit. Rev. Clin. Lab. Sci. 57(6), 365–388 (2020). https://doi.org/10.1080/10408363.2020.1783198
5. Dong, G., Li, J.: Efficient mining of emerging patterns: discovering trends and differences. In: Proceedings of the fifth ACM SIGKDD International Conference on Knowledge Discovery and Data Mining, pp. 43–52 (1999)
6. Hilderman, R.J., Peckham, T.: A statistically sound alternative approach to mining contrast sets. In: AUSDM05 (2005)
7. Huang, T.C.K., Huang, C.H., Chuang, Y.T.: Change discovery of learning performance in dynamic educational environments. Telemat. Inform. 33(3), 773–792 (2016). https://doi.org/10.1016/j.tele.2015.10.005
8. Jabbar, M.S.M., Bellinger, C., Zaïane, O.R., Osornio-Vargas, A.: Discovering co-location patterns with aggregated spatial transactions and dependency rules. Int. J. Data Sci. Analyt. 5(2), 137–154 (2018)
9. Khan, S., Parkinson, S.: Eliciting and utilising knowledge for security event log analysis: an association rule mining and automated planning approach. Expert Syst. Appl. 113, 116–127 (2018). https://doi.org/10.1016/j.eswa.2018.07.006
10. Kharsa, R., Aghbari, Z.A.: Association rules based feature extraction for deep learning classification. In: Patel, K.K., Santosh, K.C., Patel, A., Ghosh, A. (eds.) Soft Computing and Its Engineering Applications. icSoftComp 2022. Communications in Computer and Information Science, vol. 1788, pp. 72–83. Springer, Cham (2023). https://doi.org/10.1007/978-3-031-27609-5_6
11. Khedr, A.M., Al Aghbari, Z., Al Ali, A., Eljamil, M.: An efficient association rule mining from distributed medical databases for predicting heart diseases. IEEE Access 9, 15320–15333 (2021). https://doi.org/10.1109/ACCESS.2021.3052799

12. Kong, J., Han, J., Ding, J., Xia, H., Han, X.: Analysis of students' learning and psychological features by contrast frequent patterns mining on academic performance. Neural Comput. Appl. **32**(1), 205–211 (2020)
13. Larsen, J.R., Martin, M.R., Martin, J.D., Kuhn, P., Hicks, J.B.: Modeling the onset of symptoms of covid-19. Front. Public Health **8**, 473 (2020). https://doi.org/10.3389/fpubh.2020.00473
14. Mohomed Jabbar, M.S., Zaïane, O.R.: Learning statistically significant contrast sets. In: Khoury, R., Drummond, C. (eds.) Advances in Artificial Intelligence, pp. 237–242. Springer, Cham (2016). https://doi.org/10.1007/978-3-319-34111-8_29
15. Nguyen, D., Luo, W., Vo, B., Pedrycz, W.: Succinct contrast sets via false positive controlling with an application in clinical process redesign. Expert Syst. Appl. **161**, 113670 (2020). https://doi.org/10.1016/j.eswa.2020.113670
16. Saeed, M.M., Al Aghbari, Z.: ARTC: feature selection using association rules for text classification. Neural Comput. Appl. **34**(24), 22519–22529 (2022)
17. Satsangi, A., Zaïane, O.R.: Contrasting the contrast sets: an alternative approach. In: 11th International Database Engineering and Applications Symposium (IDEAS 2007), pp. 114–119. IEEE (2007)
18. Segura-Delgado, A., Gacto, M.J., Alcalá, R., Alcalá-Fdez, J.: Temporal association rule mining: an overview considering the time variable as an integral or implied component. Wiley Interdiscip. Rev. Data Mining Knowl. Discov. **10**(4), e1367 (2020)
19. Shazan, M., Jabbar, M., Zaïane, O.R., Osornio-Vargas, A.: Discovering spatial contrast and common sets with statistically significant co-location patterns. In: Proceedings of the Symposium on Applied Computing (SAC 2017), pp. 796–803. Association for Computing Machinery, New York (2017). https://doi.org/10.1145/3019612.3019665
20. Spencer, M., Takahashi, N., Chakraborty, S., Miles, J., Shyu, C.R.: Heritable genotype contrast mining reveals novel gene associations specific to autism subgroups. J. Biomed. Inform. **77**, 50–61 (2018). https://doi.org/10.1016/j.jbi.2017.11.016
21. Vandromme, M., Jacques, J., Taillard, J., Hansske, A., Jourdan, L., Dhaenens, C.: Extraction and optimization of classification rules for temporal sequences: application to hospital data. Knowl.-Based Syst. **122**, 148–158 (2017). https://doi.org/10.1016/j.knosys.2017.02.001
22. Ventura, S., Luna, J.M.: Contrast Sets, pp. 33–51. Springer, Cham (2018)
23. Webb, G.I., Butler, S., Newlands, D.: On detecting differences between groups. In: KDD 2003. Association for Computing Machinery, New York (2003). https://doi.org/10.1145/956750.956781
24. Xie, D.F., Wang, M.H., Zhao, X.M.: A spatiotemporal apriori approach to capture dynamic associations of regional traffic congestion. IEEE Access **8**, 3695–3709 (2020). https://doi.org/10.1109/ACCESS.2019.2962619
25. Zhou, H., Hirasawa, K.: Evolving temporal association rules in recommender system. Neural Comput. Appl. **31**(7), 2605–2619 (2019)

Visualization Techniques in VR for Vocational Education: Comparison of Realism and Diegesis on Performance, Memory, Perception and Perceived Usability

Eleonora Nava[✉] and Ashis Jalote-Parmar

Norwegian University of Science and Technology, Trondheim, Norway
{eleonora.nava,ashis.jalote.parmar}@ntnu.no

Abstract. Designing highly realistic and immersive virtual reality (VR) environments has become crucial for enhancing the user experience within human-machine interactions. However, there is limited research regarding the utilization, impact, and influence of different visualization techniques in VR concerning cognition and behavior. This experimental study compares two scenarios: a low-realism (LR) scenario with non-diegetic visual elements and a high-realism (HR) scenario with diegetic visual elements. The experimental context pertains to young jobseekers' training and skills development in the construction industry. The findings indicate that an HR scenario contributes to enhancing memory but may reduce perceived usability if a proper feedback system related to diegetic elements is not implemented. Furthermore, an LR scenario can foster effective immersion if the diegetic narrative remains consistent and a high level of interaction is present, enabling a more inclusive design and making learning more accessible to individuals requiring a lower level of detail. Despite the limitation of a limited number of participants, the results aim to contribute to the establishment of visualization strategies in VR for vocational education and training.

Keywords: Virtual Reality · Human-computer Interaction · Realism · Diegesis · Visualization Techniques · Cognition · Behavior · Memory

1 Introduction

The evolution of VR technologies has been primarily centered on creating virtual worlds closely resembling reality to enhance the sense of presence [1]. This focus has been facilitated by the increased accessibility afforded by game engines Unity and Unreal Engine, which provide efficient and cost-effective production capabilities. Methodologies and frameworks related to the design of VR experiences are developed and validated across various fields to establish comprehensive guidelines [2, 3]. Nowadays, applications of VR technologies span various domains, including mental health treatment and therapy [4], entertainment in gaming [5], and skill enhancements in engineering and vocational training [6, 7]. Several studies have provided evidence of the enhanced effectiveness of immersive training compared to traditional approaches, highlighting its economic, safety, and cognitive benefits [8, 9].

© The Author(s), under exclusive license to Springer Nature Switzerland AG 2024
K. K. Patel et al. (Eds.): icSoftComp 2023, CCIS 2030, pp. 104–116, 2024.
https://doi.org/10.1007/978-3-031-53731-8_9

Studies on simulation-based virtual training have highlighted that high realism can lead to enhanced performance [10] and information retention [11], although its effects may vary among age groups [12]. Another identified aspect is the possibility of individual practice without the need to interact with others, which provides greater psychological comfort [13]. With different design strategies, VR technology can improve motivation in learning and the ability to retain information [14, 15], as well as increase cognitive engagement and problem-solving [16, 17].

Vocational education typically involves hands-on training and the development of practical skills directly applicable to a particular job or industry [18, 19]. VR can be a valuable tool for exploring careers and gaining practical experience, supporting exploring career opportunities. Previously inaccessible professions that could only be experienced through volunteering, internships, or job shadowing can now be simulated accurately in VR [20, 21]. The IMTEL research group has contributed significantly to this field with the Virtual Internship [22] and VR4VET (Virtual Reality for Vocational Education and Training) [23] projects, which are exploring the concept of "Immersive Job Taste" in various sectors such as fish farming, wind turbines, construction, and dentistry [24] to provide young jobseekers with the feeling of experiencing an average working day of a professional, incorporating essential training elements.

Visualization techniques can assist in supporting the creation of immersive learning environments. Previous studies have presented various strategies pertaining to scenario complexity in terms of realism and interactions [25–27]. They provide information on performance-related aspects such as receiving instructions, performance feedback, and the incorporation of gamification elements [28, 29], such as reward systems and scores. Although different visualization techniques have been explored in VR, research on guidelines and design strategies is limited. Notably lacking is research on how these techniques facilitate knowledge and skill transfer and their impact on memory and information retention [30, 31]. Furthermore, the design of these experiences must consider usability and simulation complexity in terms of narrative and realism, as these factors can affect accessibility. Addressing the research gap, this article aims to investigate and illustrate the impact of different visualization techniques, specifically realism and diegesis, on perceptual, cognitive, and mnemonic aspects of vocational virtual education experiences. The paper first describes related work, followed by the design of the VR experiment, testing, results, and discussions.

2 Related Work

Visualization techniques in VR refer to the methods and approaches used to display and represent visual representations within the virtual environment [32, 33]. These techniques encompass a range of elements, including the design of virtual objects, environments, graphical user interfaces, and the rendering and presentation of visual elements within the virtual space. Implementing these techniques in designing a VR scenario aims to achieve various outcomes by enriching the user experience. This study specifically delves into two key facets highlighted in the state-of-the-art research: (1) display and interaction realism (or fidelity) and (2) diegesis.

Realism in VR systems concerns how accurately real-world experiences have been replicated and explored in the context of display and interaction capabilities. Display realism, also known as display fidelity, is associated with sensory realism, indicating how faithfully sensory stimuli in the virtual environment resemble those in the real world [34]. Interaction realism, or interaction fidelity, pertains to the degree to which user actions in the virtual environment mirror those in the real world in terms of biomechanics, input, and control. While several studies demonstrate that higher levels of display realism generally lead to an enhanced user experience [35], the impact of varying levels of interaction realism remains less evident. High levels of interaction realism may provide a stronger sense of being in the real world [36], whereas lower levels can promote a sense of familiarity with computer interfaces [37]. A lower interaction realism may yield comparable outcomes to higher levels of realism. Moreover, reduced realism can facilitate accessibility for individuals with impairments [38, 39], particularly in educational experiences where accessibility for all should be ensured. Furthermore, techniques employed to enhance interaction realism should adhere to user experience (UX) principles. A well-designed interaction can differentiate between an enjoyable and natural experience and a tiresome and frustrating one.

Diegesis refers to the narrative world within a fictional work, and in VR, it pertains to the narrative elements that contribute to storytelling consistency [40]. It represents the internal narrative space of the virtual environment and includes events, characters, locations, and sounds experienced by the user. Research has shown that incorporating diegetic elements enhances user engagement [41] and performance [42–45]. However, the previously cited studies argue that there is no significant difference in terms of immersion and sense of presence. Additionally, they suggest that non-diegetic interfaces may be more effective and user-friendly for individuals unfamiliar with the VR environment. Both diegetic and non-diegetic visualization patterns can be successful when designed with usability.

Furthermore, the experiment seeks to integrate the acquired knowledge and test the impact of the varying levels of realism and diegesis for educational and training use, with the intention of establishing design principles in the design virtual environment.

3 Design of VR Scenarios for the Construction Industry

This section outlines the procedure for formulating and executing the experimental study design. The selected construction case assesses the potential of VR technology in standardized instruction, particularly targeting the vocational training sector lack: the scarcity of instructors and the need for safe workplace simulations. The design process adheres to a well-established Human-centered design process [46] and encompasses the following key phases in the development and evaluation of the VR prototypes: (a) conducting user studies with experts to inform content development tailored to the needs of construction industry job seekers; (b) creating a customized VR prototypes for job seekers; (c) administering user testing of the prototype. The experiment's design compares two VR visualization scenarios: an LR scenario with non-diegetic visual elements and an HR scenario with diegetic visual elements, focusing on assessing their effects on Performance, Memory, Perception, and Perceived Usability.

3.1 Target User

The study concentrates on young job seekers in the early training stage, targeting individuals aged between 16 and 25 who opt for alternative vocational education. Their primary goal is to acquire practical experience and skills to guide their future career choices. Frequently, the target may also be tech-savvy, comfortable with digital technologies and online learning, and open to new concepts and experiences.

3.2 Experts' Interviews with Construction Professionals

Semi-structured interviews were conducted with eight construction professionals to gain insights into the construction industry's daily routines and fundamental responsibilities. Findings led to the content customization for developing VR scenarios, where the virtual environment accurately reflects the real-world site, tools and terminology. Data analysis identified three critical tasks that can provide the taste of the job: creating concrete, measuring room dimensions using laser levels, and fixing electrical cables.

3.3 Prototype Development in VR

Unity 3D software was employed to create the VR scenarios, while Figma software was utilized to design interfaces and visual elements. As shown in Table 1, the LR scenario, with non-diegetic visual components, was designed using low-poly models (refer to Fig. 1). This scenario incorporated low-realism interactions and a 2D pop-up interface system. Conversely, the HR scenario (refer to Fig. 2) was built with high-detailed models, integrating 3D objects coherent with the real environment and adopting a natural and direct interaction modality. Both scenarios underwent usability and feedback intuitiveness testing before the final experiment, ensuring a practical and engaging learning experience that inspires confidence in the young target users.

4 Experiment Design

4.1 Participants

A total of 14 participants, comprising 6 self-identifying as female and 8 as male, with ages ranging from 18 to 23 years (average age 20.5), were voluntarily recruited from the first year of the Civil and Environmental Engineering bachelor program at NTNU. Informed consent was obtained from all participants. Selection criteria prioritized participants' interest in the construction environment and the absence of direct field experience to minimize potential test result bias. None of the participants had expertise in VR, with 71.4% classifying themselves as novices and 28.6% reporting limited VR experience. The VR experiences were designed to facilitate two-handed object manipulation, obviating the need for hand-dominance-based participant selection.

Fig. 1. a) Interaction with objects via ray-cast and 2D non-diegetic interfaces; b) Wayfinding via non-diegetic arrows; c) Tasks via pop-up system of interfaces.

Fig. 2. a) Direct interaction by grasping objects and performing actions; b) Wayfinding via diegetic signals; c) Tasks via diegetic signals and animation script effects.

4.2 Set Up

The experiment was conducted over two days and encompassed two locations to accommodate all the students and consider the computation requirements of both HR and LR. Both spaces were soundproofed and isolated to minimize external noise interference. The first location, adjacent to NTNU's Civil and Environmental Engineering Department classrooms, featured a 4 m^2 room equipped with an Oculus Quest 2 for LR testing. The second location was a 3 m^2 space inside the NTNU's VR Lab, at another university facility away from the main campus. Due to its more complex and computationally intensive nature, the HR scenario's prototype was conducted in the lab, necessitating a High-Performance Computing (HPC) model computer.

4.3 Procedure

The participants were divided into two groups. One group of 7 participants tested the LR scenario, while another group of 7 participants tested the HR scenario. All participants signed a consent form, although all data remained anonymous even when collecting questionnaire answers. The experiment consisted of five stages: Introductory stage. The first phase involved a controlled process in which users were asked to enter the room individually to avoid influencing performance. Each session lasted a maximum of 15

Table 1. Experimental study to evaluate the impact of Realism and Diegesis on Performance, Perception, Perceived Usability and Memory in VR scenarios for vocational education.

Comparative parameters	Low realism (LR) and non-diegetic visual elements	High realism (HR) and diegetic visual elements
Terminology explanation	**LR:** Abstract visualizations of contextual elements and reduced realism in representing objects, buildings, and tools. Indirect interaction through point-and-click, ray-cast lasers, and wayfinding. **Non-diegetic visual elements:** 2D interfaces for providing environment interaction instructions.	**HR:** Detailed and realistic visualization of contextual elements, including objects, buildings, and tools. Direct interaction through manipulation and use of real objects. **Diegetic visual elements:** Integration of narrative realistic objects, as signal indications and billboards, offering contextual information.
Comparative VR Visualization	Fig. 3. An example of the LR scenario for the 1st task of creating concrete.	Fig. 4. An example of the HR scenario for the 1st task of creating concrete.
Tasks given to participants	Group 1 (7 participants) **1st task:** Create concrete (by using an interface to mix sand and cement) **2nd task:** Measure the dimensions of a wall (by using interfaces and ray-cast lasers) **3rd task:** Repairing an electric cable (by interacting with an interface).	Group 2 (7 participants) **1st task:** Create concrete (by using realistic objects: shovel and cement mixer). **2nd task:** Measure the dimensions of a wall (using a laser level). **3rd task:** Repairing an electrical cable (using a plier object).
Measurement criteria method	**1. During the experiment:** *Method:* Observation; *Measurement criteria:* Time measurements, Task Performance (task completion and error). **2. After experiment:** *Method:* Post-experience questionnaire (adapted from VRLEQ and combined with SUS and GEQ scales); *Measurement criteria:* Perception (immersion, presence, user flow and usability). **3. Two weeks after experiment:** *Method:* Questionnaire (customized task-based questionnaire); *Measurement criteria:* Knowledge retention.	

min and was introduced by a brief explanation of the project, followed by signing consent forms. They were provided with necessary information regarding device usage, including instructions on operating the main commands and advised to avoid questions during the test. Users were assisted in wearing the Head Mounted Display (HMD) and inquired about ergonomics issues, such as blurred vision or discomfort due to HMD misalignment. Once recording commenced, the test began. **Familiarization and tutorial stage:** All participants engaged in a customized tutorial to acquaint themselves with the VR environment and tools, explaining interactions and proper controller trigger usage. **Experiment stage:** The immersive experience involved completing all three scenario tasks, lasting 7 to 10 min on average. Testers were instructed not to seek guidance or feedback unless necessary due to VR sickness concerns. Task completion time, errors, and difficulties were noted during the experiment and later compared with video recordings. **Post-experiment questionnaire:** Participants received an online questionnaire using a 5-point Likert scale to assess cognitive engagement regarding perception (immersion, presence, and flow) and perceived usability, adapted from VRLEQ [47]. **Post-2 weeks questionnaire:** After two weeks, participants received an online questionnaire to evaluate memory and knowledge retention by recalling tasks and tools used during the experiment.

5   Results

5.1   Performance

The time to complete the tasks was longer in the LR scenario (181.3 s on average) than in the HR scenario (158 s on average). However, it should be considered that reading text in non-diegetic 2D interfaces took longer in the LR scenario. Overall, the perceived difficulty did not show significant differences between the two scenarios. As shown in Graph 1, task durations took more time to complete in the LR scenario (T1 = 64.1 s, T2 = 68.4 s, T3 = 48.8 s) compared to the HR scenario (T1 = 54.2 s, T2 = 51.4 s, T3 = 52.4 s). Slight motion sickness in the HR scenario was reported by 42.9% of participants (6 out of 14), while 57.1% (8 out of 14) reported none. No participants experienced severe motion sickness or abandoned the experience; all tasks were successfully completed.

5.2   Perception and Immersion

The post-experience questionnaire, adapted from VRLEQ [47], provided data about average participants' ratings, as shown in Graph 2, of perceived Immersion (LR = 3.85/5, HR = 4.14/5; using a 5-point Likert scale), Presence (LR = 3.97/5, HR = 4.28/5; using a 5-point Likert scale), and Flow (LR = 3.38/5, HR = 3.66/5; using a 5-point Likert scale). In addition, according to the post-2 weeks questionnaire sent via email, long-term perceived immersion and presence were rated slightly more in HR scenarios (LR = 4/5, HR = 4.42/5; using a 5-point Likert scale).

5.3   Perceived Usability

The post-experience questionnaire revealed that the LR scenario received a slightly higher perceived usability rating on a 5-point Likert scale (3.75/5) than the HR scenario (3.59/5), as shown in Graph 3. Participants in the HR scenario reported experiencing more confusion and disorientation, often seeking confirmation of task completion. In contrast, interaction with interfaces appeared more intuitive in the LR scenario. After two weeks following the test, participants in the LR scenario provided feedback indicating a lack of issues and a good understanding of the tasks. Some participants even found certain tasks overly simplistic, as expressed in their comments, such as, "Easy task. Only one click and it was over". Conversely, participants in the HR reported experiencing difficulties in determining when a task was complete. They expressed uncertainty, stating, "I didn't know when to stop" and "I wasn't sure when the task was done but just assumed based on the lack of sparkling animation".

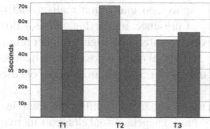

Graph 1. Comparison of the Performance duration of the three tasks for each scenario measured in seconds.

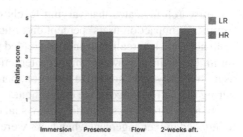

Graph 2. Comparison of the Perception of the scenarios in the post-experience and post-2 weeks after the experience.

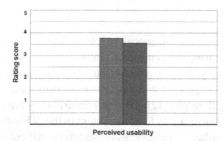

Graph 3. Comparison of Perceived Usability ratings results in the post-experience of the two scenarios.

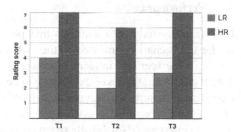

Graph 4. Comparison of participants' Memory of the tools used to complete the tasks in the scenarios after a 2-weeks period.

5.4 Memory

Two weeks after the experience, the questionnaire revealed that all 14 participants were able to recall the three tasks they performed in order. However, some participants displayed inaccuracies in recalling specific actions in the LR scenario, as shown in Graph 4. Instead of accurately recalling the act of measuring a wall, they described the second task as "pointing a laser to a yellow square". Indeed, in the HR scenario, 6 out of 7 participants were able to recall the tools used to accomplish the tasks. Conversely, in the LR scenario only 3 out of 7 participants remembered the tools employed to complete the tasks. In the LR scenario: T1 = all participants remembered mixing sand and cement, but only 57.1% recalled using a concrete mixer; T2 = all participants remembered measuring the room dimensions, but only 28.6% recalled using a level; T3 = all participants remembered fixing an electrical cable, but only 42.9% remembered using a plier. In the HR scenario: T1 = all participants remembered mixing sand and cement, and all remembered using a concrete mixer as a tool; T2 = all participants remembered measuring the dimensions, only one person did not recall using a laser level; T3 = all participants remembered fixing the electrical cable and using a plier as a tool.

6 Discussion and Limitation

The research indicates that varying levels of display and interaction realism had no significant impact on users' perception in terms of presence, immersion and user flow. Both high and low realism scenarios yielded positive results, with slightly higher ratings for higher realism. However, higher interaction and visual realism positively influenced users' memory and knowledge retention, particularly in recalling actions and tools used during the performance of the experiment.

In terms of diegesis, non-diegetic visual elements enhanced usability but could lead to reduced user engagement when tasks were too easy. Diegetic visual elements increase user engagement and presence, although clear feedback is essential to prevent confusion. Incorporating visual and audio feedback, such as animations and changes in tool states, could further enhance effectiveness.

6.1 Performance

Performance study data indicates that participants took more time to complete tasks in the LR scenario compared to the HR scenario, implying that realism and diegesis levels in the scenario influenced task efficiency. Lower realism and diegesis lead to longer completion times of the performance, in alignment with observations from similar experiments comparing varying degrees of realism [36, 42]. The LR scenario may have affected participants' information processing, particularly textual content in the 2D non-diegetic interfaces. Notably, there were no significant differences in perceived difficulty between the LR and HR scenarios, indicating that, despite longer completion times in the LR scenario, participants did not necessarily find the tasks more challenging. Some participants reported slight motion sickness in the HR scenario, although it did not affect task completion.

6.2 Perception and Immersion

Post-experience questionnaire data reveals that participants rated their perceived perception higher in the HR scenario than in the LR scenario, indicating that increased realism in the HR scenario contributed to greater immersion and presence, consistent with prior research findings cited in related works. Higher realism also fostered a more engaging and immersive experience, characterized by a sense of flow associated with deep engagement and enjoyment. The post-2 weeks questionnaire, administered via email, showed slightly higher long-term perceived immersion and presence ratings in the HR scenario compared to the LR scenario, indicating that the impact of high realism persisted even after two weeks, resulting in a stronger sense of immersion and presence for participants.

6.3 Perceived Usability

The Perceived Usability results from both post-experience and post-2 weeks question-naires show that participants perceived the LR scenario as slightly more usable than the HR scenario. The LR scenario, in fact, received a slightly higher usability rating com-pared to the HR scenario. Participants in the HR scenario reported more confusion and disorientation during task completion, often seeking confirmation or clarification. This suggests that the increased realism in the HR scenario introduced complexities that hin-dered participants' comprehension. The results may be attributed to some participants not noticing the animation indicating the task completion. Thus, the lower realism in the LR scenario offered a more straightforward and user-friendly interaction experience, enhancing participants' understanding and comprehension. However, it should be noted that users felt less perceptually involved in the immersive environment.

6.4 Memory

Memory assessment reveals that after a two weeks interval, all 14 participants could recall the sequence of the three tasks. However, in the LR scenario, some participants had trouble recalling specific actions or the tool used. Instead of remembering measuring a wall, they described the second task as "pointing a laser to a yellow square," and only 3 out of 7 participants remembered the tools used. This could be due to the use of interfaces in the LR scenario and participants finding it too simple to be memorable. In contrast, in the HR scenario, which offered a more realistic experience, 6 out of 7 participants accurately recalled the tools used. This suggests that higher realism aids in remembering task-related tools. However, it is important to note that the tasks were intentionally designed to be simple and intuitive for both scenarios. Therefore, user experiences should strive to be memorable and user-friendly. By incorporating realism into virtual scenarios, users are more likely to remember the tools used. This is particularly crucial in training scenarios that aim to prepare users for real-life situations. However, usability should not be sacrificed for realism. The user experience should balance realism and ease of use to ensure efficient task completion.

6.5 Limitation

This is a pilot study with limited users hence the reliability of the results can be further improved with increasing the number of user testing. Furthermore, during the experi-ment, task performance might have been influenced by the utilization of software ren-dering, potentially impacting the quality and efficiency of task execution. Enhancements in rendering techniques and equipment could enhance visualization and overall perfor-mance. Participants in the HR scenario experienced slight motion sickness, although they successfully completed all the tasks. Occasional oversight or lack of notice regard-ing task completion animations in the diegetic scenario was observed. Incorporating more prominent highlighting or emphasis on these animations could have significantly improved task performance and participant feedback.

7   Conclusion

Although existing studies illustrate the realm of VR and its application, few focus on the human-centered design of content, interaction and overall experiences within VR. This study aims to investigate the impact of visualization techniques, encompassing realism and diegesis influence on perception, immersion and memory of young job seekers engaged in vocational education. This study contributes to the growing body of knowledge in the area of visualization techniques in VR and how domain-specific content was designed in the customized experiment, to provide insights for practitioners and researchers. Findings are also generalizable across extent domains beyond vocational education and construction. Further research is required to investigate the role of visualization techniques and the impact on cognition in education and training.

Acknowledgements. We extend our gratitude to the IMTEL (Innovative Immersive Technologies for Learning) research group for their collaboration and support during this study, for providing us with access to the laboratory facilities and essential research instruments.

References

1. Newman, M., Gatersleben, B., Wyles, K., Ratcliffe, E.: The use of virtual reality in environment experiences and the importance of realism. J. Environ. Psychol. **79**, 101733 (2022)
2. Caputo, F., Greco, A., D'Amato, E., Notaro, I., Spada, S.: On the use of Virtual Reality for a human-centered workplace design. Proc. Struct. Integrit. **8**, 297–308 (2018)
3. Jalote-Parmar, A., Badke-Schaub, P., Wajid, A., Samset, E.: Cognitive processes as the foundation towards developing an intra-operative visualization system. J. Biomed. Inform. **43**(1), 60–74 (2010)
4. Nava, E., Jalote-Parmar, A.: Virtual reality revolution: strategies for treating mental and emotional disorders. In: IEEE International Conference on Systems, Man, and Cybernetics (SMC), Prague, pp. 3373–3378 (2022)
5. Kari, T., Kosa, M.: Acceptance and use of virtual reality games: an extension of HMSAM. Virtual Reality (2023)
6. Halabi, O.: Immersive virtual reality to enforce teaching in engineering education. Multim. Tools Appl. **79**(3–4), 2987–3004 (2019). https://doi.org/10.1007/s11042-019-08214-8
7. Xie, B., et al.: A review on virtual reality skill training applications. Front. Virt. Reality **2**, 645153 (2021)
8. Hvass, J., Larsen, O., Vendelbo, K., Nilsson, N., Nordahl, R., Serafin, S.: Visual realism and presence in a virtual reality game. In: 3DTV Conference the True Vision - Capture, Transmission and Display of 3D Video, Copenhagen, pp. 1–4 (2017)
9. McMahan, R.P., Bowman, D.A., Zielinski, D.J., Brady, R.B.: Evaluating display fidelity and interaction fidelity in a virtual reality game. IEEE Trans. Visual. Comput. Graph. **18**(4), 626–633 (2012)
10. Stevens, J., Kincaid, J.: The relationship between presence and performance in virtual simulation training. Open J. Model. Simulat. **03**, 41–48 (2015)
11. Essoe, J.K., Reggente, N., Ohno, A.A., Baek, Y.H., Rissman, J.: Enhancing learning and retention with distinctive virtual reality environments and mental context reinstatement. NPJ Sci. Learn. **7**(1), 1–14 (2022)

12. Cadet, L.B., Reynaud, E., Chainay, H.: Memory for a Virtual Reality Experience in Children and Adults According to Image Quality, Emotion, and Sense of Presence, pp. 55–75 (2022)
13. Schott, C., Marshall, S.: Virtual reality and situated experiential education: a conceptualization and exploratory trial. Comput. Assist. Learn. **34**(6), 843–852 (2018)
14. Toyoda, R., Russo-Abegão, F., Glassey, J.: VR-based health and safety training in various high-risk engineering industries: a literature review. Int. J. Educ. Technol. High. Educ. (2022)
15. Xu, Z., Zheng, N.: Incorporating virtual reality technology in safety training solution for construction site of urban cities. Sustainability **13**(1), 243 (2020)
16. Osti, F., de Amicis, R., Sanchez, C.A., et al.: A VR training system for learning and skills development for construction workers. Virtual Reality **25**(4), 523–538 (2021)
17. Howard, M.C., Gutworth, M.G., Jacobs, R.R.: A meta-analysis of virtual reality training programs. Comput. Human Behav. (2021)
18. Prasolova-Førland, E., Fominykh, M., Ekelund, O.I.: Empowering young job seekers with virtual reality. In: IEEE Conference on Virtual Reality and 3D User Interfaces, Osaka, pp. 295–302 (2019)
19. Baek, S., Gil, H., Kim, Y.: VR-based job training system using tangible interactions. Sensors, Basel, Switzerland (2021)
20. Toyoda, R., Glassey, J.: VR-based health and safety training in various high-risk engineering industries: a literature review. Int. J. Educ. Technol. High. Educ. **19**(1), 1–22 (2022)
21. Chellappa, V., Mésáros, P., Spak, M., Spisakova, M., Kaleja, P.: VR-based safety training research in construction. IOP Conf. Ser. Mater. Sci. Eng. **1252**, 012058 (2022)
22. IMTEL. https://www.ntnu.edu/imtel/virtual-internship. Accessed 25 May 2023
23. NTNU, VR4VET Project. https://www.vr4vet.eu/. Accessed 25 May 2023
24. Fominykh, M., Prasolova-Førland, E.: Immersive job taste: a concept of demonstrating workplaces with virtual reality. In: IEEE Conference on Virtual Reality and 3D User Interfaces (VR), pp. 1600–1605 (2019)
25. Chavez, B., Bayona, S.: Virtual reality in the learning process. In: Chavez, B., Bayona, S. (eds.) Virtual Reality in the Learning Process, Advances in Intelligent Systems and Computing, vol. 746, pp. 1345–1356 (2018)
26. Kwon, C.: Verification of the possibility and effectiveness of experiential learning using HMD-based immersive VR technologies. Virtual Reality **23**(1), 101–118 (2018)
27. Petersen, G.B., Petkakis, G., Makransky, G.: A study of how immersion and interactivity drive VR learning. Comput. Educ. **179**, 104429 (2022)
28. Fromm, J., Radianti, J., Wehking, C., Stieglitz, S., Majchrzak, T.A., Vom Brocke, J.: More than experience? - on the unique opportunities of virtual reality to afford a holistic experiential learning cycle. Internet High. Educ. **50**, 100804 (2021)
29. Su, C.-H., Cheng, T.-W.: A sustainability innovation experiential learning model for virtual reality chemistry laboratory: an empirical study with PLS-SEM and IPMA. Sustainability **11**(1), 1–24 (2019)
30. Strojny, P., Dużmańska-Misiarczyk, N.: Measuring the effectiveness of virtual training: a systematic review. Comput. Educ. X Reality **2**, 100006 (2023)
31. Coleman, B., Marion, S., Rizzo, A., Turnbull, J., Nolty, A.: Virtual reality assessment of classroom - related attention: an ecologically relevant approach to evaluating the effectiveness of working memory training. Front. Psychol. **10**, 1851 (2019)
32. Korkut, E.H., Surer, E.: Visualization in virtual reality: a systematic review. Virtual Reality **27**, 1447–1480 (2023)
33. Jiang, F., Haddad, D.D., Paradiso, J.: Baguamarsh: an immersive narrative visualization for conveying subjective experience. In: Kurosu, M. (ed.) HCII 2020. LNCS, vol. 12181, pp. 596–613. Springer, Cham (2020). https://doi.org/10.1007/978-3-030-49059-1_44
34. Al-Jundi, H.A., Tanbour, E.Y.: A framework for fidelity evaluation of immersive virtual reality systems. Virtual Reality **26**, 1103–1122 (2022)

35. McMahan, R.P., Bowman, D.A., Zielinski, D.J., Brady, R.B.: Evaluating display fidelity and interaction fidelity in a virtual reality game. IEEE Trans. Visual. Comput. Graph. **4**(4), 626–633 (2012)

36. Rogers, K., Funke, J., Frommel, J., Stamm, S., Weber, M.: Exploring interaction fidelity in virtual reality: object manipulation and whole-body movements. In: ACM CHI Conference on Human Factors in Computing Systems (2019)

37. McMahan, R.P., Lai, C., Pal, S.K.: Interaction fidelity: the uncanny valley of virtual reality interactions. In: Lackey, S., Shumaker, R. (eds.) VAMR 2016. LNCS, vol. 9740, pp. 59–70. Springer, Cham (2016). https://doi.org/10.1007/978-3-319-39907-2_6

38. Li, C., Ip, H.H.S., Ma, P.K.: A design framework of virtual reality enabled experiential learning for children with autism spectrum disorder. In: Blended Learning: Educational Innovation for Personalized Learning, vol. 11546, pp. 93–102. Springer, Cham (2019)

39. Zhao, Y., Cutrell, E., Holz, C., Morris, M.R., Ofek, E., Wilson, A.D.: SeeingVR: a set of tools to make virtual reality more accessible to people with low vision. In: Proceedings of the 2019 CHI Conference on Human Factors in Computing Systems (2019)

40. Gottsacker, M., Norouzi, N., Kim, K., Bruder, G., Welch, G.: Diegetic representations for seamless cross-reality interruptions. In: IEEE International Symposium on Mixed and Augmented Reality (ISMAR), Bari, pp. 310–319 (2021)

41. Saling, F., Bernhardt, D., Lysek, A., Smekal, M.: Diegetic vs. non-diegetic GUIs: what do virtual reality players prefer?. Artif. Reality Telexistence (2021)

42. Marre, Q., Caroux, L., Sakdavong, J.C.: Video game interfaces and diegesis: the impact on experts and novices' performance and experience in virtual reality. Int. J. Human-Comput. Interact. **37**, 1089–1103 (2021)

43. Beck, T., Rothe, S.: Applying diegetic cues to an interactive virtual reality experience. In: IEEE Conference on Games (CoG), Copenhagen (2021)

44. Leroux, E., Caroux, L., Sakdavong, J.C.: Diegetic display, player performance and presence in virtual reality video games. In: 11th International Conference on Applied Human Factors and Ergonomics (2020)

45. Dickinson, P., Cardwell, A., Parke, A., Gerling, K, Murray, J.: Diegetic tool management in a virtual reality training simulation. In: IEEE Virtual Reality and 3D User Interfaces (VR), Lisboa, pp. 131–139 (2021)

46. Jerald, J.: Human-centered design for immersive interactions. In: IEEE Virtual Reality (VR), Los Angeles, pp. 431–432 (2017)

47. Boletsis, C.: A user experience questionnaire for VR locomotion: formulation and preliminary evaluation. In: De Paolis, L.T., Bourdot, P. (eds.) AVR 2020. LNCS, vol. 12242, pp. 157–167. Springer, Cham (2020). https://doi.org/10.1007/978-3-030-58465-8_11

Multi-Tenant Servitization Platform-as-a-Service Model

Moses L. Gadebe$^{(\boxtimes)}$ ⓘ, Adeiza J. Onumanyi ⓘ, and Buhle Mkhize

Next Generation Enterprises and Institutions, Council for Scientific and Industrial Research, Pretoria, South Africa
{mgadebe,aonumanyi,bmkhize}@csir.co.za

Abstract. In recent times, there has been a growing effort to develop horizontal platforms that can accommodate multiple vertical systems, aiming to tackle the design, implementation, and maintenance challenges associated with vertical product-as-a-service (vPaaS) systems. The main issue faced by most vPaaS systems is the lack of interoperability and loose coupling, both of which are essential for achieving reusability of such systems. As a result, vPaaS systems have traditionally been developed in isolation, serving specific domain contexts, and following a monolithic design pattern that tightly integrates the frontend, backend, and databases into a single application. Initially, this approach seemed acceptable to developers particularly when domain experts are responsible for providing the design requirements. However, it has become evident that certain requirements often overlap with other domains. Thus, in this paper, we propose a servitization platform-as-a-service (sPaaS) model to improve reusability, efficiency, and maintainability in a multi-tenant servitization platform. Our model aims to address the challenges associated with tenant ownership, innovation, adoption, and maintenance burdens by introducing beneficial schemes as part of the sPaaS. These schemes will incentivize innovation and create new online job opportunities for artisans, plumbers, and software engineers. These professionals will then be responsible for conducting repairs, replacements, and upgrades on behalf of tenants and sPaaS providers.

Keywords: Servitization · sPaaS · Product-Service · Product-as-a-Service · Multi-Tenant · Incentivization

1 Introduction

Servitization refers to the process of transforming traditional product-oriented businesses into service-oriented models, where companies provide value to customers through integrated services alongside their core products [1–3]. These models, which facilitate the evolution of businesses from being product-oriented to service-based enterprises, are commonly referred to as servitization models [1,4,5].

Supported by Council for Scientific and Industrial Research (CSIR).

K. K. Patel et al. (Eds.): icSoftComp 2023, CCIS 2030, pp. 117–128, 2024.
https://doi.org/10.1007/978-3-031-53731-8_10

The process of developing a servitization model is depicted in Fig. 1, which presents an example of an aeroplane turbine product-as-a-service system [1,2]. The physical turbine, situated in the first layer, represents the tangible value being delivered to the customer (mobility). The second layer encompasses sensors that measure turbine status, temperature for proactive maintenance, and usage statistics [2]. The third layer involves connectivity, providing internet access for monitoring and remote control through machine-to-machine communication. Analytics forms the fourth layer, where data is collected, stored, and classified using cloud computing and data analytics [2]. The fifth layer involves the digital service interface, which enables companies to engage with customers in order to deliver the intended products or services.

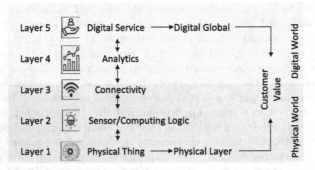

Fig. 1. IoT value creation layers of aeroplane turbine system [2]

Most importantly, data analytics plays a crucial role in the quest of manufacturers to extract value from collected data [2]. It enables them to enhance quality, boost efficiency, and optimize their products and services. In particular, the bundling layer acts as a bridge, bringing together various products and services to establish a long-term relationship between suppliers and product owners [2]. Such a service delivery model as described above thus brings about fundamental changes to the core of the business model in two key ways: (a) value creation through data analytics and information, and (b) value creation through monitoring and remote control [2,4]. In certain cases, it may also encompass (c) maintenance and asset replacement [6].

However, when it comes to a pure vertical system, consider building management systems (BMS) as an example of such a vertical system within the realm of the Internet of Things (IoT) [1]. BMS encompasses the management of various components such as heating, ventilation, and air conditioners (HVAC), light controls, power systems, security monitoring, and life systems, to name just a few. This type of a system poses several limitations that developers encounter when expanding and maintaining it. Thus, the proliferation of such product-asa-service (PaaS) systems, also known as vertical systems or silo-based service delivery models, lacks interoperability and flexibility. These qualities are often crucial for seamlessly integrating new devices and application components [7]. Moreover, these vertical systems entail higher costs in terms of hardware (including sensors), networks, middleware, and application logic, all of which are intricately linked to cater to specific contexts and domains, as illustrated in Fig. 1 and Fig. 2 [8].

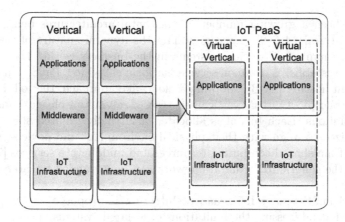

Fig. 2. Physical isolated verticals to virtual verticals [1]

Thus, to ensure the smooth operation of vertical systems, software developers will need to engage in reverse-engineering of these applications. This is because the hardware (sensors), backend application, and frontend components are tightly intertwined [7]. However, this delivery of services is inefficient due to the lack of hardware and application reuse [1,2]. Consequently, the same development process is repeated each time to create and deploy isolated vertical systems [5,6]. This approach contradicts the principles of Service Orientation Architecture (SOA) and the definition of the IoT as being "a dynamic global network infrastructure with self-configuring capabilities based on standards and communication protocols."

Hence, the authors in [1] emphasized that the fundamental concept of a horizontal architecture is to establish a domain-independent PaaS framework. This framework would provide discoverable services on the cloud for IoT solution providers, enabling them to deliver services efficiently and continuously expand their offerings [7]. Therefore, in the present paper, we propose a multi-tenant architecture that encompasses SOA principles and caters to multiple tenants. Our proposed model encourages tenants to be viewed as innovative stakeholders who can financially benefit from the business, rather than just being mere users as indicated in [5]. Additionally, our model facilitates the easy reuse of common services and simplifies maintenance processes.

The remainder of this paper is organized as follows: The related work is presented in Sect. 2. In Sect. 3, challenges and legitimization of a servitization model are presented, while the proposed multi-tenant servitization model is documented in Sect. 4. Lastly, the discussion and conclusion are presented in Sects. 5 and 6, respectively.

2   Related Work

There has been a growing number of horizontal architectures being proposed in the literature [1, 4, 6–9] to accommodate the dynamic nature of the IoT and the growing number of vertical systems in the business world. These architectures aim to establish a common service platform called IoT PaaS, which enables multi-tenant hosting [8]. By adopting a cloud-like approach, the IoT PaaS facilitates the concept of virtual verticals instead of relying on physical and isolated vertical solutions. Each tenant is allocated a virtually isolated space that can be customized to align with their physical environment and devices, while still benefiting from shared computing resources and middleware services [7,8]. Consequently, the generic IoT PaaS framework can be expanded to cater to various domains [8].

A novel multi-tenant IoT PaaS architecture is proposed and implemented in [1] aimed at addressing the limitations of isolated systems. These limitations include loose-coupling, extensibility, reuse, data cleansing, and fine-grained data analysis. Unlike isolated vertical solutions, the IoT PaaS offers a scalable and flexible approach to incorporate diverse domain-specific data models. It also provides control applications that rely on physical devices. In contrast, our proposal suggests a decentralized model based on cloud and fog computing. In this model, each tenant has their own environment where they can control applications and containerize data, eliminating latency issues associated with cloud computing. This approach also accommodates unique domain-specific differences.

Similarly, authors in [9] introduced an architecture called iTaaS for patient monitoring. This architecture enables real-time data collection, big data storage, and analytics. Devices are registered using a generic device XML with a unique ID, facilitating dynamic discovery of sensors and services. Similarly to [1], they used generic sensor XML files as a digital twin of the actual paired devices. The devices are paired using universal Bluetooth Low Energy (BLE) for on-the-fly discovery, attachment, pairing, and registration of devices within a 50m radius. When a device or sensor is detected, it is automatically registered based on its unique UUID and stored locally on the IoT side. Their architecture employs a smartphone as a gateway and a fog computing environment. This proposed architecture aligns with servitization principles prescribed in [10]. A fog computing on the smartphone ensures data control by the owner and allows sharing with cloud users, thereby promoting trust. However, it is susceptible to denial-of-service attacks and lacks data confidentiality, as personal data can be stolen from the smartphone. Although proposed as a reference model for patient monitoring based on SOA principles, the architecture lacks multi-tenancy, billing, and the ability to host multiple vertical systems. Our goal is to incorporate on-the-fly device/sensor discovery and registration within client tenant systems.

Authors in [11] proposed an architecture that enables the hosting of multiple vertical systems. In their architecture, each client can acquire and register their IoT broker on an IoT common service platform. The IoT broker, developed on a mobile ODROID-XU3 device that supports open source, allows tenants to subscribe to specific services, such as power consumption. The mobile IoT broker can then dynamically discover and capture devices and sensors at a household level, storing the information in a generic data model database. When a request is received, the IoT broker collects data from a sensor and represents it as a generic device model. It adds an additional security token and forwards the data to the database. To manage the IoT services effectively, the system invokes the IoT service management module. This module classifies client appliances (i.e., discovered devices or sensors) and registers them with a specific service agent. Consequently, the service management module receives updates and responds to the tenant's user portal.

The IoT broker operates similarly to the IoT side or fog computing introduced in [9]. In contrast, the IoT common service is hosted on the cloud. Additionally, the proposed architecture incorporates the bill generator, which charges for pay-per-use analytic services. However, a key limitation of this model is that each tenant is required to obtain their own IoT broker. Without this broker, it becomes impossible to monitor tenants and their physical environment. Moreover, this model fails to support user domain diversity. Furthermore, processing is conducted in the cloud, which can impact both Quality-of-Service (QoS) and Quality-of-Experience (QoE) due to latency concerns. To address these issues, it is necessary to expand the capabilities of the IoT broker to include intelligent computations for enhanced predictive processing and data confidentiality [12–14]. The proposed intelligent IoT for common services should be diversified to cater to different domains and improve latency. This, in turn, would contribute to enhancing trust, QoS, and QoE. As highlighted in [10], domain users (tenants) exhibit reluctance towards adopting the servitization model due to concerns regarding the confidentiality of data sharing.

Findings and Gaps of IoT-as-a-service model: While IoT-as-a-Service can support multiple tenants, it lacks essential asset management, supplier management, and customer chain capabilities crucial for implementing a successful horizontal model. Not only the disruption is not welcome, customers and employees need to be monitored and managed for effective operations, because interruption of services is not tolerated in servitization model. Ensuring uninterrupted service continuity necessitates a fundamental maintenance process involving repair, replacement, and upgrades. Unfortunately, this process is often regarded as an external and overhead expense. Additionally, the multi-tenant approach fails to account for variances in tenant domains and requirements, as Infrastructure-as-a-Service (IaaS) offers only generic services that may not suffice for diverse tenant needs. Furthermore, tenants are treated merely as users rather than stakeholders, neglecting the principles of stakeholder involvement, such as flexibility, potential revenue sources, incentive negotiation, and ambiguous social responses, as outlined in [10]. The billing system lacks incentives negotiation and benefits for tenant contributions to innovation, resulting in low adoption rates and a lack of ownership. To address these challenges, the author in [10] suggests the inclusion of a comprehensive and easily understandable servitization model that encompasses the transition and alleviates concerns about income loss. Additionally, conducting financial analyses and feasibility assessments prior to joining servitization is essential. Thus, it is the quest of the present paper to develop a model that addresses the above gaps.

3   Challenges and Legitimization of Servitization Models

The development of any proposed servitization model necessitates careful consideration of a range of critical challenges and legitimization issues [2,10]. According to [10], it is imperative to acknowledge and address these issues in order to ensure the effectiveness and viability of any servitization model. The following are some of the issues to be considered:

Flexibility: When designing a servitization model, it is crucial for developers to prioritize flexibility in order to effectively accommodate evolving service requirements. Nevertheless, this emphasis on flexibility poses challenges for customers in comprehending and distinguishing the unique value propositions offered by different suppliers. In order to address these challenges, the model should encompass diverse revenue sources, including (a) pay-per-use, (b) pay-per-value created, and (c) pay-per-zero downtime, which can be subject to negotiation. The incorporation of such revenue mechanisms can prove beneficial for organizations and owners of product-as-a-service systems, as they serve as incentivization schemes.

Feasibility Study: To ensure a smooth business transition following any servitization model, it is crucial to establish a clear and transparent business model that addresses potential concerns about income loss. servitization has wide-ranging impacts on various aspects of a company. Therefore, it is advisable for organizations to conduct thorough financial analyses or feasibility studies before committing to the model. Data analysis enables more precise monitoring of service performance, which can enhance the value proposition. However, some customers may perceive servitization as a short-term or one-time solution for optimizing their own services, which may not meet their minimum acceptable standards. Another challenge arises from the shifting risk structure, as operational risks move from customers to service providers. As a result, customers become reliant on the service without possessing the necessary technical expertise, which can disrupt their business continuity in the long run. Furthermore, the relationship between service providers and customers needs to evolve, with providers taking the responsibility of educating customers about the benefits they will gain.

Customer Involvement: However, technical service alone is insufficient. Customers need to be actively involved in defining performance-based agreements. The negotiation of incentives should be a collaborative effort between service providers and customers, treating them as partners. However, there is an unclear societal response to the servitization model [10]. The impact of the model on society and the issues related to trust in monitoring private data usage and the sense of ownership need to be supported by evidence. Therefore, conducting a feasibility study before implementing the servitization model is necessary. Additionally, the success of the servitization platform relies on societal factors such as the sense of ownership, rather than just usage.

Data Sharing Risks: Sharing corporate sensitive data poses a significant challenge when it comes to establishing legitimacy. This needs to be considered when designing a servitization model.

Role of Government: The government plays a vital role in legitimizing servitization by promoting and exemplifying best practices through policymaking. Developing such policies may pose challenges, but it is crucial for designing an effective servitization model.

Private Sector Experimentation: The legitimacy of servitization is currently evolving. Therefore, it is essential for us to continue experimenting with the servitization model in collaboration with the private sector until it gains acceptance and widespread adoption.

Data Sovereignty and Hybrid Strategies: To ensure effective data governance and organized data sovereignty, it is crucial to prioritize the value generated by data. Thus, owners should be incentivized through revenue models and contract development in order to promote its usage. Due to the rapid changes in digital technology, organizations must adopt hybrid strategies that incorporate traditional processes. These traditional processes are vital for maintaining continuity, which should be considered in a servitization model.

Long-Term Viability: It is crucial for organizations to maintain important knowledge about traditional processes. Therefore, they must inquire about the extent to which they are willing to outsource these processes. While some organizations may experience short-term benefits from servitization, others may reap advantages in the long run. Consequently, conducting a feasibility study becomes essential, as reliance on service providers can significantly impact business continuity over time.

4 Proposed Multi-tenant Servitization Model

The proposed sPaaS architecture is introduced in this section as a platform for hosting multiple vertical product-services across various domains, as depicted in

Fig. 3. Building upon the architecture proposed in [11], the sPaaS incorporates client fog-computing to address design concerns highlighted in [10,12,14,15], as explained in Sect. 3.

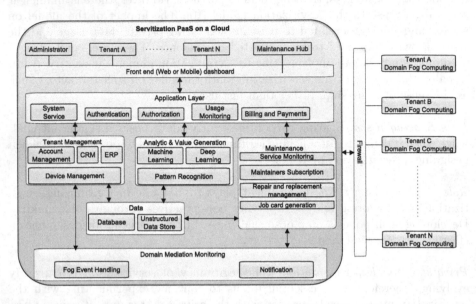

Fig. 3. Proposed multi-tenant cloud sPaaS architecture.

Within the proposed platform, the contextual and requirement specifics of tenants' domains are fundamental to each tenant vertical system. The sPaaS providers aim to gain a thorough understanding of tenants' requirements, data formats, and security needs to facilitate seamless integration. Furthermore, they seek to establish mutually beneficial billing and data usage schemes. As stated in [11], the development of vertical systems can be undertaken either by third-party developers or by the sPaaS service provider. Consequently, the provider subscribes to applications based on a predetermined tenancy agreement between the solution provider and the IoT PaaS platform provider. Once subscribed, the solution provider can configure each control application by adjusting relevant parameters through application context management.

The proposed platform offers various essential services, including tenant management, analytics and value generation, maintenance, domain mediation monitoring, and a frontend dashboard as depicted in Fig. 4. In this architecture, the administrator/provider initially gathers tenant requirements and identifies their domains. Subsequently, common services are assigned, and domain-specific contextual services are modeled, implemented, and containerized within a fog-computing environment portrayed in Fig. 4, tailored to the specific domain requirements. Any service that overlaps across domains is implemented as a common service and deployed on the sPaaS cloud, allowing other tenants to utilize it for an agreed-upon fee. The royalty fees for these services should be negotiated between the sPaaS provider and the innovator as an incentive.

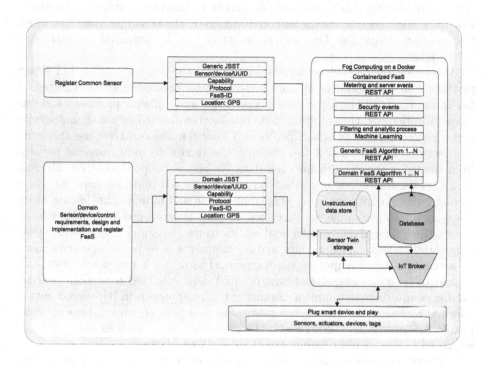

Fig. 4. Domain fog computing environment.

The account management system facilitates the tenant silo computing environment, contract agreements, and beneficiary incentives. Its purpose is to promote innovation and adoption of the servitization platform. All nodes in the tenants' fog computing environment (including sensors, devices, equipment, and FaaS services) are registered and recorded in the device management system. The enterprise resource planning (ERP) and customer relationship management (CRM) systems are used exclusively to handle processes necessary for running tenant legacy systems and managing their customers' operations based on negotiations.

Any notifications of hardware or software defects, errors, or unavailability reported by the tenant fog computing environment are captured in the repair and replacement management system. These notifications are also forwarded to the relevant tenant. The tenant has the option to use their own technicians/software developers or utilize a certified maintenance personnel from the sPaaS, subject to a fee. If the tenant chooses a certified engineer, a job card is generated. Consequently, certified technicians, engineers, artisans, and software developers receive an alert via the maintenance hub, indicating the location and problem

along with a possible diagnostic report. The first qualified personnel to respond
will be assigned the job and remunerated accordingly by either the tenant or the
sPaaS provider. It is important to note that these personnel are not employees of
sPaaS provider; they are freelancers or remote technicians/software maintainers
who have undergone screening and certification by the sPaaS provider, for which
they pay an annual fee. This unique initiative has the potential to create new
online jobs.

Value generation will be facilitated through the fog computing and sPaaS
platforms on the cloud, using data analytics and prediction tools. The tenant
fog computing will employ it to learn patterns and enhance processes and services. Additionally, the resulting value can be transformed into a shared service
and integrated into the sPaaS platform. Other tenants can then use this common service for a specified fee, allowing the tenant to gain financial rewards
from the value generated using their data. To ensure proper incentives, all common services offered by our sPaaS will be monitored and measured for usage
through digital contracts using usage monitoring and billing. The domain mediation monitoring module, as described in [11], is leveraged to handle various
events, including data transfer and notifications, to support communication,
monitoring, maintenance, repairs, and replacements in response to actions taken
by a tenant's local computing environment. The tenant's transferred data may
consist of both structured and unstructured data collected from their individual fog computing environment. Similarly to the approach in [1], shared data is
adopted to discover patterns, generate value, and predict/control tenants' computing environments. Additionally, the generated value is then enhanced and
migrated as common services to the cloud's sPaaS platform.

All services and tenants will undergo authentication to verify their authoriza-
tion for using the proposed platform, for example, by using OAuth 2.0 through
Keycloak. Data transfer from individuals' fog-computing environments will then
be accessed securely through private and public certification over the firewall.

5   Discussion

Software developers typically prefer to work on vertical domain-specific solutions
rather than horizontal solutions that are more generic. This preference stems
from their comfort in dealing with specific domains and the challenge of obtaining
sufficient details for a horizontal solution. Developing a horizontal architecture
with common features requires developers to navigate multiple domains and
thoroughly grasp their unique needs and requirements. Consequently, developers
tend to gravitate towards a vertical design approach, where they first engineer
software requirements specific to the domain.

To address this issue and improve the overall service, we propose the implementation of a software platform-as-a-service. This platform serves as a centralized hub, allowing tenants to access common services that enhance and add value
to their products through pay-per-use digital contracts. To ensure the delivery of
better solutions, our sPaaS also provides external software developers with the

opportunity to consult domain experts. This collaboration enables developers to learn and understand the specific environments and requirements of tenants, ultimately leading to improved solutions. The sPaaS will seek to eradicate the escalating costs of infrastructure such as hardware, network, and software and challenges faced by vertical systems. The primary objective of sPaaS is to offer a platform that incentivizes stakeholders, including tenants, software developers, artisans, maintainers, and engineers, to foster innovation and promote the adoption of a multi-tenant approach.

6   Conclusion and Future Work

In this paper, an architecture of a servitization platform-as-a-service (sPaaS) model has been proposed. The proposed model aims to provide a platform for both tenants and freelancing technicians, including electricians, plumbers, artisans, and software developers specializing in various domains. This platform will not only facilitate business transactions between service providers and freelancers but also create new online job opportunities, contributing to poverty alleviation and skill development. By turning tenants into co-stakeholders of business services, the platform will foster a sense of ownership and expedite its adoption. The proposed sPaaS will receive royalties and profits through the incorporated metering and billing mechanisms for hosting different vertical systems. Our future research will explore open-source multi-tenant frameworks to avoid reinventing the wheel. The sPaaS architecture will be implemented in the cloud, and software developers and technicians will undergo screening and registration processes to ensure the provision of necessary services.

References

1. Li, F., Vogler, M., Claebens, M., Dustdar, S.: Efficient and scalable IoT service delivery on cloud. IEEE Internet Things J. **7**(1), 740–747 (2020)
2. Boehmer, H.J.: The impact of the Internet of Things (IoT) on servitization: an exploration of changing supply relationships. Product. Plann. Contr. **31**(2), 203–219 (2019). https://doi.org/10.1080/09537287.2019.1631465
3. Tewari, A.: Servitization implementation in manufacturing organizations - a systematic review to identify obstacles and critical success factors. Int. J. Res. Appl. Sci. Eng. Technolo. (IJRASET) **8**(5), 851–865 (2020). https://doi.org/10.22214/ijraset.2020.5135
4. Cenamor, J.: Adopting a platform approach in servitization: leveraging the value of digitalization. Int. J. Prod. Econom. **192**, 54–65 (2016) https://doi.org/10.1016/j.ijpe.2016.12.033
5. Sklyar, A., Kowalkowski, C., Tronvoll, B., Sorhammar, D.: Organizing for digital servitization: a service ecosystem perspective. J. Bus. Res. **104**, 450–460 (2019). https://doi.org/10.1016/j.jbusres.2019.02.012
6. Amadi-Echendu, J., Dakada, M., Ramlal, R., Englebrecht, F.: Asset replacement in the context of Servitization. In: 2019 IEEE Technology and Engineering Management Conference (TEMSCON), pp. 1–7. IEEE (2019) https://doi.org/10.1109/temscon.2019.8813622

7. Pahl, M.-O.: Multi-tenant IoT service management towards an IoT app economy. In: IFIP/IEEE Symposium on Integrated Network and Service Management (IM), 2019, pp. 1–4. IEEE, Arlington, VA, USA (2019)
8. Aytac, K., Korcak, O.: Multi-tenant management in secured IoT based solutions. In: 2022 32nd Conference of Open Innovations Association (FRUCT), pp. 56–64. IEEE (2022) https://doi.org/10.23919/fruct56874.2022.9953817
9. Petrakis, G.M.E.: Internet of Things as a Service (iTaaS): Challenges and solutions for management of sensor data on the cloud and the fog. Internet of Things 3–4, 3–4 (2018) https://doi.org/10.1016/j.iot.2018.09.009
10. Langley, D.J.: Digital product-service systems: the role of data in the transition to servitization business models. Sustainability **14**(3), 1303 (2022). https://doi.org/10.3390/su14031303
11. Kim, J., Jeon, Y., Kim, H.: The intelligent IoT common service platform architecture and service implementation. J. Supercomput. **74**, 4242–4260 (2016). DOIurlhttps://doi.org/10.1007/s11227-016-1845-1
12. Anawar, M.R., Wang, S., Zia, M.A., Jadoon, A.K., Akram, U., Raza, S.: Fog computing: an overview of big IoT data analytics. Wirel. Commun. Mob. Comput. (2018). https://doi.org/10.1155/2018/7157192
13. Giannoutakis, K.M., Spanopoulos-Karalexidis, M., Filelis Papadopoulos, C.K., Tzovaras, D.: Next generation cloud architectures. In: Lynn, T., Mooney, J.G., Lee, B., Endo, P.T. (eds.) The Cloud-to-Thing Continuum: Opportunities and Challenges in Cloud, Fog and Edge Computing, pp. 23–39. Springer International Publishing, Cham (2020). https://doi.org/10.1007/978-3-030-41110-7_2
14. Soultanopoulos, T., Sotiriadis, S., Petrakis, E., Amza, C.: Internet of Things data management in the cloud for Bluetooth Low Energy (BLE) devices. In: Proceedings of the Third International Workshop on Adaptive Resource Management and Scheduling for Cloud Computing, pp. 35–39. ACM (2016) https://doi.org/10.1145/2962564.2962568
15. Al Faruque, M.A., Vatanparvar, K.: Energy management-as-a-service over fog computing platform. IEEE Internet Things J. **3**(2), 161–169 (2015). https://doi.org/10.1109/JIOT.2015.2471260

A Load-Balanced Task Scheduling in Fog-Cloud Architecture: A Machine Learning Approach

Rashmi Keshri$^{(\boxtimes)}$ (ID) and Deo Prakash Vidyarthi (ID)

School of Computer and Systems Sciences, Jawaharlal Nehru University, New Delhi, India
rashmikeshri28@gmail.com

Abstract. The swift expansion of internet-of-things (IoT) devices and the rise in the pace of task requests sent from these IoT devices to the cloud data centres led to Congestion and delays in the service. To meet the challenges, fog computing emerged as a new computer paradigm that offers services near the request-generating devices and reduces delays, particularly for real-time and delay-sensitive queries. It is crucial to consider issues like balancing the load, lowering energy consumption, and scheduling requests that impact the fog-cloud ecosystem's performance to accomplish these aims. This work suggests a Machine Learning based Task scheduling algorithm with load balancing for the fog-integrated cloud. It first deals with the task offloading to decide the layer where the service should be placed in the fog-cloud ecosystem. Then, it allocates the best available node considering the load balance of the overall ecosystem. The simulation experiments show that the suggested algorithm better balances the load and decreases reaction time compared to the state-of-art algorithms. It is also energy efficient as it minimises the number of active devices and their run time.

Keywords: Task Scheduling · Fog Computing · Cloud Computing · Load Balancing · Clustering

1 Introduction

As Internet of Things (IoT) devices advanced and the rate at which these devices send requests to cloud data centers, the network and the cloud data centers became congested, resulting in delayed performance. Fog computing emerged to resolve the problem by offering services at the network's edge to help time-critical IoT applications. Fog nodes are diverse, distributed, and resource-constrained; as a result, efficient resource management on fog nodes is crucial. Fog and Cloud, often known as Fog-integrated Cloud, provide an excellent environment for handling time-sensitive applications.

The edge-cloud-fog architecture, shown in Fig. 1, consists of three layers: edge, fog, and cloud layer. Every edge layer device connected to a fog node transmits the request to that fog node. Additionally, each cloud node makes resources available in response to requests made by the edge layer. A management node is present in each fog node, which receives task requests from the edge layer and adds them to a queue. The classifier

© The Author(s), under exclusive license to Springer Nature Switzerland AG 2024
K. K. Patel et al. (Eds.): icSoftComp 2023, CCIS 2030, pp. 129–140, 2024.
https://doi.org/10.1007/978-3-031-53731-8_11

module receives requests in the correct order, and the requests' types are identified. Non-real Time, Real-Time, and Important request categories are available. Real-time requests have deadlines and are extremely sensitive to delays. There is no deadline for important requests, although they are sensitive to delays. Requests that are not real-time are seldom affected by the delays.

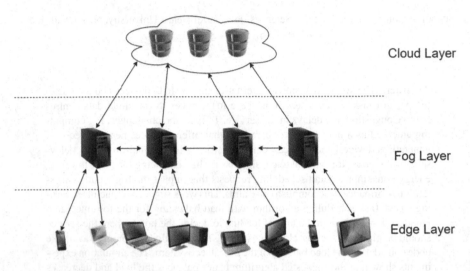

Fig. 1. Edge-Fog-Cloud Architecture

The request is sent for scheduling after the request's nature has been determined. The scheduler assigns the proper resources to handle the requests. It cannot fulfil all requests because of the limited resources in the fog layer. Therefore, non-real-time requests are sent to the cloud layer so the fog layer caters to the remaining time-sensitive demands. Doing this keeps as many as possible real-time and crucial requests away from the cloud layer. Additionally, if a fog node lacks the resources necessary to reply to a task, it will turn to the resources present in other fog nodes for assistance. If other fog nodes decline to supply the requested resources, the request is forwarded to the cloud.

Scheduling has been looked at individually and rarely received attention in the cloud-fog-edge ecosystem. The primary objectives of the task scheduling algorithms are to reduce task execution time and improve performance. To meet customer needs and improve resource efficiency, cloud providers should use effective scheduling techniques, which can promote the development of environment-friendly information technology. Load balancing is one of the most crucial objectives for task scheduling algorithms, along with other objectives, including lowering resource wastage, energy consumption, delay, cost, and reaction time. Load balancing distributes the task requests within a computing environment to improve speed and reliability among available resources. According to our best knowledge, very few academics have focused on task scheduling with load balancing in the fog-cloud ecosystem. Some of the works are presented as follows:

A task scheduling method, IEGA [1], was devised to reduce response time in a fog environment. The authors used an enhanced genetic algorithm to increase the services' standard in this study. The proposed algorithm's performance in carbon emissions, energy consumption, and fitness function efficiency was compared to previous algorithms. The simulation shows that the proposed strategy outperforms alternative strategies across the board.

A multi-agent job scheduling approach for grid computing was proposed by Fellir et al. [2]. Their proposed strategy considers waiting time, task priority, and resource status that serves the most crucial duties. The method was simulated on iFog simulator, showing that the proposed technique performs well.

Subbaraj et al. [3] attempted to use decision-making techniques like AHP and TOPSIS to address the scheduling issue. Their proposed work tried to resolve the issue using well-liked multi-criteria decision-making techniques like AHP and TOPSIS. The performance features of the fog devices are evaluated using two separate multi-criteria decision-making processes. The first technique's priority weight calculation and ranking of fog devices utilise the Analytic Hierarchical Process (AHP). The Technique for Order Preference by Similarity to the Ideal Solution (TOPSIS) algorithm ranks the fog devices in the second technique, which employs AHP for the priority weight computation. Then, the fog devices can be assigned to the duties according to their rank. Simulation results demonstrate that the proposed technique outperforms conventional scheduling algorithms in the fog environment by factoring performance, security, and cost factors into scheduling decisions.

Aburukba et al. [4] created a genetic algorithm-based scheduling solution for computing requests from fixed Internet of Things devices to reduce timeline limitations in a hybrid fog cloud environment. The work suggested a scheduling method using a three-tier fog computing architecture to accommodate maximum requests while meeting deadline criteria. To reduce missed deadlines, a mixed integer programming optimisation model is introduced in this work. With the use of an exact solution technique, the model is validated. Due to the scheduling problem's well-known NP-hardness, precise optimisation methods are insufficient for problems of a typical magnitude in fog computing. Given the difficulty of the issue, a heuristic strategy utilising a genetic algorithm (GA) is proposed. Round-robin and priority scheduling were used to test and compare the proposed GA's performance. The outcomes demonstrate that the recommended strategy has a 20%–55% better rate of deadline misses than the other techniques.

In [5] the authors extensively review the literature that uses machine learning-based methods to address resource management issues in fog computing. These difficulties include task offloading, load balancing, application placement, scheduling, resource allocation, and resource supply. The reviewed literature is comprehensively compared based on the applied tactics, objective measures, tools, datasets, and techniques. Furthermore, they also suggest future strategies for developing the discipline and identifying research needs in problems related to resource management.

In this paper, we have applied a machine-learning approach for efficient task placement in a Fog-integrated Cloud. The suggested work formulated the task placement optimisation problem as a classification and selection problem. A thorough experiment is conducted to assess the performance of the suggested method and observed that the proposed model performs much better than cutting-edge approaches.

The rest of this paper is organised as follows. Section 2 states the problem formulation and matrices considered in this work, whereas Sect. 3 deals with the methodology used and the algorithm designed. Section 4 presents the experimental verification and analyses the results followed by conclusions highlighting the technique used and results obtained in this work.

2 Problem Statement and Formulation

The task manager, often known as the broker, is responsible for collecting and managing the tasks that users have submitted. Specifically, the management process allocates incoming tasks to the available resources. The broker's primary goal is to optimise several necessary metrics, e.g., resource utilisation and energy consumption.

We take into account M IoT devices that produce data for the services. These devices are identified by group name and user ID and generate m Tasks (T) to be sent for placement. The attributes of each task include the task_id, timestamp, CPU requirements, Memory requirements, cores required and delay sensitivity. The task set T is expressed as in Eq. 1.

$$T = \{t_1, t_2, t_3, ...t_m\} \tag{1}$$

These task requests are to be placed on a processing node. The processing node may be on the cloud layer or the fog layer. Each processing node has specific attributes: node_id, CPU_rate (MIPS), number of cores, memory unit, and disk unit. The proposed method operates in two phases. The first phase is the task classification phase, as expressed in Eq. 2, in which each task is offloaded to either the cloud layer or fog layer based on its requirements.

$$Offloading = \{t_i^l : i \in [1, m], l \in \{1, 2\}\} \tag{2}$$

where $t_i \in T$ is the task request that is offloaded to l^{th} layer in the architecture.

In the second phase, the services are placed on a particular cloud or fog node as per the *Offloading*. It is presumed that each service will always be assigned to a single node and will not be preempted till it terminates. A single node can process multiple services, and each node maintains a list of Tasks allotted to it. For example, if Task t_5, t_8 and t_{12} are allotted to node I, it will be represented as given in Eq. 3.

$$Task\_Allotted_i = \{t_5, t_8, t_{12}\} \tag{3}$$

This work applies an ML-based model for the task scheduling problem.

2.1 Performance Metrics

The following Performance metrics have been considered in this work.

Makespan: The total time required to complete all tasks in the computing environment. Better computing efficiency is achieved by having the scheduler assign tasks to existing nodes per the task's and RIS information, which results in the minimum makespan value. Typically, Makespan and Utilization rates are inversely proportional. The most effective scheduling algorithm will arrange the jobs to take the least time and use the most available resources. Makespan is given as in Eq. 4.

$$Makespan^{min} = Max\{Execution\_time_{node}\} \tag{4}$$

Number of Devices Used: Fog/Edge devices are less energy-efficient than cloud data centres. This is a cause for concern, particularly for prospective battery-operated fog nodes like mobile phones, portable laptops, and cars. A node's battery drains more quickly as it operates more swiftly. A fog node is said to have failed when its battery level is zero because it can no longer perform its intended functions. Failure of a node will lower the overall availability of the fog. It could cause additional delays, eventually breaching the service-level agreement (SLA) of applications requiring extremely low latency. The energy use in access and edge network segments will surely increase due to the growing need for fog computing.

Resource Wastage: Resource wastage is calculated as given in Eq. 5.

$$R_i^w = \frac{\sum_{k=1}^r \left(\left| R_i^k - min\left(R_i^k\right) \right| \right) + \in}{\sum_{k=1}^r u_i^k} \tag{5}$$

where R_i^k given by $1\text{-}u_i^k$ is the remaining resource k in the ith device. The resource efficiency of the device is indicated by u_i^k and the lowest usage of resources in the ith device is represented by $min\left(R_i^k\right)$.

Internal Load Imbalance: Resource balancing inside a single resource and within each dimension is crucial. This equilibrium enables the resources to fulfil more requests concurrently. This causes most tasks to be resolved inside the fog node and, less frequently, sent to other fog nodes or the cloud. Equation 6 calculates internal load imbalance when using several resources.

$$disequilibrium(j) = \sqrt{\frac{1}{R} \cdot \sum_{j=1}^R \left(\frac{L_j^R - \overline{L}}{\overline{L}} \right)^2} \tag{6}$$

Since this work considers two dimensions i.e., processor and memory, R = 2. $\overline{L}$ and L_j^R are the average utilisation of all the resources and particular resource usage, respectively.

3   The Proposed Approach

The scheduling procedure is crucial for providing service to the submitted requests in the necessary levels of response times in these paradigms. However, implementing a subpar scheduling technique may result in an unintended increase in response times and, in some circumstances, catastrophes.

All non-real-time requests are forwarded to the cloud, while real-time and important requests should be handled in the fog layer due to response time constraints. To solve the optimization problem in the cloud-fog ecosystem, this work implements an ML-based intelligent admission control manager, after which service scheduling is formulated.

The service classification issue is addressed in the first phase and is covered in Sect. 4.1. An admission control manager using Fuzzy C-means is constructed to forecast the offloading layer. In the second stage, covered in Sect. 4.2, a classification and selection-based model is used to pinpoint the node with the highest convergence rate as the ideal node for the service placement.

3.1 Task Offloading

IoT services are offloaded to the cloud or fog layer based on various parameters. Proper offloading plays a crucial role in the model's overall performance. The problem is to build a suitable approach for classifying the services due to faster reaction times and the restricted processing capability of fog and edge nodes. As a result, an intelligent admission control manager is put into place.

We have used fuzzy C-means clustering, as illustrated in algorithm 1, to form three clusters of the tasks: real-time, important and non-real-time. Depending upon the cluster, the real-time and important tasks are sent to the Fog layer and the non-real time tasks which are not very delay sensitive are offloaded to the cloud layer for scheduling.

Algorithm 1: Task classification and offloading based on Fuzzy C-means clustering

> *Input: T: a set of tasks with task parameters, C: number of clusters*
> *Output: Appropriate cluster for each task*
> *Randomly select 'c' cluster centres.*
> *While the termination criterion is met*
> > *for each task t from task set T, do*
> > > *for each task parameter x do*
> > > > *Calculate membership value μ(x)*
> > > *end for*
> > *end for*
> > *for each cluster c, do:*
> > > *Compute the fuzzy centre c*
> > *end for*

3.2 Service Placement Algorithm

After deciding the layer at which the task will be offloaded, the next job is to place the task on the most suitable node. For this, the tasks are classified based on resource requirements and the nodes are classified based on resource availability. Then, a task allocation algorithm is proposed to select the most promising node for each task based on their resource availability.

Algorithm 2: Task Type decision making based on K means clustering

   Input: T: a set of tasks with task parameters

   Output: Appropriate Task classification

   Set the number of clusters K=4, one for each type of task CPU intensive, memory intensive, both CPU and memory intensive or none Select random K points or centroids from the Task dataset.

   While the termination criterion is met

      for each task t from task set T, do

         Assign t to its closest centroid

      end for

      Calculate the variance and place a new centroid of each cluster.

   If any reassignment occurs, continue else, break

   assign the final type for each task in T.

   Return type for each task

Classification of the Tasks: As given in Algorithm 2, K-means clustering has been applied to form four clusters representing CPU intensive, memory intensive, CPU and memory intensive or none. Each task belongs to one of the four mentioned clusters and is accordingly placed on the most suitable node, which leads to load balance among nodes.

Classification of the Hosts: A random dataset consisting of nodes with possible combinations of tasks allocated has been generated. Next, K means clustering, as described for task classification, has been applied to form 4 clusters of hosts. Each cluster symbolises one of the four possible states of the host, i.e., CPU overloaded, memory overloaded, both CPU and memory overloaded and underloaded. The final centroids obtained from this clustering had been saved and are used at each step to examine the host's current state before assigning the task to the host. The centroids are obtained to guide the task allocation such that the best possible host is selected for each task.

Task Allocation: As in Algorithm 3, all the tasks are checked before allocation. First, the offload layer for the task is checked. Next, the type of node is checked such that the CPU-intensive tasks are placed on memory-overloaded hosts, the memory-intensive tasks are placed on CPU-overloaded hosts, both memory and CPU-intensive tasks are placed on underloaded hosts, and overloaded hosts are avoided till it comes to a neutral state. This helps in attaining proper load balance for the nodes. In addition, the algorithm also takes into account to use minimum number of nodes in order to save energy. Since the tasks may be real-time and important, the algorithm also takes into account the waiting time required for the task placement.

Algorithm 3: Task Placement Algorithm

Input: T: a set of tasks with task parameters, N: number of nodes

Output: Appropriate Task Placement

Set a clock to track the time

For each Task t in T with arrival time ≤ current time:

Check the offload tier l of current task

Check the type of current task

Check if any suitable node on the is available to place task:

If yes place the task on the node

Else check if any suitable node will be available in near future based on the job runtime

If yes wait

else switch on a new node and place task on it

End For

4 Experimental Evaluation

The performance of the clustering-based optimal task placement is studied and compared with several other algorithms. The simulation is run on a machine of 2.40GHz 11th Gen Intel(R) Core (TM) i5-1135G7 with 16GB RAM with Spyder (anaconda3) using Python. This section describes the results of the experiments conducted for the performance evaluation. A fog-integrated cloud data centre is considered. The host configuration considered in the experiments is described in the following subsection followed by the dataset availability and results and comparative analysis.

4.1 Simulation Setup

The network environment consisted of two computation layers: cloud, and fog. The cloud layer had three data centers and 10 virtual machines are running in each of them. Additionally, the fog layer had 100 nodes (Fog nodes), each of which carried a number of resources to address requests from IoT devices. The fog nodes were linked by a network with 1 GB of capacity. Additionally, fog nodes are linked to the cloud layer via communication lines with frequently gigabyte-sized capacity.

4.2 Dataset

LCG (Large Hadron Collider Computing Grid) Dataset (Available at https://www.cs. huji.ac.il/labs/parallel/workload/l_lcg/) is used in the tests to imitate the queries from the IoT devices. The findings of the simulation are examined in four distinct scenarios that included 500, 1000, 1500, and 2000 requests from the IoT devices.

4.3 Results

Figure 2 shows the Task offloading results for 500, 1000, 1500 and 2000 tasks respectively. The points in yellow color depicts real-time requests; the green-colored points are important requests and the magenta-colored points are non-real time requests. All real time and important tasks are sensitive to the deadline and are placed on Fog nodes whereas the non-real time tasks are not as delay sensitive and are offloaded to the cloud data center. The results show that the tasks are not blindly offloaded to cloud data center. Offloading does not depend on the total number of tasks. As shown, the number of tasks offloaded to cloud is much higher for 1000 tasks as compared with the offloading at 2000 tasks.

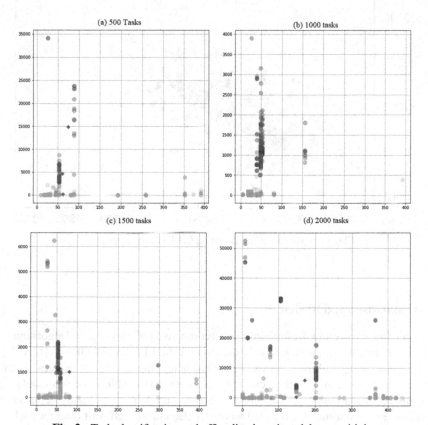

Fig. 2. Task classification and offloading based on delay sensitivity

Figure 3 shows the dataset generated for the load structure of the nodes and clustering of nodes based on its CPU and memory load. The nodes marked as red are neither CPU nor memory overloaded; the nodes pointed in green color are memory overloaded; nodes marked as blue are CPU overloaded and those marked as yellow are both CPU and memory overloaded. The centroids obtained by this clustering is used to access the node's current status every time before it is considered for task allocation. This is crucial as it leads to proper load balance during the task allocation. Tasks are allocated based on the load status of the nodes and if not done properly can negatively impact the total number of devices used for task allocation.

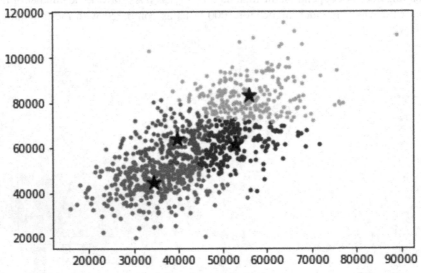

Fig. 3. Node clustering based on CPU and memory usage

4.4 Comparison and Review

In this section, we present the findings from a comparison of the proposed algorithm with the Reinforcement learning fog scheduling (RLFS) [6], LBSSA [7], DRAM [8], GA [9], and PSO-SA [8] algorithms in terms of load balancing, the number of devices required, the average response time, and execution time. As shown in Table 1, the proposed model performs better than the other models. It is because the load is internally balanced and the resources are optimally utilized. The model also considered wait time for each task during scheduling and attained a balance between wait time and number of devices used. Task offloading and host clustering is done beforehand to reduce the overall runtime of the task scheduling.

Table 1. Comparison analysis of proposed model

Number of Tasks	Proposed Approach	RLFS	LBSSA	DRAM	GA	PSO-SA
Load Balance Variance						
500	0.15	0.1	0.15	0.2	0.7	0.75
1000	0.48	0.5	0.8	1.45	1.8	1.9
1500	0.9	1.25	1.7	2.25	2.6	2.75
2000	1.2	1.4	2.2	2.9	3.45	3.5
Average Response Time (s)						
500	0.28	0.33	0.38	0.65	0.35	0.42
1000	0.3	0.35	0.41	0.71	0.49	0.53
1500	0.42	0.45	0.53	0.78	0.72	0.73
2000	0.48	0.51	0.58	0.83	0.79	0.82
Number of Devices Used						
500	35	40	44	60	105	100
1000	37	43	47	67	118	115
1500	40	45	50	75	145	142
2000	45	52	55	81	168	162
Execution Time (s)						
500	2.3	2.5	2.5	2.7	12	5
1000	2.8	3	2.5	2.7	17.5	6
1500	3.2	3.5	3	5	27	7
2000	3.8	4	3.9	6	37.5	8

5 Conclusion

In this work, a Machine Learning based model has been proposed to attain a proper load-balanced task scheduling considering various parameters. The proposed model first classifies the tasks as real-time, important, and non-real time. All the real-time and important tasks are offloaded to Fog layer and the non-real time tasks are offloaded to the cloud layer. Next at the time of scheduling the tasks are again classified as CPU intensive, memory intensive, both CPU and memory intensive and none. Similarly, all the nodes are also classified as CPU overloaded, memory overloaded, both CPU and memory overloaded or underloaded. All the tasks are assigned the most promising node available for it. The proposed model is simulated in Python environment on a real set data and compared with existing models, proving it to be better than others. The results provided good outcomes in reducing the load balance variance, average response time, number of devices used and execution time.

References

1. Abdel-Basset, M., Mohamed, R., Chakrabortty, R.K., Ryan, M.J.: IEGA: an improved elitism-based genetic algorithm for task scheduling problem in fog computing. Int. J. Intell. Syst. **36**(9), 4592–4631 (2021)
2. Fellir, F., El Attar, A., Nafil, K., Chung, L.: A multi-Agent based model for task scheduling in cloud-fog computing platform. In: 2020 IEEE International Conference on Informatics, IoT, and enabling technologies (ICIoT), pp. 377–382. IEEE (2020)
3. Subbaraj, S., Thiyagarajan, R.: Performance oriented task-resource mapping and scheduling in fog computing environment. Cogn. Syst. Res. **70**, 40–50 (2021)
4. Aburukba, R.O., Landolsi, T., Omer, D.: A heuristic scheduling approach for fog-cloud computing environment with stationary IoT devices. J. Netw. Comput. Appl. **180**, 102994 (2021)
5. Fahimullah, M., Ahvar, S., Agarwal, M., Trocan, M.: Machine learning-based solutions for resource management in fog computing. Multimedia Tools Appl. 1–27 (2023). https://doi.org/10.1007/s11042-023-16399-2
6. Ramezani Shahidani, F., Ghasemi, A., Toroghi Haghighat, A., Keshavarzi, A.: Task scheduling in edge-fog-cloud architecture: a multi-objective load balancing approach using reinforcement learning algorithm. Computing **105**(6), 1337–1359 (2023). https://doi.org/10.1007/s00607-022-01147-5
7. Alqahtani, F., Amoon, M., Nasr, A.A.: Reliable scheduling and load balancing for requests in cloud-fog computing. Peer-to-Peer Network. Appl. **14**, 1905–1916 (2021)
8. Xu, X., et al.: Dynamic resource allocation for load balancing in fog environment. Wireless Commun. Mobile Comput. **2018** (2018)
9. Aburukba, R.O., AliKarrar, M., Landolsi, T., El-Fakih, K.: Scheduling internet of things requests to minimize latency in hybrid fog–cloud computing. Futur. Gener. Comput. Syst. Comput. Syst. **111**, 539–551 (2020)

Using Machine Learning Techniques and Algorithms for Predicting the Time Length of Publishing a Law (TLOPL) in the Domain of e-Parliament

Safije Sadiki Shaini[(✉)], Majlinda Fetaji, and Fadil Zendeli

South East University, Tetovo, North Macedonia
ss29749@seeu.edu.mk

Abstract. The analysis of legislative data using machine learning has the potential to greatly enhance parliamentary policymaking. With a focus on estimating when laws and amendments will be published in Parliament, this literature review seeks to learn more about how machine learning can be used in legislative decision-making. However, more than a basic platform with integrated tools and software applications is required for the legislative workspace, particularly during the policy development stage, where many users, such as parliamentary actors and/or stakeholders, are frequently active. To find studies about machine learning application to parliamentary research, a thorough search of electronic databases was performed. To better understand how machine learning is being used in the legislative branch, we conducted a Systematic Literature Review (SLR) of 35 primary papers. The objective of this study is to examine the use of machine learning in legislative decision-making. In addition, we pointed out research needs and gaps and predicted developments in this area.

Keywords: Machine learning · Parliament · Legislative Data · Decision-making processes

1 Introduction

Machine learning uses algorithms to automate the creation of models and discover concealed information from data. It is an iterative process, which enables machines to adapt to new data and scenarios without explicit instructions on where to search [1]. New uses for artificial intelligence and machine learning are added to the body of work on a weekly basis [2]. As the availability of accessible data sources continues to grow, predictive models are increasingly utilized to make more accurate estimations and classifications of data than ever before [3]. Machine learning has the potential to have a major impact in the legislative arena by helping to analyze data, predict outcomes, and automate a number of processes. The legislative process as a whole would be improved by this. This review aims to critically examine the existing literature on ML in Parliament, point out research gaps, and make recommendations for future study.

K. K. Patel et al. (Eds.): icSoftComp 2023, CCIS 2030, pp. 141–154, 2024.
https://doi.org/10.1007/978-3-031-53731-8_12

Techniques that utilize artificial intelligence (AI) and machine learning (ML) are now undergoing revolution in a number of scientific and industrial domains, including computer vision, autonomous driving, natural language processing, and speech recognition [4, 5]. Research on the application of machine learning within the context of Parliament is insufficient, despite its potential benefits. Therefore, the objective of this literature review is to examine the current state of machine learning in Parliament and to determine its benefits and drawbacks.

The following questions were developed as guidelines for this paper:

RQ1: What findings have been addressed in existing Machine Learning Techniques and Algorithms in different research domains?

RQ2: What are the existing solutions in the field of politics, specifically in the parliament for predicting the length of publishing a law or amendment?

RQ3: How can we predict the time length of publishing a law or amendment using machine learning Techniques and Algorithms?

RQ4: How can we improve the management of resources in the parliament using Machine Learning Techniques and Algorithms?

RQ5: What are the potential benefits of using ML techniques prediction to the parliament systems?

RQ6: What methods can be employed to evaluate the key factors that impact the precision of predicting the time it takes to approve a law using machine learning techniques?

RQ7: How to evaluate the importance of factors in the predictive model?

RQ8: Which machine learning techniques are commonly employed for prediction in legislation?

The structure of this paper is as follows: In Sect. 2, we present the methodology employed for carrying out the Systematic Literature Review (SLR), which encompasses the process of data extraction. Section 3 provides an overview of the background and related works in this paper. The results, addressing the research questions, are detailed in Sect. 4. Section 5 contains the paper's conclusion, offering the final reflections on the SLR and outlining directions for future work.

2 Methodology

The methodology used here is based on Kitchenham's [6] recommendations for conducting a systematic literature review. Articles published between 2010 and 2022 were the focus of the search strategy, which involved a thorough search of the major databases such as the ACM Digital Library, IEEE Xplore, Springer Link, ScienceDirect, Elsevier, and Google Scholar. Machine learning, Parliament, government, politics, legislation, voting, and electoral politics were among the most popular search terms. This review only includes English-language articles that specifically address the application of ML in Parliament.

2.1 Research Questions and Scope of the Study

This section presents the research questions and their motivations. In Table 1, we provide the motivations, demonstrating the importance of exploring the research questions.

Table 1. Research questions and motivations

Research Questions	Motivation
RQ1: What findings have been addressed in existing Machine Learning Techniques and Algorithms in different research domains?	To give a broad overview of the innovation in using machine learning in several fields today
RQ2. What are the existing solutions in the field of politics, specifically in the parliament for predicting the length of publishing a law or amendment?	To give some solutions if there are any, in the field of parliament, using machine learning and algorithms
RQ3: How can we predict the time length of publishing a law or amendment using machine learning Techniques and Algorithms?	To assist in optimizing and streamlining the legislative process by estimating the required approval time
RQ4: How can we improve the management of resources in the parliament using Machine Learning Techniques and Algorithms?	To assist legislative managers in allocating resources effectively and efficiently
RQ5: What are the potential benefits of using ML techniques prediction to the parliament systems?	Determine the possible benefits of applying machine learning for prediction in parliament systems
RQ6: What methods can be employed to evaluate the key factors that impact the precision of predicting the time it takes to approve a law using machine learning techniques?	To discover the important parameters that influence the accuracy of approval time prediction
RQ7: How to evaluate the importance of factors in the predictive model?	To analyze the relative relevance of the predictive model's variables in order to enhance its accuracy
RQ7: Which machine learning techniques are commonly employed for prediction in legislation?	To discover which machine learning techniques are more used in legislative branches

2.2 Conduct a Comprehensive Literature Search

In our research, we conducted a systematic search across reputable academic databases, including IEEE, ACM, Springer Link, Science Direct, Elsevier, and Google Scholar (Fig. 1).

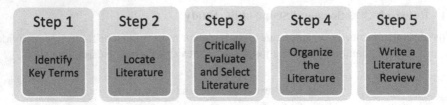

Fig. 1. The steps to conduct the SLR (Systematic Literature Review)

Sources of selected papers	Number of Selected Papers	Search terms
IEEE	6	machine learning AND prediction AND parliament OR legislation
ACM	1	machine learning AND prediction in parliament OR government, AND voting in parliament
Springer Link	8	ML in parliament OR machine learning for prediction in parliament
Science Direct	3	ML in parliament OR government, AND politics
Elsevier	1	ML in parliament OR government, AND politics
Google Scholar	30	machine learning in parliament OR government, AND prediction in parliament laws

2.3 Data Extraction

Abstracts and the titles, also the key terms regarding the papers were originally considered in the screening process. We then developed the inclusion and exclusion criteria in order to perform the selection procedure on the collected papers. This systematic literature review included a total of 35 publications. Data were extracted from each article, including the author(s), year of publication, research methodology, benefits, challenges, and applications of machine learning in Parliament.

3 Related Work

3.1 Background

From 1991 to 2022, 18,267 laws were proposed in the Assembly of the Republic of North Macedonia by the deputies and the government.

The legislative authority of the Republic of Macedonia is placed in the Assembly, which is a body that represents the population. The majority required for adoption is outlined in the Constitution of the Republic of Macedonia. The Assembly sets the legislation governing sessions. Any individual, group of individuals, organization, or society may propose a law to the appropriate authorities [7].

From 1991 to 2022, 18,267 laws were proposed in the Assembly of the Republic of North Macedonia by the deputies and also the government. From these proposed laws, 5,798 were accepted and published in the official gazette, while 12,469 proposed laws were not voted on in the plenary session.

The Rules of Procedure of the Assembly contain specifics about the adoption of laws. The authorized instances must present the law proposal to the Assembly's President in accordance with the rules of procedure. The legislative process starts when the President of the Assembly distributes the proposal to the Members of Parliament in writing or electronically, no later than three (3) working days after the proposal is submitted. According to the Rules of Procedure, the legislative procedure consists of three readings. The legislative process in the Assembly of North Macedonia goes through four stages: the first reading, the second reading, and the final reading, after which the law is published in the Public Gazette.

3.2 Related Works

In the field of politics and parliament, several different works have been done using machine learning techniques for prediction.

Cavalieri et al. [8] have presented a machine-learning approach to automate the task of answering parliamentary questions, a process that can take parliamentary staff a lot of time. To determine the optimal classification models for the particular task at hand, the authors conducted experiments comparing various classical machine learning and deep learning-based text classification models.

In their paper [9] the authors describe an interdisciplinary study that aims to investigate whether an SVM classifier can use recorded speeches from plenary debates of the European Parliament to accurately predict the party affiliations of the participants. The way to find the party affiliation using SVM classifier, accurately through the debates made in the 6th plenary session of the European Parliament and the 5th plenary session, the final results show that such predictions can be made.

Abercrombie et al. [10] created a sizable corpus of UK parliamentary debate speeches that are labeled at the speech level, and this corpus is meant for evaluating the effectiveness of supervised speech-level sentiment classification systems. They conducted initial experiments using different machine learning techniques and methods of text representation to classify speeches from a subset of the corpus, and reported the findings. The authors come to the conclusion that this is a complex work, which requires more analysis and work in the future.

In their paper [11], the authors explore the use of machine learning techniques for analyzing the negativity of parliamentary speeches. Their approach involves conducting supervised sentiment analysis of political speeches in German from Austria, which are referred to as application data in their analysis. Each sentence in the application dataset will receive a predicted negative score according to their system. They used supervised machine learning and created a classification model for sentiment analysis. They achieved good results, but there are a lot of limitations, which are considered as future works.

By favoring data and machine learning above human previous knowledge, De Luca et al. [12] examine the issue of analyst bias when analyzing political discourse through a comparative political lens. It created a machine-learning system that can identify the key themes in Croatian political debate on immigration. The authors used the Decision Tree algorithm, and it gave the best results for comparative political analysis.

Salah [13] in his thesis determined the best approach to using sentiment analysis in the context of public policy discussions. Classification-based methods and lexicon-based methods are both examined as potential solutions. We offer two methods—direct generation and adaptation—to build sentiment lexicons tailored to certain domains. A study comparing several sentiment mining methods found that it is possible to accurately anticipate the speakers' emotions using this technique.

Budhwar et al. [14] analyze hundreds of hours of recorded floor debates from the 2015–2016 session of the California State Legislature to create individual voting behavior models, which are then used to train classifiers and predict future votes. They utilize each lawmaker's words spoken throughout the hearing to forecast the final fate of the bill and then investigate any links between a member's words and their vote. Many machine learning methods and characteristics collected from the voice text are used to evaluate our performance. As a result, our prediction system takes an innovative method to predicting the final outcome of a bill by analyzing data from lawmakers' speeches.

One relevant paper is by John J. Nay et al. [15] where Machine learning was used to forecast the likelihood of a bill becoming law, with a language model used to assign a score and a global sensitivity analysis to identify key factors. They used five models and compared them. From the results obtained, if the bill is proposed by the largest party in power, then the chances are greater that the bill will be voted and passed. Also, if the law is proposed by smaller parties, then the probability is much lower.

Their paper [16] offered a straightforward baseline for predicting the likelihood that a new bill would be voted on and can effectively estimate the prediction that a measure filed into the Kenyan parliament will not be voted on. Logistic Regression and Support Vector Machine models were used to gain their results, then employed meta-learners for achievement of better results. They came to the conclusion that in order to achieve accurate results through machine learning, to see how likely it is that a bill will become law, the text within the law is not as important as the date and year when the law was presented.

Nona Naderi et al. [17] compared the honesty categorization of statements in the PolitiFact dataset with analysis of the classification of facts, falsehoods, dodges, and stretches in the Canadian Parliament. They used machine learning to understand if there is truth in the debates held in the Canadian parliament. Their work was done by training the neural net, and including its probabilities into their model.

Their paper [18] measured the policy stances and their causes using written inquiries from MEPs to the European Commission. The results of their study demonstrate that the tone of the MEP-written questions of Russia is considerably influenced by nationality and EPG affiliation. They used a Sentiment dictionary approach and Inverse text regression approach for human classification.

To anticipate the tone of Austrian parliamentary speeches and news stories, [19] this study used a number of supervised machine learning approaches. And compared their efficacy (German language). They used a phrase tokenizer to predict sentiment scores at the sentence level for all parliamentary speeches.

In [20], authors have introduced the idea of prominence, which relates to the necessity for organizations to be taken seriously by political elites. Additionally, they have created a prominence metric that considers the linguistic context of legislative discussions. Their findings demonstrate that the significance given to organizations by Australian Parliamentarians varies significantly. They observe that many groupings just do not become prominent among elites.

In their paper [21], the authors intend to investigate the significance negative linkages have in relations of graph partitioning further. To achieve this, they make use of a collection of signed networks derived from voting information representing the behavior of the European Parliament members. Their key contribution is a novel algorithm for performing partitioning algorithms that accounts for negative linkages in the networks, hence increasing the amount of available space for data processing.

Therefore, according to [22] the algorithms that were used to develop a model that detects irregularities and outliers in the compensation for the quota for the performance of the parliamentary work. The two successful uses of Autoencoders in Brazilian government fraud detection swayed their decision.

The authors [23] used machine learning algorithms to examine the correlation between countries with high income GDP growth and the proportion of women in parliament, controlling for baseline GDP per capita, population growth, and foreign direct investment as a fraction of GDP. They used advanced machine learning techniques to gain their results. This paper presents results that show that if we only have a 10% increase in the number of women in parliament, then we have an increase in economic development.

The authors in [24] built the network and the timeline using data on countries and organizations taken from discussions in the British Parliament. Those interested in international relations may find this study useful because it used scientific metrology to develop the correlation network and the evolution of subjects to demonstrate international relations from the British perspective.

Meanwhile, our work differs in a few key ways from previous research. These articles don't focus on how long it takes to publish a law, but they do show how machine learning can be utilized to forecast other aspects of the legislative process. In the realm of e-Parliament, similar methods could be used to predict the length of the publishing of laws or amendments in Parliament. There appears to be a gap in the research literature when it comes to knowing the time when a law or amendment will be published in parliament, despite the potential benefits of machine learning in increasing the efficiency of parliamentary decision-making processes. Some studies have been conducted, but there needs to be more in order to fill in the gaps.

Predicting when a law or amendment will be published in parliament is important because it allows interested parties to schedule their activities in advance. Lobbyists, activists, and other interest groups, for instance, may want to modify their advocacy strategies in light of the anticipated publication date of the law. In a similar vein, parliamentary staff may need to rearrange their timetables and allocate additional resources in preparation for the publication of the law.

Use of historical data on the time it took to publish similar laws or amendments in parliament, as well as incorporating factors such as the complexity of the law or amendment, the political climate, and the level of public interest as predictors in a machine learning model, is currently available solutions for predicting the time of publishing a law or amendment in parliament. However, more study is needed to enhance and perfect the current methods that have been developed.

Predicting when a law or amendment will be published in parliament using machine learning techniques might help stakeholders save time and make more informed decisions about their lobbying efforts.

4   Results

In the next section, we will discuss the findings of our SLR in relation to each of the study questions.

1. What findings have been addressed in existing Machine Learning Techniques and Algorithms in different research domains?

The methodologies and algorithms of machine learning have been implemented in a variety of study disciplines, such as image and speech recognition, medical and bioinformatics, and encompassing the domains of financial services [25]. First, we must constantly strike the right balance between data-driven methods and human judgment, and this balance will vary from issue to problem [26]. In the domain of text processing, an assessment was conducted on four core machine learning models: logistic regression, Naive Bayes, decision tree, and support vector machine [27]. It was evident that the support vector machine (SVM) was better than the other models in terms of precision. Recurrent neural networks (RNNs) have become the customary norm for modeling the temporal dependence of audio characteristics in streaming speech recognition models [28]. In natural language processing, pre-trained language models such as BERT and GPT presented significant improvements in tasks such as text classification and question answering [29]. In bioinformatics, deep-learning models have been used to predict molecular evolution, protein structure analysis, systems biology, and disease genomics [30]. In finance, machine learning algorithms have been applied for cost reduction, improved productivity and improved risk management. [31]. Utilizing machine learning methods applied to marketing to detect client demands from internet data, anticipate consumer behavior accurately, customize pricing, and provide product suggestions. [32].

Medical data sources are mined for patterns using machine learning, which has great potential for disease prediction [33].

2. What are the existing solutions in the field of politics, specifically in the parliament for predicting the length of publishing a law or amendment?

Although no research has been conducted thus far to estimate the time required for publishing a law or amendment in parliament, there are other options that can be explored. For instance, leveraging machine learning methods in parliament to predict the likelihood of a bill passing, prioritize bills, enhance decision-making, and reduce the time and energy required to make decisions can streamline the legislative process [16, 34].

Included in a machine learning model prediction like identifying influential members in a parliamentary network, sentiment analysis of public policy discussions, prediction of party affiliation, negativity of parliamentary speeches forecasting the likelihood of a bill becoming law [10, 11, 16]. Along with political scientists, legal specialists, and other stakeholders, developing a new technique or methodology to anticipate the length of publishing a law or amendment will call for substantial knowledge and resources [35].

3. How can we predict the time length of publishing a law or amendment using machine learning Techniques and Algorithms?

The following steps can be done to estimate how long it will take to publish a law or amendment using machine learning techniques and algorithms:

- Collect and preprocess data on historical laws or amendments, including their publication time and relevant factors such as complexity and public interest.
- The data to choose and train a machine learning model, such as a regression or time series model.
- Mean absolute error and mean squared error are two measures that Could be utilized for evaluating the model's efficacy [36].
- A trained model may be applied to estimate how long it will take to publish a new law or amendment based on those parameters.

4. How can we improve the management of resources in the parliament using Machine Learning Techniques and Algorithms?

Machine learning techniques and algorithms can be used to enhance the management of resources in the parliament in the following ways: Predicting the demand for resources such as staff, equipment, and meeting rooms based on historical data and current trends, and adjusting the allocation of resources accordingly. Analyzing the workload and performance of staff members, and identifying areas for improvement or optimization [37]. Automating routine tasks such as document classification and data entry frees up staff time for more complex and high-value tasks [38].

5. What are the potential benefits of using ML techniques prediction to the parliament systems?

In many legislatures throughout the world, the use of artificial intelligence, machine learning, and vast amounts of data is quickly evolving to a crucial part of the legislative process. The number of national legislatures that have considered or experimented with the use of AI in the conduct of their deliberations has increased, and several nations have recently joined this expanding trend [39].

Prioritizing artificial intelligence and machine learning as essential tools in the legislative process is crucial. Due to the complexity of the legislation currently being presented to parliaments and the legislators' limited capacity to comprehend the bill and its implications in-depth [40, 41].

The use of ML approaches for prediction in the parliament systems may have a number of possible benefits, one of which is an improvement in the efficiency and accuracy of resource allocation and management, which may result in cost savings and an increase in productivity [37]. Providing timely and accurate information on critical aspects, such as the amount of time needed to publish a law or modification (amendment), in order to facilitate improved decision-making.

6. What methods can be employed to evaluate the key factors that impact the precision of predicting the time it takes to approve a law using machine learning techniques?

To evaluate the key factors that influence the precision of the prediction of the length of approving a law using ML techniques, feature importance analysis can be performed on the trained machine learning model. This analysis can help identify which features or factors have the most impact on prediction accuracy. Various techniques can be used for feature importance analysis, such as permutation importance, feature importance from decision trees, or SHAP(SHapley Additive exPlanations) values [42–44].

7. How to evaluate the importance of factors in the predictive model?

The importance of components in the predictive model can be evaluated using feature importance analysis, as was already discussed. Different techniques can be employed depending on the specific machine learning algorithm and type of features used. It is also possible to do a sensitivity study to determine how modifications to each attribute influence the model's predictions. Another strategy is to evaluate the model's performance with and without each feature to ascertain how it affects prediction accuracy. [45, 46]. In previous instances, businesses have employed machine learning to forecast customer behavior or results, such as determining which items customers prefer based on past purchases [47]. Several methods exist for creating predictive algorithms utilizing various prediction analytic tools/software. Neural networks, support vector machines, and decision trees are a few examples of these varieties of systems. To provide accurate forecasts of a number of possible outcomes, decision trees, for instance, employ statistical approaches like classification and regression trees, boosting, and random forest. [48].

The first stage in creating a prediction model is to think about the study issue and perform a preliminary data analysis [49]. However, taking into account the overall sample size, the fraction of events, and the number of predictors will help to better estimate predictive performance [50].

8. Which machine learning techniques are commonly employed for prediction in legislation?

There are different techniques used in different fields, but the most basic techniques used are: reinforcement learning, unsupervised learning, and supervised learning.

a) Supervised learning involves generating functions to map new instances of the provided attribute after first employing a collection of training data to train algorithms to carry out analytical tasks [51]. This type of learning requires the user to construct functions to map new instances of the given attribute. These algorithms generate the appropriate functions in order to link the information that is input with the results that are wanted. The resolution of classification issues is a common application of supervised learning, in which an algorithm learns to map input vectors to specific classes by analyzing several examples of input and output pairs.

b) Unsupervised learning - Unsupervised learning is the process of recognizing patterns in data without the assistance of a target attribute. Because all variables are used as inputs in this scenario, the method is suitable for activities such as association mining and clustering, which locate linkages and groups among data points without the use of preconceived categories [52]. These are examples of tasks that are suited for the technique.

c) Reinforcement Learning - When an agent learns from interacting with an environment, reinforcement learning has served its purpose [53]. An RL framework's principal function is to facilitate learning through practice and experience. An RL framework will attempt to learn an optimal sequence of actions to carry out in the environment in order to achieve the objective that it has set for itself [54].

According to the previous research presented [8–11], it would seem that studies from legislative and parliamentary procedures more frequently refer to and use supervised machine learning techniques. Using labeled data, these supervised techniques are applied to problems including sentiment analysis, text categorization, party affiliation prediction etc.

The paper largely emphasizes the use of supervised approaches for specific prediction and analysis tasks within legislative and parliamentary research, while unsupervised techniques like clustering or topic modeling may also have their applications in this context. In summary, in this research, supervised machine learning techniques are more usually used in the applications mentioned. However, the choice between supervised and unsupervised procedures depends on the objectives of a given study or application.

5 Conclusion

The systematic literature review presented in this text examines the prospective utilization of machine learning to analyze and predict parliamentary activities such as voting behavior, sentiment analysis of political speeches, identification of influential members in a parliamentary network, and the likelihood of a bill becoming law. Also, GDP growth in high-income countries and the proportion of women in parliament have been studied using machine learning techniques, and anomalies in the reimbursement of the quota for the performance of legislative proceedings have been uncovered. However, it also identifies a gap in the research literature when it comes to predicting the time of publishing a law or amendment in parliament. In conclusion, this comprehensive literature analysis offers an enlightening summary of how machine learning might be used to the functioning of legislative systems. The SLR shows the potential benefits of utilizing methods and algorithms for machine learning across a variety of study disciplines, including politics, as one of such domains. The time it takes to publish a bill or amendment may be estimated by utilizing machine learning, which has the possibility of useful applications for many stakeholders, including lobbyists and legislative staff. In addition to this, using machine learning in parliament may help enhance the administration of parliamentary resources. In order to address these objectives and provide a prediction that is more accurate, we need to do further study. The effectiveness of legislative decision-making processes, as well as the planning and lobbying strategies used by a variety of stakeholders, may be improved through more research in this area.

References

1. Bhardwaj, R., Nambiar, A.R., Dutta, D.: A study of machine learning in healthcare. In *2017 IEEE 41st annual computer software and applications conference (COMPSAC)* (Vol. 2, pp. 236–241). IEEE July (2017)
2. Fluke, C.J., Jacobs, C.: Surveying the reach and maturity of machine learning and artificial intelligence in astronomy. Wiley Interdiscip. Rev.: Data Mining Knowl. Disc. **10**(2), e1349 (2020)
3. Mersy, G., Santore, V., R., I., Kleinman, C., Wilson, G., Bonsall, J., Edwards, T.:, . A comparison of machine learning algorithms applied to Aamerican legislature polarization. In: 2020 IEEE 21st International Conference on Information Reuse and Integration for Data Science (IRI), (pp. 451–456). IEEE August (2020)
4. Galbusera, F., Casaroli, G., Bassani, T.: Artificial intelligence and machine learning in spine research. JOR spine **2**(1), e1044 (2019)
5. Kononenko, I.: Machine learning for medical diagnosis: history, state of the art and perspective. Artif. Intell. Med. **23**(1), 89–109 (2001)
6. Kitchenham, B.: Procedures for performing systematic reviews. Keele, UK, Keele University **33**(2004), 1–26 (2004)
7. https://www.sobranie.mk/
8. Cavalieri, A., Ducange, P., Fabi, S., Russo, F., Tonellotto, N.: An Intelligent system for the categorization of question time official documents of the Italian Chamber of Deputies. J. Inform. Technol. Politics 1–22 (2022)
9. Høyland, B., Godbout, J.F., Lapponi, E., Velldal, E.: Predicting party affiliations from European Parliament debates. In: Proceedings of the ACL 2014 Workshop on Language Technologies and Computational Social Science, (pp. 56–60), June (2014)
10. Abercrombie, G., Batista-Navarro, R.T.: ParlVote: a corpus for sentiment analysis of political debates. In: Proceedings of the 12th Language Resources and Evaluation Conference, (pp. 5073–5078) May (2020)
11. Rudkowsky, E., Haselmayer, M., Wastian, M., Jenny, M., Emrich, Š., Sedlmair, M.: Supervised sentiment analysis of parliamentary speeches and news reports. In: 67th Annual Conference of the International Communication Association (ICA), Panel on Automatic Sentiment Analysis (2017)
12. De Luca, G., Beck, M.: Natural language processing for the analysis of the political characterisation of migration in the croatian political discourse. RUDN J. Political Sci. **22**(3), 517–532 (2020)
13. Salah, Z.: Machine learning and sentiment analysis approaches for the analysis of Parliamentary debates (Doctoral dissertation, University of Liverpool) (2014)
14. Budhwar, A., Kuboi, T., Dekhtyar, A. and Khosmood, F.: May. Predicting the vote using legislative speech. In: Proceedings of the 19th Annual International Conference on Digital Government Research: Governance in the Data Age, (pp. 1–10) (2018)
15. Nay, J.J.: Predicting and understanding law-making with word vectors and an ensemble model. PLoS ONE **12**(5), e0176999 (2017)
16. Babafemi, O., Akinfaderin, A.: Predicting and Analyzing Law-Making in Kenya. arXiv preprint arXiv:2006.05493 (2020)
17. Naderi, N., Hirst, G.: Automated fact-checking of claims in argumentative parliamentary debates. In: Proceedings of the First Workshop on Fact Extraction and VERification (FEVER), (pp. 60–65) Nov (2018)
18. Dekalchuk, A., Khokhlova, A., Skougarevskiy, D.: National or European Politicians? Gauging MEPs Polarity When Russia is Concerned. Gauging MEPs Polarity When Russia is Concerned (May 13, 2016). Higher School of Economics Research Paper No. WP BRP, 35 (2016)

19. Rudkowsky, E., Haselmayer, M., Wastian, M., Jenny, M., Emrich, Š., Sedlmair, M.: Supervised sentiment analysis of parliamentary speeches and news reports. In: 67th Annual Conference of the International Communication Association (ICA), Panel on Automatic Sentiment Analysis (2017)
20. Fraussen, B., Graham, T., Halpin, D.R.: Assessing the prominence of interest groups in parliament: a supervised machine learning approach. J. Legislative Stud. 24(4), 450–474 (2018)
21. Mendonça, I., Trouve, A., Fukuda, A.: Exploring the importance of negative links through the European parliament social graph. In: Proceedings of the 2017 International Conference on E-Society, E-Education and E-Technology (pp. 1–7) Oct (2017)
22. Gomes, T.A., Carvalho, R.N., Carvalho, R.S.: Identifying anomalies in parliamentary expenditures of brazilian chamber of deputies with deep autoencoders. In: 2017 16th IEEE International Conference on Machine Learning and Applications (ICMLA), (pp. 940–943). IEEE Dec (2017)
23. Khorsheed, E.: Women parliamentarians impact on economic growth: a cross-country analysis evidence. In: 2019 8th International Conference on Modeling Simulation and Applied Optimization (ICMSAO), (pp. 1–5). IEEE Apr (2019)
24. Wang, J., et al.: Analyzing international relations from British parliamentary debates. In: Proceedings of the ACM/IEEE Joint Conference on Digital Libraries in 2020 (pp. 463–464) Aug (2020)
25. Haldorai, A., Murugan, S., Ramu, A.: Evolution, challenges, and application of intelligent ICT education: aAn overview. Comput. Appl. Eng. Educ. 29(3), 562–571 (2021)
26. Petropoulos, F., et al.: Forecasting: theory and practice. Int. J. Forecast 38(3), 705–871 (2022)
27. Sengupta, S., Dave, V.: Predicting applicable law sections from judicial case reports using legislative text analysis with machine learning. J. Comput. Soc. Sci. pp.1–14 (2021)
28. Sak, H., Senior, A.W., Beaufays, F.: Long short-term memory recurrent neural network architectures for large scale acoustic modeling (2014)
29. Topal, M.O., Bas, A., van Heerden, I.: Exploring transformers in natural language generation: Gpt, bert, and xlnet. arXiv preprint arXiv:2102.08036 (2021)
30. Auslander, N., Gussow, A.B., Koonin, E.V.: Incorporating machine learning into established bioinformatics frameworks. Int. J. Mol. Sci. 22(6), 2903 (2021)
31. Leo, M., Sharma, S., Maddulety, K.: Machine learning in banking risk management: a literature review. Risks 7(1), 29 (2019)
32. Proserpio, D., et al.: Soul and machine (learning). Mark. Lett. 31, 393–404 (2020)
33. Shailaja, K., Seetharamulu, B., Jabbar, M.A.: Machine learning in healthcare: A review. In: 2018 Second International Conference on Electronics, Communication and Aerospace Technology (ICECA), (pp. 910–914). IEEE March (2018)
34. Pal, A.: DeepParliament: A Legal domain Benchmark & Dataset for Parliament Bills Prediction. *arXiv preprint* arXiv:2211.15424 (2022)
35. Jabeur, S.B., Ballouk, H., Arfi, W.B., Khalfaoui, R.: Machine learning-based modeling of the environmental degradation, institutional quality, and economic growth. Environmental Modeling Assessment, pp.1–14 (2021)
36. Bouktif, S., Fiaz, A., Ouni, A., Serhani, M.A.: Optimal deep learning LSTM model for electric load forecasting using feature selection and genetic algorithm: comparison with machine learning approaches. Energies 11(7), 1636 (2018)
37. Zekić-Sušac, M., Mitrović, S., Has, A.: Machine learning based system for managing energy efficiency of public sector as an approach towards smart cities. Int. J. Inf. Manage. 58, 102074 (2021)
38. Anagnoste, S.: Robotic Automation process-the next major revolution in terms of back office operations improvement. In: Proceedings of the International Conference on Business Excellence (Vol. 11, No. 1, pp. 676–686) July (2017)

39. Saari, M.: IR 4.0 in Parliament: Conceptualising the application of artificial intelligence and machine learning in the Parliament of Malaysia's parliamentary questions. In: International Journal of Law Government and Communication, **5**(20), pp.124–137 (2020)
40. Furst, K.: Why the legislative and election process needs artificial intelligence (2018)
41. Reis, J., Santo, P.E., Melão, N.: Impacts of artificial intelligence on public administration: A systematic literature review. In: 2019 14th Iberian conference on information systems and technologies (CISTI) (pp. 1–7). IEEE June (2019)
42. Gómez-Ramírez, J., Ávila-Villanueva, M., Fernández-Blázquez, M.Á.: Selecting the most important self-assessed features for predicting conversion to mild cognitive impairment with random forest and permutation-based methods. Sci. Rep. **10**(1), 1–15 (2020)
43. Lundberg, S.M., Erion, G.G., Lee, S.I.: Consistent individualized feature attribution for tree ensembles. arXiv preprint arXiv:1802.03888 (2018)
44. Guo, M., Yuan, Z., Janson, B., Peng, Y., Yang, Y., Wang, W.: Older pedestrian traffic crashes severity analysis based on an emerging machine learning XGBoost. Sustainability **13**(2), 926 (2021)
45. Ghorbani, R., Ghousi, R.: Comparing different resampling methods in predicting students' performance using machine learning techniques. IEEE Access **8**, 67899–67911 (2020)
46. Popovici, V., et al.: Effect of training-sample size and classification difficulty on the accuracy of genomic predictors. Breast Cancer Res. **12**(1), 1–13 (2010)
47. Martínez, A., Schmuck, C., Pereverzyev, S., Jr., Pirker, C., Haltmeier, M.: A machine learning framework for customer purchase prediction in the non-contractual setting. Eur. J. Oper. Res. **281**(3), 588–596 (2020)
48. Waljee, A.K., Higgins, P.D., Singal, A.G.: A primer on predictive models. Clin. Transl. Gastroenterol. **5**(1), e44 (2014)
49. Steyerberg, E.W., Vergouwe, Y.: Towards better clinical prediction models: seven steps for development and an ABCD for validation. Eur. Heart J. **35**(29), 1925–1931 (2014)
50. van Smeden, M., et al.: Sample size for binary logistic prediction models: beyond events per variable criteria. Stat. Methods Med. Res. **28**(8), 2455–2474 (2019)
51. Libbrecht, M.W., Noble, W.S.: Machine learning applications in genetics and genomics. Nat. Rev. Genet. **16**(6), 321–332 (2015)
52. Nasteski, V.: An overview of the supervised machine learning methods. Horizons. b **4**, 51–62 (2017)
53. Berry, M.W., Mohamed, A., Yap, B.W. eds.: Supervised and unsupervised learning for data science. Springer Nature (2019)
54. Mousavi, S.S., Schukat, M., Howley, E.: Deep reinforcement learning: an overview. In: Proceedings of SAI Intelligent Systems Conference (IntelliSys) 2016: Vol. 2, (pp. 426–440). Springer International Publishing (2018)

Deep Metric Learning with Music Data

Vignesh Bhat$^{(\boxtimes)}$ and J. Angel Arul Jothi⬤

Department of Computer Science, Birla Institute of Technology and Science Pilani, Dubai Campus, Dubai International Academic City, Dubai, UAE
bhatvignesh17@gmail.com, f20190088d@alumni.bits-pilani.ac.in

Abstract. Music Information Retrieval has often relied on supervised learning with handpicked features. These supervised models are often trained with a fixed set of labels and require to be retrained whenever new labels are introduced to the dataset. In this paper we develop a distance metric between a database of songs provided to it and use it to mine deeper and more complex relationships between songs that may not be easily quantified. We achieve this by the usage of Deep Metric Learning using a Siamese network. We have developed a metric between multiple songs of artist. Our developed model is successful in maintaining a healthy separation between artists and able to also encode deeper relations such as genre. The model performs well even when new labels are introduced to the dataset and boasts comparable accuracy to previous research done in this field as it is well generalized due to hard triplet mining and can be used to handle multiple tasks like artist prediction, genre auto-tagging and content-based retrieval.

Keywords: Deep Metric Learning · Siamese Network · Triplet Mining · Mel Spectrograms · Music Data

1 Introduction

Traditional supervised learning has been used extensively to build deep learning models. But recently there has been a surge in interest in adapting deep metric learning (DML). DML is a subfield of deep learning focused on learning meaningful feature representations for similarity-based tasks. In DML, the objective is to train neural networks to map input data into a high-dimensional embedding space, where similar samples are closer together, and dissimilar ones are farther apart. Siamese or triplet network architectures are widely employed for this by optimizing loss functions to preserve relative distances between data points. By learning a discriminative metric directly from the data, it enables the network to handle complex and high-dimensional data effectively. It is commonly used in tasks like face recognition, image retrieval, and content-based recommendation systems. It has also demonstrated significant improvements in various applications, particularly in cases with limited labeled data, few-shot learning scenarios, and large-scale similarity-based tasks [1].

Extracting useful information from huge amounts of music data is music mining which involves tasks like music retrieval, music similarity detection for music recommendation, artist/genre classification etc., to name a few. Music information retrieval

K. K. Patel et al. (Eds.): icSoftComp 2023, CCIS 2030, pp. 155–167, 2024.
https://doi.org/10.1007/978-3-031-53731-8_13

(MIR) is a field of research that focuses on developing algorithms, techniques, and systems to extract meaningful information from music signals and enable efficient retrieval, organization, and analysis of music content [2]. Music similarity detection is the process of quantifying the degree of similarity between different music tracks. It aims to find songs with similar musical characteristics, aiding in tasks like music recommendation, content-based music retrieval, and playlist generation. Over the years, significant advances have been made in these tasks, driven by the availability of large music databases, advancements in signal processing, and the rise of artificial intelligence techniques.

Traditional methods for music mining relied on handcrafted features, while modern approaches leverage deep learning techniques such as convolutional neural networks (CNNs) and recurrent neural networks (RNNs) to learn high-level representations from raw audio data. But these models might fail to extract the hidden features from the enormous amounts of music data. Thus, the requirement is to extract the inherent, hidden and useful representation from this data, that could be used for several other downstream tasks.

The goal of this paper is to learn a metric (a metric satisfies the properties of non-negativity, symmetry, and transitivity) between songs on the basis of artists. For this, a Siamese network which uses three identical CNNs is trained to derive a meaningful representation (embedding) of the input audio data using deep metric learning. The n-dimensional embedding received is further reduced to 2-dimensions and plotted to view the artist classes and genre classes captured. Additionally, the trained model is also used for artist prediction. The trained Siamese network can be used to generate an embedding (a n-dimensional point) for any raw audio data fed into it and can be positively used for genre tagging and content-based music retrieval.

The following sections of this research paper will discuss in further detail the literature review, the dataset used, the proposed system, the implementation details, the evaluation metrics, the experimental results obtained, the recommended use cases and finally the conclusion.

2 Literature Review

Music Information Retrieval has seen multiple works that have utilized Siamese networks before for feature representation and others have explored supervised approaches.

Jiyoung Park et al. [3] used audio features to represent with artist labels and compared performances between a standard CNN for multiclass artist classification and Siamese Neural Network for maximizing similarity based on artist labels. They found best performance on song retrieval and genre classification task and that using larger number of artists increases model performance.

Jaehun Kim et al. [4] explored multiple different networks architectures with multiple learning sources. They found that it was difficult to achieve generality in the music domain. A single source was found to be effective for specific cases but for general cases, models, with less shared information from different sources were found to outperform base models.

Siamese networks also have seen usage in several papers such as [5] which introduces Siamese neural networks for one-shot image recognition, where a model must recognize objects with very limited training examples. The Siamese network consists of twin sub-networks with shared weights, learning to map images into a feature space that places similar images close together and dissimilar images apart. The contrastive loss function encourages the network to optimize the distances between image pairs. The approach shows promise in learning meaningful image representations with limited data, making it valuable for few-shot learning tasks.

Jongpil Lee et al. [6] propose a method for music similarity detection using deep convolutional embeddings. The approach involves transforming music audio into compact representations through CNN. These embeddings capture the salient features of the audio, enabling effective comparison and similarity measurement between music pieces. The method is evaluated on various datasets, demonstrating its ability to accurately assess music similarity and outperforming traditional techniques. The research contributes to the field of music information retrieval by providing a deep learning-based solution for music similarity analysis and content-based music retrieval tasks.

Rui Lu et al. [7] propose an innovative approach called "Triplet MatchNet" for music metric learning using deep ranking. Metric learning involves learning a similarity function to rank data instances based on their similarities. It demonstrates its efficacy in enhancing music metric learning and recommendation systems. By training on large-scale music datasets, Triplet MatchNet can effectively capture intricate relationships between songs, leading to more accurate music similarity assessments and improved music recommendations.

This research contributes to music information retrieval and content-based recommendation systems by providing a robust method for audio similarity learning.

Another paper on using the concepts of Deep Metric Learning [8] introduces a triplet network for deep metric learning. The paper addresses the challenge of learning effective feature representations for comparing images in a similarity-based manner. The triplet network learns to map images into a common space, where the distance between anchor-positive pairs is minimized, while the distance between anchor-negative pairs is maximized. The proposed approach outperforms traditional metric learning methods and achieves significant improvements in various tasks, such as face verification and image retrieval.

Triplet loss was first used in face recognition as seen in the groundbreaking approach [8] for face recognition and clustering using deep metric learning. The paper proposes FaceNet, a CNN that learns a compact and discriminative embedding space for face images. It can also be seen in other papers dealing with facial recognition such as [9] which focus on handling subtle differences between visually similar images in fine-grained categories. The proposed method uses a deep ranking loss function to train a neural network that learns a discriminative feature representation. The network's objective is to rank similar images higher than dissimilar ones in the learned feature space.

3 Dataset

The dataset used for this project is the Free Music Archive [11], which is an open-source dataset for the purpose of MIR. It provides 917 GB of music data comprising 106,574 tracks from 16,341 artists and 14,854 albums, with a total of 161 genres. The dataset is available as mp3 files and is divided into multiple subsets, namely - fma_full (917 GB of unaltered audio), fma_large (93 GB of 30 s of 106,576 songs), fma_medium (25,000 songs of 30 s each of 22 GB) and fma_small (8,000 songs of 30 s each, 7.2 GB). The dataset also includes a set of metadata files which is sub-divided into 4.csv files as follows:

a) tracks.csv - per track metadata such as ID, title, artist, genres, tags and play counts, for all 106,574 tracks.
b) genres.csv – 163 genres with names
c) features.csv – features extracted with the librosa library
d) echonest.csv – additional audio features provided by echonest/spotify

In this paper, only the subset fma_small is utilized which comprises of songs distributed over 8 balanced genres. It should be noted that the dataset has an artist imbalance with some artists having over 100 songs while many others only have a single song. Figure 1 shows the distribution of the songs in the dataset.

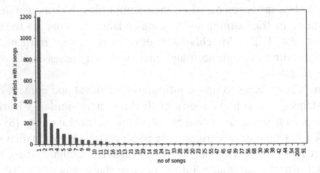

Fig. 1. Distribution of songs in the dataset.

4   Proposed Method

This section details the various steps of the proposed method like preprocessing, the Siamese model used and triplet mining.

4.1   Data Preprocessing

The raw audio files require two steps of preprocessing: first conversion to an image format (Mel spectrograms), and then subset creation.

Mel Spectrogram
Mel spectrograms are widely utilized as input representations [12, 13] for convolutional neural networks (CNNs) in audio analysis. By transforming the linear frequency scale into a logarithmic scale that aligns with human auditory perception, Mel spectrograms capture the relevant spectral information of an audio signal. This representation enables

CNNs to effectively learn and recognize patterns in audio data. To learn a suitable representation from the audio data as raw as possible, a Mel Scale magnitude spectrum on the dB scale is utilized using a 128 bin Mel filter Short-Time Fourier Transform (STFT). The input shapes of the resulting graphs are 396 × 469 in grayscale as color is only used to represent decibel intensity. Figure 2 shows sample Mel spectrograms.

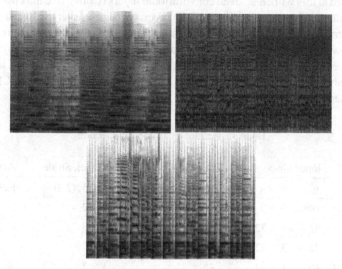

Fig. 2. Spectrograms generated.

Subset Creation

For training by using a Siamese network, our final dataset will be several triplets consisting of an anchor (song of an artist), a positive sample (another song of said artist), and a negative sample (song of a different artist). So, for any given list of songs of length n. the total theoretical dataset size would be of size C^n_3. In addition, running three simultaneous CNNs bears a high computation and memory cost. Considering this the following datasets were created.

A subset of 354 artists who had at least 6 songs were selected to be used as the training set. The train set only includes 6 songs from each artist, while the remaining songs of the same artists totaling 1276, belonging to these artists were used as the test set. A total of 3400 songs were derived as the first subset. Each anchor in this dataset has 5 options for a positive sample and 353 options for a negative sample. Total theoretical triplets in the training set here come to 624,810.

This will be referred to as balanced-large. Another subset of 100 artist classes each with at least 10 songs was later selected to assist in triplet mining. The remaining songs of these artists totaling 455 were used as the test set. This subset was comprised of a total of 1455 songs. Total theoretical triplets here come to 89,100.

4.2 The Siamese Network

Siamese networks are a type of neural network architecture designed for similarity-based learning tasks. They were first introduced in the 1990s and have gained popularity in various domains, including image similarity, text similarity, and audio similarity. This work uses Siamese neural network with a CNN feature extractor. The CNN feature extractor used in this work has seven convolutional layers (Conv), five max pooling layers (MaxPool), a global average pooling (GAP) layer, two dense layers with a dropout layer between them and an output layer. The first layer is the only layer that has a stride of (2, 3), to lower the dimensionality, this is also the only layer that does not have same padding. The final layer in the model is an L2 normalization layer. The final layer has an alterable embedding size (64 or 128). Table 1 shows the architecture of the CNN feature extractor used.

Table 1. Feature extractor Architecture

Layer	Input Shape	Out Shape	Kernel, Stride	Activation
Conv1	$369 \times 496 \times 1$	$186 \times 166 \times 16$	5×5, (2,3)	Relu
MaxPool1	$186 \times 166 \times 16$	$92 \times 83 \times 16$	2×2, -	
Conv2	$92 \times 83 \times 32$	$92 \times 83 \times 32$	3×3	Relu
Maxpool2	$92 \times 83 \times 32$	$46 \times 41 \times 32$	2×2,-	
Conv3	$46 \times 41 \times 32$	$46 \times 41 \times 64$	3×3	Relu
Maxpool3	$46 \times 41 \times 64$	$23 \times 20 \times 64$	2×2,-	
Conv4	$23 \times 20 \times 64$	$23 \times 20 \times 64$	3×3	Relu
Maxpool4	$23 \times 20 \times 64$	$11 \times 10 \times 64$	2×2,-	
Conv5	$11 \times 10 \times 128$	$11 \times 10 \times 256$	3×3	Relu
Maxpool5	$11 \times 10 \times 128$	$5 \times 5 \times 128$	2×2,-	
Conv6	$5 \times 5 \times 256$	$5 \times 5 \times 256$	3×3	Relu
Conv7	$5 \times 5 \times 256$	$5 \times 5 \times 256$	1×1	Relu
GAP	$5 \times 5 \times 256$	256	–	
Dense	256	256	–	Relu
Dropout	256	256	–	
Dense	256	64	–	None
Lambda	64/128	64/128	–	

Our Siamese network as shown in Fig. 3 consists of triplet network with shared weights of the CNN feature extractor. Each sub-network processes one input, and all three sub-networks have the same architecture and parameters. For each input it embeds a unique output which is trained to maximize the similarity scores of similar songs. The architecture is set up to accept three inputs namely – the anchor (any song of an artist), the positive (a song of the same artist as that of the anchor), and the negative (a song of a different artist). The goal of the network is to minimize the distance between the embedding of the anchor sample and the positive sample and maximize the distance between the anchor sample and the negative sample as shown in Fig. 4. This can be achieved by using triplet loss as given by Eq. 1.

$$\|f(x_i^a) - f(x_i^n)\|^2 + \alpha < \|f(x_i^a) - f(x_i^p)\|^2 \tag{1}$$

where x_i^a represents an anchor sample (an artist's song), x_i^p represents a positive sample (another song of the same artist), x_i^n represents a negative sample (a song by a different artist), $f(x_i^a)$ represents the output of the CNN embedding network for the

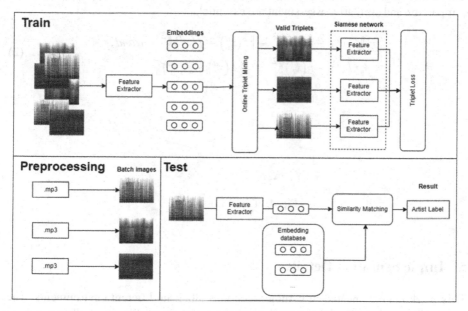

Fig. 3. Overall architecture

anchor sample, $f(x_i^p)$ represents the output of the CNN embedding network for the positive sample, $f(x_i^n)$ represents the output of the CNN embedding network for the negative sample, $\|.\|^2$ denote the Euclidean norm, and α is the margin variable used to limit the minimum difference between the positive and negative sample distance. If we were to generate all possible samples, it would have too many easy samples that would not lead to a significant loss, inhibiting learning.

4.3 Triplet Mining

To assure faster convergence it is required to select triplets [7] that violate Eq. 1, (i.e.,) for a given anchor, a positive pair should be in such a way that we have the hardest negative samples for it. This can be achieved in two ways, one is by generating all possible hard combinations as a batch before training, also known as offline triplet mining. Although this would require very high memory overhead that makes it infeasible for even moderately sized datasets. For this we use an alternative known as online triplet mining. In online triplet mining we compute the triplets that satisfy the criteria within a mini batch of a thousand examples. In this case as the dataset used itself is small, the mini batch comprises all the 1000 samples. For selection, all anchor positive pairs are utilized as it causes faster convergence early on. Selection of hardest negative of each positive pair was avoided as it led to model to stagnate at a local minimum early on, so we select negative sample that satisfies the criteria given by Eq. 2. The negative samples obtained are semi-hard, as they are closer to the positive samples within the margin value but also close to the anchor variable. This training approach helps the network learn a meaningful similarity metric, which can be useful for various applications where comparison and similarity assessment are essential.

$$
\begin{aligned}
&\left\| f\left(x_i^a\right) - f\left(x_i^n\right)\right\|^2 - \left\| f\left(x_i^a\right) - f\left(x_i^p\right)\right\|^2 < \alpha \, and, \\
&\left\| f\left(x_i^a\right) - f\left(x_i^p\right)\right\|^2 <= \left\| f\left(x_i^a\right) - f\left(x_i^n\right)\right\|^2
\end{aligned}
\tag{2}
$$

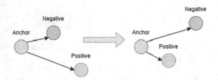

Fig. 4. Triplet Loss Goal

5   Implementation Details

This work is implemented in python notebooks albeit in different environments. The music data was initially preprocessed into images and numpy files in locally on an Intel i7 8th gen CPU, in a Python 3.8 environment. The processed data was made available via Google Drive to Python Notebooks powered by GPU accelerated Google Colab. The training and predictions were done on Tesla T4 GPU with 16 GB RAM machines. Librosa library was used for raw audio extraction along with pillow for image conversion. The converted images are then read and processed into NumPy to save and hold the data. The Tensorflow library was used for all deep learning modelling and training. The sklearn library used testing and predictions.

For model evaluation, the area under the receiver operating characteristic (ROC) curve (AUC) is used as it is particularly useful when dealing with imbalanced datasets, where the classes are unevenly distributed. It measures the classifier's ability to distinguish between positive and negative instances across various thresholds, providing a more comprehensive assessment of the model's discrimination power compared to single-point metrics like accuracy. It plots the true positive rate (sensitivity) against the false positive rate (1-specificity). The AUC value ranges from 0 to 1, with a higher value indicating better classifier performance.

Two models were trained, one with an embedding size of [64, 128] with optimizers [Adam, SGD], for 100 classes and another with an embedding size of 128 with the Adam optimizer for 354 classes. The model used for further performance testing and prediction is the single CNN model that the Siamese network comprises of. For training the models, the dataset utilizes the spectrograms as input and the artist of each song, derived from the tracks.csv file is taken as the anchor value for the Siamese network. Artist, in particular was chosen as semantic labels instead of genre and mood because they are usually ambiguous and become difficult to annotate by a crowd. High quality music annotation by music experts is expensive and very time consuming while artist names are labels that are strongly associated with the music releases. This becomes a powerful metric as each artist has their own unique style of music. Often prolific artists tend to define the genre they make music of. This would act as a secondary genre representation derived from a model trained on a different label set.

6 Results and Discussion

6.1 Model Evaluation

The model was fed the balanced subset of 1455 songs, and its corresponding embedding was generated. These embeddings were used for a binary classification task, where songs of the same artist were positive and songs of two different artists were considered negative. From these embeddings, an input array was generated as the distance between the embedding of each song and all other songs, and an equivalent array of labels were created. For predictions, if the distance between two songs fell below a threshold of 0.15, it was considered positive. The Area Under the Receiving Operating Characteristic (AUC) was calculated and the results can be seen in Table 2 and Fig. 5. It can be noted that the best result was provided by the 64 feature extractor trained on SGD and shows the best generalizability. Prediction was also done on the 3400-song dataset of 354 classes with the 128-feature extractor, it provided similar results to that of the earlier models.

Table 2. Performance of the models

Classes	Embedding size	Optmizer	Train AUC	Test AUC
100	64	SGD	0.954	0.826
100	128	Adam	0.969	0.796
354	128	Adam	0.963	0.808

6.2 T-SNE Plots

The dataset of 1455 songs were taken and passed through the 64-embedding Adam optimized model to get 1455 datapoints. To visualize these data points, their dimensionality

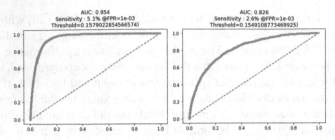

Fig. 5. ROC Curves for the 64-embedding SGD model (a) Training data (b) Testing data

is reduced with T-distributed Stochastic Neighbor Embedding (t-SNE). The finals datapoints are plotted as shown in Fig. 6. In Fig. 6 each dot represents a song embedding reduced to 2 dimensions by t-SNE. The graph above on its own does not provide any useful data but it can be viewed through different lenses to extract further meaningful relationships as follows.

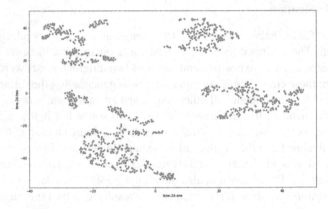

Fig. 6. 1455 songs plotted in 2-dimensions using T-SNE

Artist Class Visualization
From Fig. 6, each song was categorized by artist, to get 100 artists across the 1455 songs which is shown by Fig. 7a. It can also be seen from Fig. 7a that most artists form clusters. It should be noted that though the model was only trained and has seen a maximum of 10 songs per artist, but the clusters seem to group all songs of an artist in most cases.

Genre Class Visualization
If the songs plotted by Fig. 6 are then categorized by genre, we obtain 1455 songs split into 8 genres. Figure 7b shows the same plot but colored by each song's genre. It could be seen from Fig. 7b that the songs are grouped according to genre fairly well despite the model being trained for artists. The feature extractor seems to have successfully learnt a complex relationship between the artists and genre of songs too. From Fig. 7b, it is apparent that 'International' category and 'Folk' categories have distinct signatures that

both form distinct clusters while 'Instrumental' is mostly its own cluster and 'Hip-Hop' and 'Electronic' seem to merge into each other. 'Pop' and 'Experimental' seem scattered with no specific cluster. These models also seem unable to fit different genres perfectly.

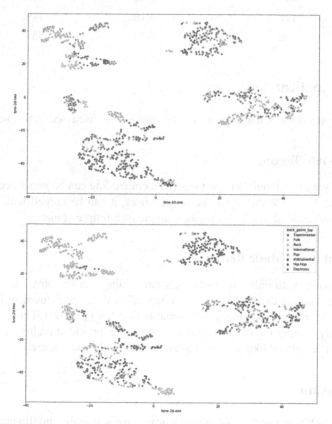

Fig. 7. a.) 1455 songs split into 100 artist classes b.) 1455 songs classified by genre

6.3 Artist Prediction

To predict an artist, the dataset is reformatted to the songs as the input and each songs' artists as the label. This is then fed to the trained CNN to produce an n-dimensional embedding. To simplify the prediction, the n-dimensional embedding is brought down to a 2-dimensional map by Neighborhood Component Analysis. For each embedding generated by the model, its nearest neighbors are calculated, and the most common artist is used as the predicted label. The performance of testing is shown in Table 3.

Table 3. Artist Prediction Accuracy

Classes	Embedding size	Optimizer	Accuracy
100	64	SGD	70.6%
100	128	Adam	74%
354	128	Adam	56%

7 Other Applications

The other applications of the trained Siamese model are discussed in this section.

7.1 Genre-Auto Tagging

For any new song that requires to be tagged, an embedding can be generated by passing it through the model. With the use of K-neighbors, it can be tagged with the K most common genres around it, which may be a single or multiple value.

7.2 Content-Based Music Retrieval

For a given song, with only raw audio, it's embedding can be retrieved by feeding it to the model. Then, a distance metric (say Euclidian distance as used in [10]) can be used to find the nearest song using a reasonable threshold (<0.05). This could be used to successfully retrieve the song's meta-data and name, provided it already exists in the database. This would be like apps like Shazam or Google audio search.

8 Conclusion

This paper exhibits the success of metric learning for generating intelligent representations and for learning patterns in data that otherwise would not fully be captured from a simple classification model. The goal of the paper to have a model learn deeper relationships beyond what was taught has been achieved as the results showed that the trained model was also able to map song genres, despite only being trained on artists. The models trained surely have deeper relations that can be fully explored by training on larger and more extensive datasets. Future work on this can incorporate more music metadata like lyrics, user data [12, 13] and tags to the model to enhance graph representation.

References

1. Kaya, M., Bilg, H.Ş: Deep metric learning: a survey. Symmetry **11**(9), 1066 (2019)
2. Nasrullah, Z., Zha, Y.: Music artist classification with convolutional recurrent neural networks. In: 2019 International Joint Conference on Neural Networks (IJCNN). IEEE (2019)
3. Jiyoung, P., Lee, J., Park, J., Ha, J.-W., Nam, J.: Representation learning of music using artist labels. arXiv preprint arXiv:1710.06648 (2017)

4. Kim, J., Urbano, J., Liem, C.C.S., Hanjalic, A.: One deep music representation to rule them all? A comparative analysis of different representation learning strategies. Neural Comput. Appl. **32**(4), 1067–1093 (2020). https://doi.org/10.1007/s00521-019-04076-1
5. Koch, G., Richard, Z., Ruslan, S.: Siamese neural networks for one-shot image recognition. In: ICML Deep Learning Workshop, vol. 2, no. 1 (2015)
6. Nam, J., Choi, K., Lee, J., Chou, S.-Y., Yang, Y.-H.: Deep learning for audio-based music classification and tagging: teaching computers to distinguish rock from bach. IEEE Signal Process. Mag. **36**(1), 41–51 (2019)
7. Lu, R., Wu, K., Duan, Z., Zhang, C.: Deep ranking: triplet MatchNet for music metric learning. In: IEEE International Conference on Acoustics, Speech, and Signal Processing (ICASSP) (2017)
8. Hoffer, E., Ailon, N.: Deep metric learning using triplet network. In: Feragen, A., Pelillo, M., Loog, M. (eds.) SIMBAD 2015. LNCS, vol. 9370, pp. 84–92. Springer, Cham (2015). https://doi.org/10.1007/978-3-319-24261-3_7
9. Schroff, F., Dmitry, K., James, P.: Facenet: a unified embedding for face recognition and clustering. In: Proceedings of the IEEE Conference on Computer Vision and Pattern Recognition (2015)
10. Sakti, S.M. Laksito, A.D., Sari, B.W., Prabowo, D.: Music recommendation system using content-based filtering method with euclidean distance algorithm. In: 6th International Conference on Information Technology. Information Systems and Electrical Engineering (ICITISEE), Yogyakarta, Indonesia, pp. 385–390 (2022)
11. Defferrard, M., Benzi, K., Vandergheynst, P., Bresson, X.: FMA: a dataset for music analysis. arXiv preprint arXiv:1612.01840 (2017)
12. Cai, D., Qian, S., Fang, Q., Hu, J., Xu, C.: User cold-start recommendation via inductive heterogeneous graph neural network. ACM Trans. Inf. Syst. **41**(3), 64 (2023)
13. La Gatta, V., Moscato, V., Pennone, M., Postiglione, M., Sperlí, G.: Music recommendation via hypergraph embedding. IEEE Trans. Neural Netw. Learn. Syst. **34**(10), 7887–7899 (2023)

Online Health Information Seeking in Social Media

Maureen Olive Gallardo[1,2(✉)] and Ryan Ebardo[2]

[1] Ateneo de Zamboanga University, Zamboanga City, Philippines
maureen_gallardo@dlsu.edu.ph
[2] De La Salle University, Manila, Philippines
ryan.ebardo@dlsu.edu.ph

Abstract. Communication platforms including social media have become a resource for various information for diverse people. This includes health information for individuals who seek advice, explore symptoms, and learn about various health conditions. As it significantly impacted how people seek and consume health information, this paper aims to appraise current scholarship on how health information seeking or HIS behavior on social media has been recently investigated and portray the characteristics of social media platforms that facilitate HIS. The search in the Scopus, Science Direct, and PubMed databases returned 252 records, of which 24 studies published from 2013 to 2023 met the eligibility criteria after reviewing the full-text documents. Results of the study indicated that the prominent users of social media for HIS are generally younger, more educated, have lower health conditions, and are female. Influenced by varied motivations, the most used social media platforms are social networking sites (SNS), video-sharing platforms, and social Q&A websites. The different features of these platforms allowed information seekers to engage with other users, receive context-specific information, access varied information sources, and inquire anonymously. However, despite its advantages, HIS in social media raises concerns about unreliable data, the spread of misinformation, and anxiety. This review presented how far the existing literature has gone on HIS in social media, but it also highlighted the dearth of research on information-related factors and anxiety leading to cyberchondria, and inequality in the distribution of studies in some social media platforms and for some social groups.

Keywords: health information · information seeking · social media · systematic review · social networks

1 Introduction

The continuous evolution of social media platforms has revolutionized how individuals seek and share health information. Social media use in healthcare includes searching for health information, sharing personal health experiences, connecting with a support group, and posting reviews about healthcare and its allied professionals [10]. Searching for possible diagnoses of current medical conditions appears to be the most prevalent among its use. Due to the recent COVID-19 pandemic, online HIS has increased specifically in social media. For example, 76% of American respondents [20] relied on social media, and 88.4% of Southeast Asian respondents [12] utilized social media for information seeking.

In 2017, the research of Zhao and Zhang [35] synthesized studies on health information seeking or HIS in social media providing an overview of its characteristics explaining why it is popular, the discussion topics, the practice of seeking medical information from close social networks and acquiring social as well as emotional support from other users of social media. On the other hand, prior studies have also highlighted that concerns may arise in accessing health information that may not necessarily validated by a medical expert. However, the review is limited to studies conducted from 2011 to 2016. A more recent review was conducted by Gupta et al. [10] mapping existing literature from the viewpoints of patients in the use of social media and HIS. Nevertheless, the scope of their study is on the general use of social media and not specifically on HIS. Thus, this study would like to address the gap in those related reviews by reviewing recent studies on HIS in social media.

With the popularity of social media use and the increasing prevalence of HIS within these platforms, it is crucial to conduct a systematic review in the field to gain an overview of how HIS in social media progressed and to investigate the characteristics of social media platforms that facilitate HIS. Specifically, this review is guided by these questions:

- Who uses social media for HIS?
- What platforms are utilized by individuals in their HIS behavior?
- What are the features of social media that facilitate HIS?
- What are the motivations of individuals in using social media for HIS?
- What are the possible concerns in using social media for HIS?

The next section explains the methodology of this paper in identifying relevant studies included in the review. Section 3 explains the study results based on the research questions. Section 4 discusses the overall findings while Sect. 5 presents the implications and limitations of this review. Lastly, Sect. 6 presents the conclusion of this study.

2 Methodology

This study follows a systematic review method. The databases selected for the literature search are Scopus, Science Direct, and PubMed. Scopus is the largest database of peer-reviewed articles in several reviews on information seeking [6, 16]. Science Direct and PubMed are databases commonly used in health-related reviews [4, 6, 18, 30, 36]. Since the common phrase found in the literature on HIS is the phrase "online health information seeking", the keyword search used in the three databases is: "online health information seeking" AND ("social media" OR "social network").

The search was conducted in May 2023. The extracted records from Scopus and Science Direct included the abstract of the studies. On the other hand, PubMed did not have the option to include the abstract in the extraction. Thus, the abstract of the studies was manually collected. Since a study may be found in multiple databases, duplicate studies were identified and removed from the list.

During the first screening of the extracted records, irrelevant studies were filtered out through keyword searches on the title and abstract. Records that did not include "social media", "social network", or "health information" were excluded from the list. Reviews, commentaries, editorials, and conceptual articles were also excluded.

The full text of the studies resulting from the first screening was downloaded for further assessment. Full-text records that were not accessible from the sources were retrieved from various online sources. The eligibility of the study was assessed by both authors based on its relevance to HIS in social media and whether it answers at least one of the guide questions for this review or users and their interactions on social media platforms in seeking health-related information. This selection process is presented in Fig. 1.

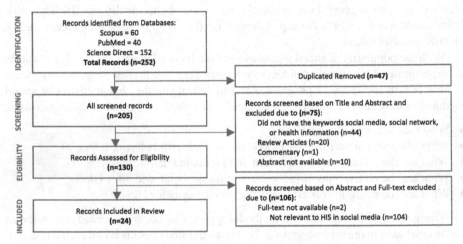

Fig. 1. Literature selection process

A total of 252 records were identified (Scopus = 60; Science Direct = 152; PubMed = 40) based on the search keyword used in the three databases. After deleting the duplicates (n = 47), the 205 article titles and its respective abstracts were screened. In the first screening, there were 75 records that were excluded based on the exclusion criteria. These excluded papers are review articles (n = 20), commentary articles (n = 1), the abstract is not available (n = 10), and did not have the keywords "social media", "social network", or "health information" (n = 44) in either the title or abstract. Furthermore, from the remaining 130 records, the abstract and full text of the records were carefully analyzed to determine their eligibility based on the consensus agreement of the researchers. There were 106 records excluded since the full text is not accessible (n = 2) and irrelevant to HIS in social media or does not answer any of the research questions (n = 104), ultimately resulting in 24 records selected for review.

3   Results

From the 24 included studies, the themes and details answering the guide questions of this review were determined from the methods, results, and discussions of the studies. The succeeding sections describe the 24 studies that resulted from the eligibility stage and are followed by discussions on how these papers answer our research questions.

3.1   HIS and Social Media in the Last 10 Years

The articles considered in this review represent published investigations on how individuals use social media in their HIS behavior from 2013 up to 2023 and are summarized in Table 1. Sixteen of the studies applied the quantitative approach (n = 16) followed by the qualitative approach (n = 7), and mixed method approach (n = 1). In terms of the source of data, five studies collected posts from social media platforms (Facebook, Facebook Group, Weibo, WebMD, and Yahoo! Answers) for data analysis. Among the studies that collected data from respondents (n = 19), most of the studies collected data from educational institutions (students = 8; faculty = 1) followed by the general population (n = 6; 5 is through an organization's survey), outpatients (n = 2), and gender-specific (women = 1; men = 1).

Looking at the context of the included studies, a significant number (n = 15) discussed how social media facilitates HIS. Eight of these studies focused on specific social media platforms (Webpages in the Baidu Ecosystem, Facebook, Yahoo! Answers, WebMD Answers, Weibo, Facebook Group, Facebook Page, WhatsApp, and Twitter). The next highest number of studies (n = 6) is on the factors affecting HIS such as the factors motivating online HIS and the relationship between chronic diseases, variables in online HIS, and SNS use. There are studies (n = 5) that focused on specific social groups (women; people with multiple chronic diseases; lesbian, bisexual, and queer women; adolescents after seeing a physician; college men in Latino fraternities). Moreover, among the studies that focused on the selection of health information sources (n = 4), three of these looked at the preference of the respondents between social media platforms, mobile applications, and internet-based sources which include search engines, health, and news webpages. Studies that examined information-related topics (n = 3) include information validation, information usefulness, information overload, and information trust. Also, there are studies that are focused on specific health conditions (n = 2) which include maternity healthcare and hypertension. Other contexts of the studies (n = 6) are on the social support dimensions, engagement in the social media platform, health decision-making, different search tasks, and HIS during the recent COVID pandemic. Other studies included the investigation of cultural differences in how users consume and understand digital health information from social media.

3.2   Social Media Users Seeking Health Information

Based on the studies that discussed the characteristics of individuals who have a higher tendency to do HIS on social media, HIS is influenced by the individual's age, education, family members with health conditions, gender, having health care provider, health condition, health efficacy, health literacy, income, marital status, race/ethnicity,

Table 1. Summary of Included Articles

Authors and Year	Method	Participants / Source of Data
Abdoh (2022)	Mixed	Undergraduate Students at Taibah University, Saudi Arabia
Briones (2015)	Quali	University Students
Chavarria et al. (2017)	Quali	College Men in Latino Fraternities in 2 Florida Universities
Deng & Liu (2017)	Quanti	Two University Hospitals in Central China
Gazibara et al. (2021)	Quanti	High School Students in Belgrade
Hassan & Mosoud (2021)	Quanti	Non-medical Students at Alexandria University, Egypt
Iftikhar & Abaalkhail (2017)	Quanti	Outpatients at King Abdulaziz University, Jeddah, Saudi Arabia
Kim (2013)	Quanti	210 Pew Internet & American Life Project
Lama et al. (2022)	Quanti	2017–2020 HINTS
Maslen & Lupton (2018)	Quali	Australian Women (21–74 yo)
Mitsutake et al. (2023)	Quanti	Japanese Individuals; 2020 INFORM Study
Oh et al. (2013)	Quanti	Undergraduate College Students at Midwestern University and Individuals recruited through Facebook
Park er al. (2020)	Quali	Yahoo! Answers: Questions
Rahim et al. (2019)	Quanti	Facebook Posts: Ministry of Health Malaysia
Rashid et al. (2022)	Quanti	Faculty in Universiti Purta Malaysia
Ruppel et al. (2017)	Quali	Facebook Group: LBQ Conception and Parenting
Sbaffi & Zhao (2020)	Quanti	Post Graduate Students in UK University
Song et al. (2016)	Quanti	College Students in US, South Korea, and Hong Kong
Song et al. (2021)	Quanti	Chinese College Students
Thackeray et al. (2013)	Quanti	US Residents (18 yo and above); 2010 Health Tracking Survey
Zhang et al. (2020)	Quanti	2012 HINTS-China
Zhang & Wang (2016)	Quali	WebMD Answers: Messages
Zhao et al. (2020)	Quali	Weibo Posts
Zheng et al. (2023)	Quanti	Chinese Individuals

reporting insurance coverage, and urban/suburban location. The characteristics that are supported by the highest number of studies are age (n = 5), health condition (n = 5 gender (n = 5), and education (n = 5). Other studies have characteristics that include marital status (n = 3) and race/ethnicity (n = 3).

Individuals who have a higher tendency to seek health information on social media platforms have one or more of these characteristics [7, 14, 31]: younger age, female, higher education, higher income, living in an urban or suburban location, higher health literacy, higher health efficacy, having family members with a health condition, having a regular health provider, and reporting health insurance coverage.

As for the marital status and health condition of the individual, studies have mixed results. According to Zhang et al. [32], married individuals tend to acquire information from social media less. However, other studies showed that more married individuals use SNS for health [19] and watch health videos on YouTube [15]. As for the health condition of the individual, those with lower health conditions [15, 32], chronic diseases [19, 31], and depression or anxiety disorder [19] are more engaged in HIS. But Oh et al. [21] concluded that health concerns are not significantly associated with seeking health information.

Moreover, the race or ethnicity of an individual is a factor in the tendency to engage in HIS in social media [15, 27, 31].

3.3 Social Media Platforms Used for HIS

The commonly used platforms for HIS are SNS (e.g., Facebook, Twitter), YouTube, Q&A websites with social features (e.g., Yahoo! Answers, WebMD Answers, Baidu Knows), blogs (e.g., Weibo), and forums. Other social media platforms mentioned in the studies are wiki websites (e.g., Baidu Experience) and photo-sharing platforms (e.g., Instagram).

SNS and YouTube are the two most popular platforms. SNS is the top preference [1, 11]. However, among adolescents, YouTube appears to be the most widely used to search for information [9] due to its rich media content.

It is also notable that messaging applications such as WhatsApp [13, 24] and WeChat [27, 37] are widespread for HIS. These studies considered these applications as social media, which may be because of their social media features despite being primarily used for communication.

The choice of social media platform depends on the type of search task [26] and the kind of information [27] the individual seeks. For factual and personal experience search tasks, social Q&A websites and SNS are most likely used by individuals with higher channel experience. For accurate search tasks, individuals who value the usefulness of information will most likely choose social Q&A websites, while those who prefer reports recommended by others will most likely select SNS. And for exploratory search tasks, age is a factor in selecting social Q&A websites. In the study conducted by Song et al. [28] that analyzed the choice of websites in the Baidu ecosystem, results showed that for receptive tasks ("seek to understand, remember, and reproduce what is taught"), respondents primarily used wikis and for critical tasks ("seek to criticize and evaluate ideas from multiple perspectives"), most of the respondents selected and used the social Q&A websites. In terms of the type of information, experienced-based health information is more sought by a holistic culture like the Koreans and Hongkongers, which are usually found in blogs, SNS, and social support groups, while expertise-based health information is more sought by analytic culture like the Americans [27].

3.4 Features of Social Media Facilitating HIS

The design and features of social media facilitate information seeking [32], which allows health information seekers to engage with other users, receive context-specific information, access various information sources, and inquire anonymously.

Social media platforms allow users to post and receive comments or replies from other users, thus encouraging engagement. Individuals use SNSs and social Q&A websites to search for health-related personal issues because of the chance to interact with other users [26]. When posting, users can put their information needs into context [22, 32]. And when users ask questions, there is an expectation of drawing responses from others [22]. The study conducted by Zhang et al. [33], where the questions and answers in WebMD Answers were examined showed that responses can reach up to 20 answers. In using a Facebook group, many of the questions for medical advice received several responses, and some were even unsolicited [25].

Hashtags provide an efficient way to find information on social media. For example, users can use a hashtag on Twitter to filter the fields [1, 34]. On Weibo, a microblogging website, patients with COVID-19 during the pandemic used the hashtag "#COVID-19 Patient Seeking Help" to seek health information [34]. Another way is to create or join groups on Facebook that provide group members with expert advice or specific health-related recommendations [25, 26]. In social Q&A websites, the platform allows users to ask specific health-related questions that are personalized [28] from healthcare institutions, health experts, and other users [33]. Moreover, social media users can see related posts based on their interests and attitudes through the algorithms used in social media [1].

Through the follow and subscribe functions and sponsored posts, users can access posts of different entities in social media such as celebrities, organizations, and close ties such as family members, peers, and friends [1]. Furthermore, information seekers who are asking private questions can use the settings in social Q&A websites that let them ask questions without divulging their identity. For example, 92.43% of social media users, using the anonymity feature of the platform, post questions on dieting topics encouraging lively discussions online [33].

3.5 Motivations for Using Social Media for HIS

Health information seekers turn to social media for social support, convenience, and access to trusted information sources.

Among the different types of social support, the predominant among Facebook users is emotional support [21]. Also, one possible reason why social Q&A websites are popular is because of the emotional support provided to users [28]. However, during the COVID-19 pandemic, when there were lockdowns, using Weibo to seek help checking older adults, taking them to the hospital, or helping them with their medications provided users more with tangible support and connected them to a more empathetic crowd [34]. Being able to use social media anytime and anywhere also increased the convenience of getting real support [7].

The convenience that motivates people to use social media is being able to get the information they need immediately [1, 3, 17] when they don't have the time to visit the doctor [1] or when there is difficulty or long travel time to access health care resources [34].

Lastly, people contributing to social media are perceived as trusted sources who are not driven by the agenda [17]. During COVID-19, social media users relied on their preferred platforms for medical information as there was mistrust in businesses such as pharmaceuticals [1]. Moreover, social media users who seek medical information interact with health professionals from WebMD [33] and YouTube [1] who speak about illnesses and how to handle them.

3.6 Challenges in Social Media for HIS

While using social media for HIS brings several advantages, it is not overlooked that using these platforms to search for critical health-related information raises issues of unreliable information, widespread misinformation, and anxiety.

The quality of health information in social media varies depending on the topics discussed [25]. This could also be attributed to inefficient regulation [26]. However, not everyone verifies the health information that they see. In the study of Iftikhar & Abaalkhail [13], only less than half of the respondents ascertain the credibility of the data. Misinformation can quickly spread on social media [5], as only about one-quarter of the respondents share health information without verifying if the notification is accurate [13]. One way to beat the spread of misinformation is for health organizations to spread more credible health information on social media. However, health organizations are not able to exploit the use of social media to its optimum. The engagement rate on the Facebook Page of a health organization showed that their posts have only average and poor engagement rates than reasonable engagement rates [23].

Moreover, HIS in social media may be beneficial or detrimental to the individual since the relationship between anxiety and the use of social media is bidirectional. Zheng et al. [37] demonstrated that HIS in social media platforms increases information trust. However, the perception of social media as a trustworthy source of health information may trigger more searches and intensify a sense of distress and confusion that will eventually lead to cyberchondria, which is an excessive online search that is accompanied by increased feelings of anxiety [37]. On the other hand, the study conducted by Mitsutake et al. [19] showed that users with undiagnosed psychological disorders such as anxiety and depression are prone to rely on social media platforms for their unmet informational needs.

4 Discussion

The objective of this review is to synthesize research in the last decade on how various social media platforms facilitated HIS behavior. Key findings include: a) existing literature is concentrated on the usage of social media platforms and the factors influencing or motivating the health information search; b) HIS in social media is influenced by the characteristics of health information seekers, type of search task, type of information sought, motivations of information seekers, and affordances of social media platforms for HIS; and, c) concerns about HIS in social media is on the unreliability and appraisal of health information in social media and the possibility of anxiety that is triggered by repeated searches as information trust is increased.

Having more studies on the usage and factors influencing or motivating HIS in social media provides a broad understanding of this behavior. At the same time, that gave room for more research to be conducted on other aspects of HIS such as on the assessment of health information quality and management of anxiety triggered by repeated searches on social media platforms.

The choice of using social media for HIS depends on the motivations and perceptions of the individual. Older literature says that health information in social media is perceived as unreliable and deters information seekers from searching for health information from these platforms. However, the perception of health information in social media has changed, as seen in recent literature. Information seekers think that they can tap trusted sources on social media platforms. During the COVID-19 pandemic, people prefer seeking information from social media because the sources of information would not have an agenda compared when searching from other platforms. However, according to Soroya et al. [29], there is also a cohort of social media users who prefer other sources of medical information aside from popular social media platforms. As such, there is a mixed sentiment on HIS in social media during the pandemic.

HIS is mainly done on Facebook, YouTube, and social Q&A websites. These different platforms each have their unique characteristics, which can be based on their design or features affecting the preference of information seekers. Understanding the technological affordances of social media platforms and how they affect human behaviors when searching online may provide fresh insights into how health information on these platforms is sought and evaluated.

5   Implications and Limitations

One of the main concerns in using social media for HIS is information reliability, mainly because misinformation can quickly spread on these platforms. However, the number of included studies in this review that examined information-related activities and characteristics is limited. Another concern is the anxiety caused by HIS. This context is less studied and is an essential concern as it may lead to cyberchondria [37], an emerging topic in online HIS [36]. With the increasing information trust in health information in social media [37] and less verification of health information [13], this topic could be one of the future directions of studies on HIS.

Social media is a broad category of internet-based applications with an elaborate environment accommodating the participation of many users for multi-way interaction and the creation and sharing of multimedia content [8]. Different types of social media offer unique features that make it a preference for specific types of search tasks and health information. Having Facebook, YouTube, and social Q&A websites as popular social media platforms, analyzing the contents and interactions in these platforms is a promising research direction that can give more insights into the actual use of these platforms for seeking health information.

The practical implication of this review is how health organizations and health professionals can maximize the use of the appropriate social media platforms including emerging platforms such as TikTok [2], to spread verified health information and maximize engagement considering the media type, information types, and target audience.

Like other reviews, the interpretation of our study's results should consider its methodical limitations. First, the choice of the search keyword might not be able to retrieve all studies related to the use of social media for HIS. Second, although this study retrieved records from three databases, there might still be other relevant studies that were not found on these databases. Third, the selection process might have excluded relevant articles that may provide different results.

6 Conclusion

This study systematically reviews existing research on online HIS in social media, providing the landscape of what has already been done and insights on what could still be done. The 24 studies included in this review provided details of the perceptions of information seekers in choosing social media platforms and their characteristics that facilitate HIS. However, more studies still need to be done by employing qualitative or mixed method approaches for in-depth understanding, analyzing social media contents and interactions, focusing on information characteristics and validation, investigating cyberchondria in social media, or getting the perspective from different social groups such as adolescents, older adults, and those with specific health conditions among others to provide more insights on the adoption and non-adoption of social media for HIS. Online HIS is not a new occurrence but conducting it on social media that continuously evolves entails continuous research for new knowledge and perceptions.

References

1. Abdoh, E.: Online health information seeking and digital health literacy among information and learning resources undergraduate students. J. Acad. Librarianship **48**, 102603 (2022)
2. Basch, C., Hillyer, G., Jaime, C.: COVID-19 on TikTok: harnessing an emerging social media platform to convey important public health messages. Int. J. Adolesc. Med. Health **34**(5), 367–369 (2022)
3. Briones, R.: Harnessing the Web: how e-health and e-health literacy impact young adults' perceptions of online health information. Medicine 2.0 **4**(2), e5 (2015)
4. Brown, R., Skelly, N., Chew-Graham, C.: Online health research and health anxiety: a systematic review and conceptual integration. Clinical Psychology: Science and Practice **27**(2), e12299 (2019)
5. Chavarria, E.A., Chaney, E.H., Stellefson, M.L., Chavarria, N., Dodd, V.J.: Types and factors associated with online health information seeking among college men in Latino fraternities: a qualitative study. Am. J. Men's Health **11**, 1692–1702 (2017)
6. Daei, A., Soleymani, M.R., Ashrafi-Rizi, H., Zargham-Boroujeni, A., Kaleshadi, R.: Clinical information seeking behavior of physicians: A systematic review. Int. J. Med. Inform. **139**, 104144 (2020)
7. Deng, Z., Liu, S.: Understanding consumer health information-seeking behavior from the perspective of the risk perception attitude framework and social support mobile social media websites. Int. J. Med. Inform. **105**, 98–109 (2017)

8. Duong, C.T.P.: Social media: a literature review. J. Media Res. **13**(3), 112–116 (2020)
9. Gazibara, T., Cakic, M., Cakic, J., Grgurevic, A., Pekmezovic, T.: Patterns of online health information seeking after visiting a physician: perceptions of adolescents from high schools in central Belgrade, Serbia. Family Practice **38**(3), 231–237 (2020)
10. Gupta, P., Khan, A., Kumar, A.: Social media use by patients in health care: A scoping review. I. J. Healthc. Manage. **15**(2), 1–11 (2020)
11. Hassan, S., Masoud, O.: Online health information seeking and health literacy among non-medical college students: gender differences. J. Public Health Theory Pract. **29**, 1267–1273 (2021)
12. Htay, M.N.N., et al.: Digital health literacy, online information-seeking behaviour, and satisfaction of Covid-19 information among the university students of East and South-East Asia. PLoS ONE **17**(4), e0266276 (2022)
13. Iftikhar, R., Abaalkhail, B.: Health-seeking influence reflected by online health-related messages received on social media: cross-sectional survey. Journal or Medical Internet Research **19**(11), e382 (2017)
14. Kim, Y.-M.: Does online searching cause or enforce health information disparity? J. Inf. Knowl. Manage. **12**(4), 1350032 (2013)
15. Lama, Y., Nan, Z., Quinn, S.C.: General and health-related social media use among adults with children in the household: findings from a national survey in the United States. Patient Educ. Couns. **105**(3), 647–653 (2022)
16. Lim, H.M., Dunn, A.G., Lim, J.R., Abdullah, A., Ng, C.J.: Association between online health information-seeking and medication adherence: a systematic review and meta-analysis. Digital Health **13**, 20552076221097784 (2022)
17. Maslen, S., Lupton, D.: "You can explore it more online": a qualitative study on Australian women's use of online health and medical information. BMC Health Serv. Res. **18**(1), 916 (2018)
18. Menon, V., Kar, S.K., Tripathi, A., Nebhinani, N., Varadharajan, N.: Cyberchondria: conceptual relation with health anxiety, assessment, management and prevention. Asian J. Psychiatr. **53**, 102225 (2020)
19. Mitsutake, S., et al.: Chronic diseases and sociodemographic characteristics associated with online health information seeking and using social networking sites: nationally representative cross-sectional survey in Japan. J. Med. Internet Res. **25**, e44741 (2023)
20. Neely, S., Eldredge, C., Sanders, R.: Health information seeking behaviors on social media during the COVID-19 pandemic among American social networking site users: survey study. J. Med. Internet Res. **23**(6), e29802 (2021)
21. Oh, H.J., Lauckner, C., Boehmer, J., Fewins-Bliss, R., Li, K.: Facebooking for health: an examination into the solicitation and effect of health-related social support on social networking sites. Comput. Hum. Behav. **29**, 2072–2080 (2013)
22. Park, M.S., Oh, H., You, S.: Health information seeking among people with multiple chronic conditions: contextual factors and their associations mined from questions in social media. Library Inf. Sci. Res. **42**, 101030 (2020)
23. Rahim, A.I., Ibrahim, M.I., Salim, F.N., Ariffin, M.A.I.: Health information engagement factors in Malaysia: A content analysis of Facebook use by the ministry of health in 2016 and 2017. Int. J. Environ. Res. Public Health **16**, 591 (2019)
24. Rashid, A.A., Devaraj, N.K., Xuan, L.Z., Selvanesan, K., Noorazalan, A.A.: Social media use and hypertension knowledge among undergraduate students during the COVID-19 pandemic in faculty of medicine and health sciences, Universiti Putra Malaysia. Malays. J. Med. Health Sci. **18**, 1–9 (2022)
25. Ruppel, E., Karpman, H., Delf, C., Merryman, M.: Online maternity information seeking among lesbian, bisexual, and queer women. Midwifery **48**, 18–23 (2017)

26. Sbaffi, L., Zhao, C.: Modeling the online health information seeking process: information channel selection among university students. J. Am. Soc. Inf. Sci. **71**(2), 196–207 (2020)
27. Song, H., et al.: Trusting social media as a source of health information: online surveys comparing the United States, Korea, and Hong Kong. J. Med. Internet Res. **18**(3), e25 (2016)
28. Song, X., Liu, C., Zhang, Y.: Chinese college students' source selection and use in searching for health-related information online. Inf. Process. Manage. **58**, 102489 (2021)
29. Soroya, S.H., Farooq, A., Mahmood, K., Isoaho, J., Zara, S.-E.: From information seeking to information avoidance: understanding the health information behavior during a global health crisis. Inf. Process. Manage. **58**(2), 102440 (2021)
30. Starcevic, V., Berle, D., Arnaez, S., Vismara, M., Fineberg, N.: The assessment of cyberchondria: instruments for assessing problematic online health-related research. Curr. Addict. Rep. **7**, 149–165 (2020)
31. Thackeray, R., Crookston, B.T., West, J.H.: Correlates of health-related social media use among adults. J. Med. Internet Res. **15**(1), e21 (2013)
32. Zhang, L., Qin, Y., Li, P.: Media complementarity and health information acquisition: a cross-sectional analysis of the 2017 HINTS-China survey. J. Health Commun. **25**, 291–300 (2020)
33. Zhang, Y., Wang, P.: Interactions and user-perceived helpfulness in diet information social questions & answers. Health Info. Libr. J. **33**, 295–307 (2016)
34. Zhao, X., Fan, J., Basnyat, I., Hu, B.: Online health information seeking using "#COVID-19 patient seeking help" on Weibo in Wuhun, China: descriptive study. J. Med. Internet Res. **22**(10), e22910 (2020)
35. Zhao, Y., Zhang, J.: Consumer health information seeking in social media: a literature review. Health Info. Libr. J. **34**, 268–283 (2017)
36. Zheng, H., Sin, S.-C.J., Kim, H.K., Theng, Y.-L.: Cyberchondria: a systematic review. Internet Res. **31**(2), 677–698 (2020)
37. Zheng, H., Chen, X., Jiang, S., Sun, L.: How does health information seeking from different online sources trigger cyberchondria? The roles of online information overload and information trust. Inf. Process. Manage. **60**, 103364 (2023)

Blockchain Segmentation: An Industrial Solution for Large Scale Data

Anooja Ali[1]([✉]) [iD], Nisha Joseph[2], and TousifAhamed Allabksha Nadaf[3]

[1] School of CSE, REVA University, Bengaluru 560064, India
anooja.ali@reva.edu.in
[2] Department of CSE, Saintgits College of Engineering, Pathamuttam, Kottayam 686532, India
nisha.joseph@saintgits.org
[3] Architect, Wipro Arabia Ltd., Dhahran, Saudi Arabia
ahamedpapers@gmail.com

Abstract. Blockchain technology finds diverse applications, encompassing the security and managing the vast amounts of data generated by IoT devices, with secure and transparent communication between them. In Supply chain management, blockchain can track products and goods from their origin to destination. A blockchain is an autonomous digital ledger employing Distributed Ledger Technology (DLT) to securely and transparently record transactions in a decentralized manner. It accomplishes this by using a network of computers (nodes) and these components collaborate to verify and register transactions within a communal database. This research proposes a new approach called Segment Blockchain that divides a blockchain into smaller segments and enables nodes to retain only one segment in place of the entire blockchain. This approach can potentially reduce the storage requirements for participating nodes, facilitate the incorporation of addional nodes into the network and maintain a copy of the blockchain. Our proposed methodology aims to address the concern of the risk of a singular vulnerable point, wherein a malicious entity keeps all copies of a specific segment and leaves the system, causing in the irreversible deletion of that segment. The proposed blockchain system can handle big data while reducing storage space demands while ensuring heightened security for user data. Theoretical evidence shows that this is achieved by limiting the number of blocks a malicious entity has the capability to both store and distribute every segment over a cluster of cloud-based blocks, the storage burden is significantly reduced compared to conventional designs. The system was successful in reducing storage space by 33% for large scale data. This makes the proposed segmentation approach more practical for processing and managing large volumes of data.

Keywords: Blockchain · Decentralization · Distributed Ledger Technology · Network · Segmentation · Shared Databases

1 Introduction

Blockchain systems are renowned for its anonymity and independence, hence, it is essential to ensure that each transaction can be authenticated to build trust and ensure security. Therefore, it is crucial to keep a full and accurate record of all transactions.

K. K. Patel et al. (Eds.): icSoftComp 2023, CCIS 2030, pp. 180–192, 2024.
https://doi.org/10.1007/978-3-031-53731-8_15

However, as the blockchain size has increased over time, the cost of storing all the blocks of the mainchain has also risen sharply, making it more expensive to sustain a complete node within the Bitcoin network. Blockchain can be used to securely verify and authenticate identities, which can help prevent fraud and protect sensitive information [1]. Blockchain-based finance allows decentralized financial transactions and services. It is essential to have a secure and decentralized data sharing and collaboration while ensuring data privacy and ownership.

When the size of the blockchain grows, computational requirements and storage regarding the preservation of a full node also increase, making it more difficult and expensive for users to engage in the network. This potentially causes centralization of the network as only large and well-funded players can afford to maintain a full node. In order to tackle this, various proposals have been put forward to reduce the storage and computational requirements of maintaining a full node while still ensuring the security and trustworthiness of the blockchain. Even a small transaction in Bitcoin or other Distributed Ledgers occupies only hundred bytes, but nodes with limited resources are gradually leaving the mining game, leading to more devices operating in lightweight mode or participating in mining pools.

The common approaches to increase the blockchain performance include weighted models [2], pruning, blockchain sharding [3] and off-chain [4]. Pruning removes the old and unnecessary data from the blockchain without compromising its integrity. Sharding refers to the process of dividing the blockchain into smaller, more manageable segments, allowing different nodes to process them in parallel. Additionally, there are also efforts to develop off-chain solutions, such as the Lightning Network, which enable faster and cheaper transactions by conducting them off the main blockchain.

Decentralization becomes a major issue for large data. Blockchain holds the potential to distribute complex and data-intensive tasks through leveraging smart contracts, to unidentified nodes across the network. The expansion of blockchain size and associated storage costs poses a challenge for maintaining a decentralized and secure network. Segments, or also known as blocks, are a fundamental component of blockchain technology. The use of segments is crucial for several reasons including security, scalability and consensus [5]. This guarantees the integrity and immutability of transaction data by maintaining efficiency and speed. A distributed ledger is a database variant that is spread out across multiple locations, institutions, or nodes in a network. This decentralization allows for greater transparency, security, and resilience than traditional centralized databases.

This paper discusses segmentation of blocks within their assigned segment without the need to access the entire blockchain. Each segment would have its own set of validators that are responsible for validating transactions and blocks within that specific segment. This approach can improve the scalability of blockchain networks by reducing the time required to propagate blocks across the network [6]. Since nodes only need to transmit blocks to other nodes within their segment, the network can potentially handle more transactions per second without overwhelming individual nodes.

The proposed research, organizes a defined quantity of blocks into segment, which is subsequently stored by multiple nodes. The approach put forth addresses the concern of a sole vulnerability point, wherein a malicious entity could obtain all instances of a blockchain segment and exit the system, causing irrevocable loss. The theoretical proof states by restricting the storage capacity for malicious entities and distributing each segment across a group of cloud-based blocks, the storage requirements are significantly decreased in comparison to standard designs. Thus, the proposed system establishes a blockchain capable of managing big data by minimizing storage demands while ensuring data protection for users. The system achieved a 33% reduction in storage space during the experiment. This renders the suggested approach capable of effectively handling substantial volumes of data and its processing.

The next section of the paper is literature survey, later methodology is explained and is followed by implementation and results. Later the paper is concluded.

2 Literature Survey

Blockchain is a type of decentralized ledger technology that was first described in literature. It was initially used primarily in the field of cryptocurrency, with Bitcoin and Litecoin [7], Monero [8], and Zcash [9] being the most prominent examples. As blockchain technology has developed rapidly, it has become an effective means of ensuring the genuineness, safety, and dependability of data. It has been applied in a wide range of areas, including medical data, safeguarding personal information, and strategies for distributing data. Blocks serve as the fundamental components of a blockchain, comprising a segment header with primary data and transactional information as a block body.

Block data is utilized for establishing a connection to the preceding block and for indexing data using the hash value of the current block. Transactions within the blockchain involve engagement with a hash function, thereby ensuring security. The security of the blockchain system is of utmost importance and encompasses various aspects such as data security, security of smart contracts, safeguarding privacy, and mitigating application risks. In order to guarantee the integrity of data immutability, constant enhancements are required in the foundational data structure, cryptographic techniques, and communication networks of the blockchain. These improvements are essential to foster the robust advancement of blockchain applications.

The decentralized and unidentified characteristics of blockchain guarantee that correct results are recognized by the majority as far as the security threshold is upheld [10]. The issue of excessive growth raises the storage demands for participants, creating challenges for handling data-intensive tasks such as training AI models within a decentralized system, particularly when the system maintains a universally open membership approach. Few researchers aim to alleviate individual load and address the predicament of balancing the capacity to handle all aspects, sustaining a decentralized, and high-performance architecture.

Weighted models distribute the responsibilities of a node based on their weights [11]. An instance of a weighted model is the lightweight node system, where a lightweight node refrains from storing blocks; instead, it operates as a client to certain full nodes. These nodes verify a new transaction through Simple Payment Verification (SPV) queries. Lightweight nodes consume a maximum of 4.2 megabytes, irrespective of the overall blockchain dimensions, they cannot verify new blocks. Delegated Proof of Stake (DPoS) exhibits superior performance since representative nodes often possess advanced computational capacity, storage capabilities, and network bandwidth [12, 13].

A co-signed contract involving both parties is often made in off-chain approaches and this marks the beginning of the deal [14]. Off-chain channels perform secure trading, without publishing transactions through blockchain. Transactions are published to the blockchain only when there is a violation in off-chain transactions., There are few non-financial applications for these approaches. Instead of broadcasting the assignment and outcome to the network, entities employing off-chain techniques need to ensure confidential communication, by undermining the inherent anonymity of the blockchain.

Methods for blockchain sharding involve the allocation of nodes across distinct shards, dividing storage and assigning tasks to various shards that run in parallel, guaranteeing that the workload for individual nodes remains manageable with the global increase in transactions [15]. The primary focus of sharding design is to decrease the likelihood of an adversary exerting control over a majority of the inside shard positions, even though they haven't acquired the majority of nodes across the entire network. If an adversary controls a shard but does not meet security thresholds globally, the integrity of the entire system's security is jeopardized. Therefore, to maintain system security, there are stringent requirements on the quantity of shards and nodes within each shard.

Clients utilizing cloud computing across diverse application domains seek assurance regarding the accuracy and reliability of their data [16]. To establish a tamper-proof cloud computing environment, blockchain, ledger, is employed in conjunction with cloud technology. The distributivity of blockchain means the absence of central governing body or single point of failure. Instead, multiple nodes in the network work together to validate and record transactions, which helps to ensure the integrity and security of the ledger.

Blockchains utilize technologies and techniques to maintain the integrity and security of the ledger. Distributed (peer-to-peer) networks allow for the validation and sharing of information across the network. Encryption and cryptography methods are also crucial components of blockchain technology, as they are used to secure the data on the blockchain and prevent unauthorized access or tampering [17]. DLT is decentralized, immutable, transparent, secure, trustless, and relies on a consensus mechanism to validate transactions. Every newly appended block becomes immutable and irreversible, with transactions being documented in a publicly accessible ledger, visible to every participant. Transactions are validated by the network participants, and trust is not required between participants [18]. Consensus among all network participants is necessary to validate a transaction before incorporating it into the Blockchain.

Segment Blockchain has a primary focus on reducing the size of the blockchain. In blockchain sharding systems, the security depends on the majority of nodes in every shard being honest. However, in terms of storage, if an adversary fails to possess segment copies, their attack will fail to achieve any objective. Hence, within the Segment Blockchain, the honest nodes don't need to constitute the majority among those storing a specific segment. The only requirement is that each segment has at least one faithful keeper. With this more lenient security, it is possible to assign a reduced count of nodes to store a segment securely, as opposed to blockchain sharding systems, which require the majority of nodes to be honest. Hence, the storage can be significantly reduced.

3 Methodology

This section discusses the basic techniques in a segmented blockchain system including the implementation of a Cryptography system with AES algorithm. It has Logical Block Addressing (LBA), and leveraging cloud storage. In distributed ledger, data is stored in multiple copies across different nodes in the network, and each node has a copy of the entire ledger. Changes to the ledger are made through a consensus mechanism, in which network nodes collaborate to collectively validate and authenticate transactions. Once a transaction is validated and appended in ledger, it is replicated across all nodes ensuring that all copies of the ledger are kept in sync [19].

Segment Blockchain is appropriate for many of the current blockchain applications, such as notation, and identity control. These applications do not necessitate high rates of transaction throughput but do advantage from decentralization. Additionally, Segment Blockchain can facilitate integrating blockchain into an IoT ecosystem where edge devices might have limited storage to store a complete record, and the systems do not demand substantial transactions per second. Blockchain is a method of recording data in a manner that prevents alteration, unauthorized access, or manipulation of the system [20]. Every block within the chain holds a group of transactions, and when a fresh transaction transpires on the blockchain, a notation of it gets appended to the ledger of each participant. In Blockchain, transactions are logged using an unalterable cryptographic signature known as a Hash.

3.1 Implementation of Cryptography with AES Algorithm

Cryptography involves converting plain text into an encoded form referred to as ciphertext which is understood by authorized individuals who possess a decryption key. Cryptography provides secure communication over insecure networks, such as the Internet. AES (Advanced Encryption Standard) stands as a symmetric-key block cipher algorithm created to supersede the outdated Data Encryption Standard (DES) algorithm [21]. The size of the block and the length of the key can vary, but AES typically uses 128-bit blocks and keys of either 128, or 256 bits. The algorithm uses a series of permutation operations to transform the plaintext into ciphertext.

Additionally, AES is a symmetric key algorithm, it is easy to implement and use, which makes it a popular choice for many security applications. AES is considered to be a very secure encryption algorithm and is resistant to attacks, including brute-force and known-plaintext attacks. The relationship between the plaintext and the ciphertext is highly complex and difficult to reverse-engineer without knowledge of the encryption key.

Hashing is a process in which an input (such as a file, message, or password) is passed through a mathematical function that produces a hash with a constant-length output. Hash functions are one-way, which means that generating a hash from an input is a straightforward process, but very difficult to generate the actual input from its hash. This property makes hash functions useful for a wide range of applications, including data integrity checks, digital signatures, and password storage. One of the primary benefits of using hashing is that it allows for efficient and secure storage and retrieval of data.

3.2 Hashing with Message Digest Algorithm (MDA-5)

MD5 algorithm is a hash function generating digest, 128-bit for any input, irrespective of its length. The resultant digest is commonly displayed as a hexadecimal sequence with 32 digits. MD5 is extensively employed as a one-way function, implying it as simple to compute hash value from the message, but intricate to reverse-engineer the initial message from the hash. MD5 is commonly used for data integrity checks, changing the input message results in variant hash value, making it useful for verifying that the original data has not been tampered with. It is also used for storing passwords by hashing them with MD5 and storing the new hash instead of password. Hence, if the stored hash value is modified, the original details cannot be directly obtained, since reversing the hash function is very difficult.

3.3 Map Reduce

MapReduce is a programming paradigm and implementation that enables parallel and distributed processing of large data. The key idea behind MapReduce is to divide a large input data set into smaller chunks, process each chunk independently in parallel, and then combine the results of these independent processes. The map function applies a transformation to each data element in the input data set and produces key-value pairs. Subsequently, the reduce function merges the values linked with each intermediate key to generate a conclusive output.

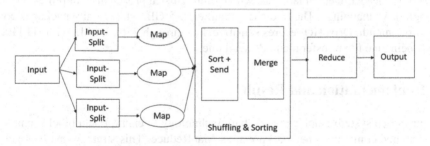

Fig. 1. MapReduce process flow

MapReduce is a flexible programming model and can be used for a wide range of applications, including data analysis and warehousing and implemented on a distributed system, making it a versatile tool for large-scale data processing [22]. MapReduce can distribute data processing across a large number of nodes, enabling efficient processing of large amounts of data and improving scalability in blockchain. MapReduce process flow is in Fig. 1. On the other hand, a reducer class handles the reducing phase, which involves aggregating and reducing the output of various data nodes to produce the final output. The data to be processed using MapReduce is stored in HDFS, and it can be a single or multiple files. The input determines the specification and specifies how files will be separated and read. During reducing phase, reducer processes the map-generated data by applying operations through a reducer function. The final output, is a smaller set of tuples, is stored in HDFS.

3.4 LBA

LBA is used in MapReduce to allow the operating system to address blocks of data on the disk using logical block addresses, rather than physical block addresses. This helps to simplify the process of accessing and managing data on the disk. In MapReduce, data is typically stored in a distributed file system across multiple nodes, such as HDFS. Data is typically partitioned into blocks of a fixed size, such as 64 or 128 MB. In HDFS, each block is assigned block ID, used to track the location of the block across the distributed file system.

When data is processed using MapReduce, the MapReduce framework assigns a task to a node to process a specific block of data. The node retrieves the block from the distributed file system, processes it using the MapReduce task, and then writes the output back to the distributed file system. In MapReduce, the processing of data is typically handled by a large number of worker nodes, while a separate set of master nodes are responsible for coordinating and managing the overall job. The worker processes data by executing Map and Reduce functions on the data blocks assigned to them.

3.5 Cloud Storage

In the proposed solution, involving cloud storage, the platform utilized is Drive HQ, which happens to be the pioneer in cloud IT solutions. Its inception in 2003 was aimed at providing a one-stop-shop for all Cloud IT solutions. Over time, DriveHQ has grown to become a leader in essential areas, with a minimalist approach that simplifies services through drive mappings. The free basic Plan, offer 5 GB, while family packages come for $4 per month. DriveHQ secures sensitive data through 256-bit SSL, with HTTPS or FTPS being the file transfer methods available.

4   Implementation and Results

The proposed system employs blockchain technology to encrypt data blocks, enabling a distributed computing method known as MapReduce. This strategy involves generating blocks prior to uploading them, while the database stores the metadata for these blocks. Moreover, Segment Blockchain includes a secure deduplication feature that utilizes an ownership protocol. This prevents any side-channel information exposure in deduplication.

The proposed system aims to offer a secure and efficient means for users to upload their files to the cloud. This is accomplished by employing both blockchain technology and sophisticated data deduplication techniques that ensure the confidentiality and integrity of data. User picks up the file from the local device. Upon selection, the file is divided utilizing LBA. This ensures that every block has a predetermined size with data alignment in a specific boundary. To guarantee distinctiveness, the system generates an MD5 hash per block as a block-level, 128-bit fingerprint, produced by encoding the input.

Users choose a file from their local system and divide into smaller blocks using LBA later, MD5 hash is generated for the block. Figure 2 is file upload method. System validates the distinctiveness of block by contrasting their MD5 values with those of already present blocks in the database (DB). Upon detecting distinct block, metadata is uploaded in DB, followed by encrypting the block and depositing it in the cloud using a blockchain structure. Thus, the system provides immutability to the stored data.

Duplicate blocks identified in the verification process, will not be allocated to the private cloud, only, the block instance is updated. The system eliminates the necessity for storing multiple duplicates of identical data, resulting in lowered storage expenses and efficiency. The proposed file upload method has advantages, including capability to create blocks before data upload, enhancing both efficiency and security. The Proof of Ownership protocol has the potential to facilitate secure deduplication, leading to additional reductions in storage expenses and enhanced efficiency. Overall, these attributes render the suggested system well-suited to contemporary requirements for data storage and security. Through the eradication of the necessity to retain multiple replicas of identical data, the system concurrently lessens the potential for data loss. It remains crucial to rigorously assess any novel system, prior to its deployment to guarantee alignment with essential criteria and intended functionality.

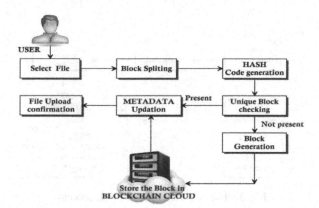

Fig. 2. File upload process

Figure 3 describes the process for downloading a file and partitioning it into smaller blocks with LBA. The user selects the file, the file is transferred from the server to the user's device in chunks. The file is divided into smaller blocks using LBA. Each block is assigned a unique address that indicates its location within the file. Overall, this process is designed to make it easier and more efficient to download and manage large files, by breaking them down into smaller, more manageable pieces that can be accessed and transferred more quickly and reliably.

> Algorithm 1: Method of Block Development

> Input: File Data Block
> Output: Block of Block Chain

    1 for N blocks in upload queue
    2   Create the Root-Hash-code
    3   Fetch Hash code for the Previous Block (PBC)
    4   Generate Header PBC-TimeStamp-Nonce
    5   Data encryption and the creation of the Block Body
    6   Combine Body and Header for block formation
    7 otherwise
    8   wait and check

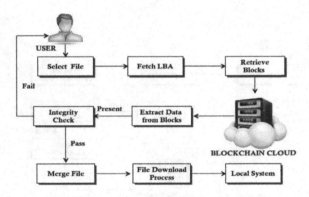

Fig. 3. The proposed file download process

To ensure that each block retains its uniqueness and integrity, a unique MD5 hash per block is created, serving as a digital signature that verifies the block's data. This hash is produced by applying the MD5 algorithm to the block's input data, resulting in a 128-bit output that remains the same for identical input data. The merge file feature combines multiple blocks into a single file, which users can download to their local systems. The auditor verifies whether the blocks are intact or not. For verification, Algorithm 2 is used.

In comparison to traditional blockchain systems, the proposed system has achieved optimized storage. Block storage optimization reduces the amount of storage while maintaining the confidentiality and integrity of the system. Storage optimization is important because blockchain requires a significant storage space to store data. These findings suggest that the suggested system is a very secure and effective way to store transaction in a hybrid cloud by offering faster uploading and downloading times and optimised storage space utilization.

> Algorithm 2: Block Verification

> Input: Block ID
> Output: Block Status

1   if Block count > 0 then
2   List the details
3   Retrieve block, Test and Extract
4   Generate Hash for transaction
5   Detect Previous Hash Code of the Block
6   Compare the Hash Codes
7   Print Result
8   else
9   Display – Verification Completed

The suggested system has been implemented and assessed within a hybrid cloud setup, confirming the proper functionality of all four user capabilities. A comparison has been conducted regarding the process of uploading blocks and downloading times to evaluate the system's performance. Based on the results presented in Fig. 4, it appears that the system demonstrates quicker block upload and download times in contrast to conventional blockchain systems. This is a positive indication that the system is meeting the requirements and specifications that were set out for it.

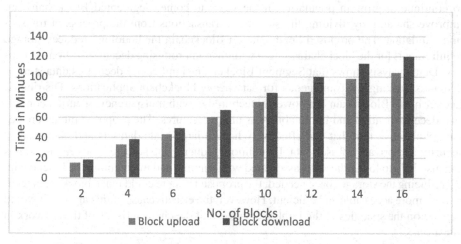

Fig. 4. Time is taken Vs No of blocks for File upload and download graph

Initially, the count of files uploaded is 30 files and later divided, and saved in cloud without the map-reduce algorithm with a total block of 850. However, the block count was decreased to 415 using the map-reduce approach. This is mentioned in Table 1. Hence, the findings indicate that the proposed method is effective and secure for hybrid cloud data storage. In comparison to existing blockchain, it is capable of offering faster uploading and downloading times and optimized storage space utilization.

Table 1. Total and Unique Blocks

File Count	Total Block	Unique Block
5	150	88
10	320	174
15	450	209
20	580	318
25	720	378
30	850	415

5 Conclusion

Blockchain creates a highly secure, transparent, and decentralized system for recording and verifying transactions. A blockchain is a digital ledger, that uses cryptography and distributed computing to record transactions transparently. Distributed ledgers are often used in applications such as cryptocurrencies, supply chain, and voting systems, with significant benefits in terms of security, transparency, and efficiency, and is likely to continue to grow in popularity in the years to come. Segmented Blockchain can improve sharding by dividing the storage of transactions from the process of transaction validation. The proposed Segmentation Blockchain methodology reduces storage requirements for the blockchain system without compromising decentralization or security. Data analysis shows that Segment Blockchain significantly decreases data storage demands, making it advantageous for data-heavy blockchain applications. Distributed Ledger based Blockchain is a powerful technology with transparency, security, consensus, decentralization and immutability as unique features. The proposed model creates a blockchain system that can effectively handle large-scale datasets and also ensures security and privacy of user data. In addition, reducing the storage space requirements can also help to lower the costs associated with running and maintaining the blockchain. By reducing the storage space needed, the proposed system could make blockchain technology more accessible and efficient. However, the effectiveness of this approach would depend on the specifics of the implementation and the characteristics of the network in question.

Funding.   There is no funding for this research publication.

Conflict of Interest.   The authors declare the absence of any conflicting financial interests. They also assert that the study was conducted without any involvement of financial affiliations that could be perceived as a possible source of conflicts of interest.

Contributions. Each of the authors made an equal contribution to the preparation of the manuscript.

References

1. Biryukov, A., Tikhomirov, S.: Security and privacy of mobile wallet users in Bitcoin, Dash, Monero, and Zcash. Pervasive Mob. Comput. **59**, 101030 (2019)
2. Mahapatro, R.K., Ali, A., Ramakrishnan, N.: Blockchain segmentation: a storage optimization technique for large data. In: 2023 8th International Conference on Communication and Electronics Systems. ICCES. IEEE (2023)
3. Luu, L., et al.: A secure sharding protocol for open blockchains. In: Proceedings of the 2016 ACM SIGSAC Conference on Computer and Communications Security (2016)
4. Divakaruni, A., Zimmerman, P.: The lightning network: turning bitcoin into money. Finan. Res. Lett. **52**, 103480 (2023)
5. Jere, S., et al.: Recruitment graph model for hiring unique competencies using social media mining. In: Proceedings of the 9th International Conference on Machine Learning and Computing (2017)
6. Sharon Priya, S., Ali, A.: Localization of WSN using IDV and Trilateration Algorithm. Asian J. Eng. Technol. Innov. **4**(7) (2016)
7. Tu, Z., Xue, C.: Effect of bifurcation on the interaction between bitcoin and litecoin. Finan. Res. Lett. **31** (2019)
8. Dong-Her, S., et al.: Verification of cryptocurrency mining using ethereum. In: IEEE Access 8, 120351–120360 (2020). 2018 Crypto Valley Conference on Blockchain Technology, CVCBT. IEEE (2018)
9. Akcora, C.G., Gel, Y.R., Kantarcioglu, M.: Blockchain networks: data structures of bitcoin, Monero, Zcash, Ethereum, ripple, and iota. Wiley Interdisc. Rev. Data Min. Knowl. Disc. **12**(1), e1436 (2022)
10. Tschorsch, F., Scheuermann, B.: Bitcoin and beyond: a technical survey on decentralized digital currencies. IEEE Commun. Surv. Tutorials **18**(3), 2084–2123 (2016)
11. Palm, E., Olov, S., Ulf, B.: Selective blockchain transaction pruning and state derivability. In: 2018 Crypto Valley Conference on Blockchain Technology, CVCBT. IEEE (2018)
12. Patil, S.S., Ali, A., Ajil, A.: Approaches for network analysis in protein interaction network. Int. J. Hum. Comput. Intell. **2**(2), 47–54 (2023)
13. Vasin, P.: Blackcoin's proof-of-stake protocol v2, **71**. https://blackcoin.co/blackcoin-pos-protocol-v2-whitepaper.pdf (2014)
14. Burchert, C., Decker, C., Wattenhofer, R.: Scalable funding of bitcoin micropayment channel networks. Roy. Soc. Open Sci. **5**(8), 180089 (2018)
15. Ali, A., Sumalatha, D.P.: A survey on balancing the load of big data for pre-serving privacy access in Cloud. Asian J. Eng. Technol. Innov. (AJETI), 176 (2018)
16. Varshney, T., et al.: Authentication & encryption-based security services in blockchain technology. In: 2019 International Conference on Computing, Communication, and Intelligent Systems, ICCCIS. IEEE (2019)
17. Kuo, T.-T., Kim, H.-E., Ohno-Machado, L.: Blockchain distributed ledger technologies for biomedical and health care applications. J. Am. Med. Inform. Assoc. **24**(6), 1211–1220 (2017)
18. Ølnes, S., Ubacht, J., Janssen, M.: Blockchain in government: benefits and implications of distributed ledger technology for information sharing. Gov. Inf. Q. **34**(3), 355–364 (2017)
19. Morkunas, V.J., Paschen, J., Boon, E.: How blockchain technologies impact your business model. Bus. Horiz. **62**(3), 295–306 (2019)
20. Song, X.-S., et al.: A markov process theory for network growth processes of DAG-based blockchain systems. arXiv preprint arXiv:2209.01458 (2022)

21. Ali, A.: Analytical study on fast and secure authenticated key agreement protocol for low power networks. IJCST **6**(4) (2015)
22. Ali, A., Hulipalled, V.R., Patil, S.S.: Centrality measure analysis on protein interaction networks. In: 2020 IEEE International Conference on Technology, Engineering, Management for Societal impact using Marketing, Entrepreneurship and Talent (TEMSMET). IEEE (2020)

Non-linear Finite Element Restorative Analysis of Low Shear Resistant RC Beams Strengthened with Bio-Sisal and GFRP

Tara Sen[(✉)]

Department of Civil Engineering, National Institute of Technology, Barjala, Jirania,
Agartala 799046, India
tarasen20@gmail.com

Abstract. Restoration is strengthening of structural components to its original strength characteristics or with an even better strength characteristic. Damaging events such as earthquakes incepts various degradation processes in structures, restoration aids in making the structures overcome these damaging effects and enhances its characteristics thereby leading to structural life enhancement. Non-linear finite element method (NLFEM) is a mathematical tool by means of which various strengthening or restorative methodologies can be evaluated through virtual platform and thus it results in economic savings and aids in decision making before actually being implemented with accuracy and optimization in the field works. In this paper the restorative analysis by means of NLFEM is carried out for evaluating the restorative potential of sisal fibre made fibre reinforced polymer (FRP) and a comparative analysis with that of GFRP has been done, in enhancing the shear resistance of RC beams (designed for shear-oriented failure) with these respective bonded FRPs. Sisal fibres have huge environmental positive contributing factors over artificially fabricated glass fibres and its composites, as glass is mainly obtained from silica, which is again obtained from non-renewable resources. Concrete crushing parameter alongside concrete cracking parameters, small deflection, static analysis with appropriate boundary conditions, and suitable sub-steps were all considered in the NLFEM analysis of bio-sisal FRP affixed RC beam and also the GFRP affixed RC beam. FRP composite and RC beam affixation was virtually designed buy using contact elements. The results obtained demonstrated huge potential in shear resistance enhancements in RC beams using both sisal FRP and GFRP affixation process. The paper ended with a recommendation to practicing engineering for utilizing natural sustainable bio based FRPs over artificial ones for restorative purpose.

Keywords: Analytical verification · finite element analysis · shear · FRP · Sisal · Beams

1 Introduction

Restoration is the science of parametrical enhancements of various building components with strength, stiffness and lateral stability, so that it can again be brought back to life with improved functionalities. The process of restoration can be done with fibre reinforced polymer composites for strength parametrical enhancements using different fibres of artificial nature such as carbon based fibre reinforced polymer composite, glass fibre based reinforced polymer composite, and aramid fibre based reinforced polymer composites, successfully as past research has divulged [1–10]. There are various advantages of using the fibre reinforced polymer (FRP) restorative methodology over other methodologies such as reinforced concrete affixation on top of or surrounding the various structural components, steel plate affixation, ferrocement affixation, supplemental damper affixation, wing-wall affixation etc. FRPs are completely light weight than reinforced concrete layer affixation over a particular structural component or addition of wing wall component. FRPs have higher flexibility in the affixation process over all other processes. FRPs possess very high ultimate tensile strength and tensile modulus as compared to other restorative methodologies. Affixation process with FRPs are very simple over other methodologies. Non-linear finite element method (NLFEM) is a mathematical tool by means of which various restorative methodologies can be evaluated through virtual platform and thus it results in economic savings and aids in decision making before actually being implemented with accuracy and optimization in the field works. In reality, experimental procedures are intensively labour dependent and bear heavy economic liabilities, hence before practically implementing any new methodology with newer materials, numerical simulation aids us in proper justification of the methodology to be implemented. In this paper the restorative analysis by means of NLFEM is carried out for evaluating the restorative potential of sustainable and renewable based material which is bio-sisal obtained from local resources and using completely organic fibres which are connected with rural economy and a comparative study with the restorative potential of glass fibre reinforced polymer composite (GFRP) which bear non-renewable primary pre-cursors and are a processed industrial product, have been studied.

Effectiveness and accuracy of various linear and non-linear analysis with static pre and post deflection models, has been strongly established by various past researchers, for simulating the effect and restorative potential of various forms of FRP attached to reinforced concrete surfaces [11–32]. Predominantly non-linear analysis of RC beams designated with shear failure characteristics are of important aspect-oriented study, the reason being, shear, associated with damage which is completely brittle in nature. And can result in sudden, non-deflecting, highly catastrophic failures without giving out prior signs or warnings to the building occupants [1, 3, 6, 9, 10]. The study incorporates reinforced concrete RC beams designated with shear failure characteristics and understanding the restorative potential of the same, when affixed with bio-sisal FRP vis-à-vis GFRP respectively. Understanding the crushing and cracking features and the maximum load at shear before the RC beams goes into the plastic deformation phase, of these respective models, with the aid of the NLFEM of analysis forms the core of the paper.

The organization of the article (research as presented) basically deals with the following sub-steps: Firstly the design of the reinforced concrete (RC) beams for shear failure criteria (so that bending failure takes place after the onset of shear dominated cracks and failure), then preparation of the RC beam reinforcement detailing, followed by the generation of the finite element model in the Ansys virtual platform considering the reinforcement detailing and the RC beam properties. Then, considering all the evaluated values of the various components such as Concrete, reinforcement steel, FRP composites of Bio-sisal and Glass and the steel plates for loading and support, and assigning respective properties in the Ansys virtual environment. Thereafter crack modelling parameters and failure criteria parameters and properties being set in the Ansys virtual environment, also ensuring proper inter-elemental connectivity for monolithic effect, thereafter the loadal application and the support system modeling in the Ansys virtual environment, then the setting of the convergence criteria in the Ansys virtual environment for solution obtainment. And finally obtaining the Ansys based crack pattern and detailed analysis-based output and presentation of the Ansys generated results and thereby giving a comparative statement of the maximum ultimate shear load being carried by bio-sisal FRP wrapped RC beams and glass FRP wrapped RC beams.

2 Models for Non-linear NLFEM

IS- 456:2000 [33] has been referred to for the design of the reinforced concrete beam designated for shear failure. The reinforced concrete beam design was such that shear failure was predominant and the beam board higher flexural resistance in-order to generate shear cracks before the onset of bending cracks. Model ShearRC1 is that finite element model where no FRP was bonded on the face of the beam. Model Bio-Sisal2, is that finite element model where sisal FRP composite was bonded over the three faces of the RC beam, and finally, Model GFRP3 is that finite element model where GFRP composite was bonded over the three faces of the RC beam.

2.1 Design for Predominant Shear Failure in the NLFEM of Analysis

Shear brittle failure is characterized by shear cracks which appear to be almost 45^0 inclined cracks near the support regions of the beams. These come without any prior deflections so that sufficient warning signs can be vetted out. As these cracks are sudden and brittle in nature without any display of post bending deformations, hence study of the same becomes extremely important to avoid and prevent any catastrophic damage to the structures. The design in the simulative environment i.e. NLFEM of analysis using the Ansys 12.0 software was done considering the provision of additional flexural reinforcements i.e. longitudinal tensile reinforcements considering double bending effect, thereby providing in higher tensile reinforcement of 159.435 mm^2. It was observed that considering only single bending effect, the tensile reinforcement requirement was of 106.29 mm^2, whereas by considering the double bending effect, the total area of tensile reinforcement almost increased by 50%. In this way, determinant shear cracks can be obtained in the simulative environment. In total 4 numbers of 8 mm dia bars was provided in the pure bending zone, as the tensile reinforcement. Considering single shear effect, such that shear resistance of the RC beam is lesser than the bending resistance, 2 legged 8 mm dia steel stirrup bars at a clear spacing.

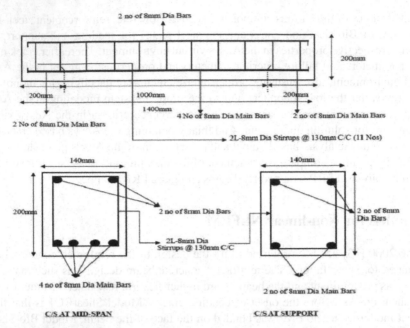

Fig. 1. Reinforcement detailing of Model ShearRC1, Model Bio-Sisal2 and Model GFRP of 130 mm C/C was provided throughout the beam. For all models i.e. Model ShearRC1, Model Bio-Sisal2 and Model GFRP3, the same reinforcement design was carried out and followed while modeling in the simulative environment. Figure 1 Shows the detailing of the steel reinforcement of all the beams.

3   Non-linear Restorative Analysis Using Non-linear Finite Element Method

3.1   Methodology of the Modeling and Meshing in the Simulative Environment

Finite element method of analysis is an elemental level analysis which basically starts with the discretization technique. Using the elemental discretization procedure, various composite elements in the model to be built up is broken down into elements connected at nodes, having individual characteristics or properties. In order to discretize concrete the SOLID-65 element has been very widely accepted with good accuracy of results. LINK-8 element has been widely accepted for simulating the steel reinforcement in discreet analysis using the FEM platform. SOLID-45 element has been widely used by past researchers for simulating the effect of steel loading plates and steel supports, and these have provided very good results without causing any deformities and have provided monolithic effect without any in-compatibility within the various elements in the structural model. FRPs of glass fibre reinforced polymer and bio-degradable sisal fibre reinforced polymer composites have been discretized using the SHELL-99 element. SHELL-99 element has been successfully used for discretizing reinforced polymer composites in various engineering simulative problems by various past researchers [11–25]. This elemental discretization of various concrete constituents such as hardened cement, steel reinforcement, loading and support plating system utilizing these stated elements have been widely studied and results with good ac curacy has been established by past works by various researchers [11–25]. After assigning the elemental properties, material properties were assigned to each of these elements for the generation of the RC beam model. The concrete element which is basically done elemental discretization using the Solid65 element in Ansys uses Material Model Number 1 for its property classification. Concrete elemental discretization basically uses both characteristics of linear and multi-linear isotropic models. The Willam and Warnke model [26] aids in defining the multilinear characteristics of the failure criterion of reinforced concrete in conjunction with the Von Misses failure criterion. The modulus of elasticity of the concrete (Ec) is defined in the Ansys platform by the parameter EX, and PRXY is the parameter which defines the Poisson's ratio (v) of reinforced concrete. The modulus of elasticity of concrete can be evaluated as per the equation as defined by (Clause. 6.2.3.1 of IS 456: 2000), $Ec = 5000\sqrt{fck}$, with the characteristic compressive strength value of fck being considered as 20 MPa. Poisson's ratio of reinforced concrete is considered as 0.2 (Kachlakev et al., [16] for the said analysis. SOLID-65 was considered for multilinear isotropic values and the following stress-strain graph as shown in Fig. 2 was considered to model the bending characteristics of concrete. For modeling the crack-crush characteristics following characteristics for concrete have been used, shear transfer between elemental interfaces plays a vital role in the crack modeling of the non-linear finite element analysis of reinforced concrete structures, the coefficient of transfer of shear between elemental interfaces is considered in the range of 0.0 to 1.0. Basically coefficient of 0 describes smooth elemental crack in the SOLID65 concrete model and coefficient of 1 describes rough elemental crack in the SOLID65 concrete model. Kachlakev et al. [16] has carried out massive research for determining the shear transfer coefficients for crack modelling of reinforced concrete using finite element method. Their study ranged from determining coefficients of shear transfer from smooth to rough cracks and also cracks falling in the intermediate sections, i.e. semi rough cracks. In non-linear finite element analysis convergence of the solution and obtainment of results is the primary criterion. It has been reported that non-convergence of solution (related to crack modeling of concrete) takes place when the shear transfer coefficient for cracks which are considered "open" falls below 0.2. Considering the said the coefficient of shear-transfer for open crack is considered 0.3 here, for the carried out analysis.

The reinforcement in the concrete is described by the parameter, Material Model Number 2. This parameter describes the elemental properties of the reinforcement defined by the Link8 element. Link8 element has been considered for elemental discretization of the longitudinal reinforcements at the beam bottom and top (reinforcement bars) and also the stirrup reinforcements, used for taking care of the shear forces. The said material, as defined in Material Model Number 2, is considered as bilinear-isotropic and has been derived from the failure criterion of "Von Mises failure criteria". This bilinear model has a requisite of two parameters being defined, one is the yield stress designated by fy, which is considered here as 415 MPa, and the other is the hardening modulus of the steel reinforcement, which is also required to be defined. For reinforcement steel modeling, the parameter of the tangent modulus (mainly considered in the plastic zone of the stress-strain curve) is considered to be zero. The modulus of elasticity for the link8

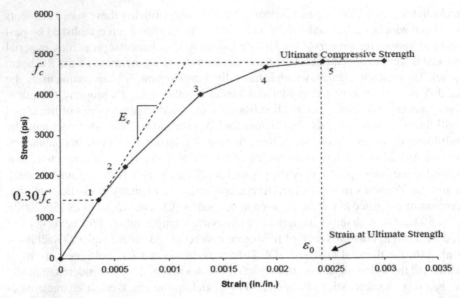

Fig. 2. Stress-strain curve for concrete as presented by by Ibrahim and Mahmood (2009), Wolanski [27] and Kachlakev et al. [16].

model is considered to be 2,00,000 MPa, which is designated as EX. The Poisson's ratio, which is designated as PRXY is considered as 0.3. Figure 3 shows the stress-strain curve for steel reinforcement, i.e. modeled using the Link8 element. Same characteristics have been used for describing the real constant properties of Solid45 element used for elemental discretization of the loading and the supports.

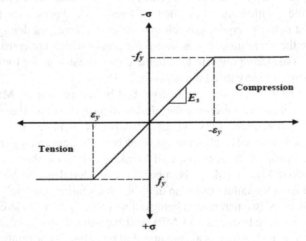

Fig. 3. Tension-compression (bending) curve for steel-reinforcement and loading as well as support roller system.

For assigning material properties to the FRP composites, SHELL-99 element was used for simulating Bio-sisal FRP composites, modulus of elasticity as designated by EX is considered as 234 MPa and the characteristic Poisson's ratio as designated by PRXY is considered as 0.32, with a material thickness of 3.98 mm for the bio-sisal FRP composites. SHELL-99 element was also used for elemental discretization of the GFRP composites, modulus of elasticity as designated by EX is considered as 678 MPa and the characteristic Poisson's ratio as designated by PRXY is considered as 0.21, with a material thickness of 1.4 mm for the GFRP composites.

Volumetric modeling of all individual elements was carried out, and all volumes were overlapped. Any orphaned nodes, which basically loose connection with the adjacent nodes, were carefully removed. All precautions were taken such that the nodes which are orphaned are removed, and all volumes are "glued" together so as to result in all the good properties to be satisfied by the displacement function, i.e. the convergence and the compatibility criteria are all fulfilled. Rectangular mesh of concrete element has been recommended by Wolanski, [27] and Kachlakev et al., [16] and the same has been carried out. All volumes of concrete, and steel rollers for loading and the steel support were meshed using the rectangular "mapped" meshing option. Even the composites of bio-sisal FRP and GFRP and their respective volumes were mapped mesh in the Model Bio-Sisal2 and Model GFRP3 for understanding the evaluating the restorative effect on reinforced concrete beams. Only half length of the RC beam alongside the composite face has been modeled due to symmetricity. NLFEM of analysis makes use of symmetrical configuration of loading and geometry and automatically projects the output for the complete beam. Symmetrical BC option was utilized for activating this feature.

Concentrating nodes, defined on a single plane of the bottom steel support (modeled at the support location) was given the constraint to model or simulate the support condition, which is "simply-supported" in this case of analysis. (degree of freedom to be constrained) restrainment was generated in the global UY and the global UZ directions, by applying 0 constraint. By this method of restrainment generation, the "simply supported" beam will be allowed to rotate at the support. The loadal force, P, is basically applied at top location where the steel plate for load application, has been modeled. The application of the vertical force is done by dividing the total force by the total number of nodes present along the plane of force application, so that all the nodes can balance the entire applied force equally and can result in equal force distribution at the center-line of nodes present on the steel plate for load application. Small deflection alongside static analysis was carried out with a total number of sub-steps being considered as 100. The convergence criteria was based on the structural load value, which was set at 200 KN (Figs. 4 and 5).

3.2   Finite Element Restorative Analysis and Conclusionary Studies

A lot of research has been undertaken by the past researchers in understanding the crack and crush characteristics of reinforced concrete [2, 6, 12–14, 28–32]. Red colour of vector plots in the form of red lines represents cracks in reinforced concrete with lesser width, whereas the blue-green colour of the vector plots in the form of blue-green lines represents critical failure criteria resulting in cracks of larger width which ultimately results in failure. The blue-green colour crack pattern basically results in the ultimate failure of the

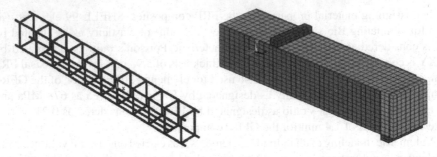

Fig. 4. NLFEM of analysis and its constructive model generation of: (a) steel reinforcement inside the RC beam; (b) The beam with the FRP composite in a three sided wrapping configuration.

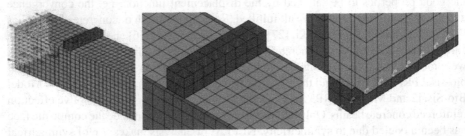

Fig. 5. NLFEM of analysis and its constructive model generation of: (a) symmetricity in the X longitudinal direction; (b) Loading generation at the top of the loading plate; (c) Support reaction at the support end.

specimens. So, in order to understand the NLFEM of analysis generated failure patterns, understanding the trajectory of blue-green vector plots becomes extremely important. Also the non-linear FEM of analysis stores and saves the cracking pattern generated by all the individual load step and finally presents a cumulative crack-crush vector plot at the ultimate or the failure load of the respective specimen. So, a composite load crack and crush characteristic vector plot of all the load sub steps is presented at the end of the iteration when the convergence criteria is met.

We can see well from the Fig. 6(a) that the beam where no FRP was bonded on the face of the beam, i.e. Model ShearRC1, the failure pattern is that of diagonal shear cracking, indicating weakness in the shear resistance over the bending resistance. From Fig. 6(b) i.e. Model Bio-Sisal2, where sisal FRP composite was bonded over the three faces of the RC beam, the failure pattern is a mixed composite failure of flexure-shear tensile cracking and shear - diagonal cracking, the overall shear-cracks in the model is considerably lower than that of the Model ShearRC1 in the shear zone, indicating the strength-stiffness contribution of the Bio-Sisal FRP to the RC beam. From Fig. 6(c) i.e. Model GFRP3, where GFRP composite was bonded over the three faces of the RC beam, the failure pattern mild shear-tensile cracking and mild shear - diagonal cracking, the overall shear-cracks in the model is considerably lower than that of the Model ShearRC1 as well as Model Bio-Sisal2, in the respective shear zones, indicating the strength-stiffness contribution of the GFRP to the RC beam. As GFRP has higher ultimate tensile strength over Bio-Sisal FRP, so strength-stiffness contribution is also higher in terms of restorative effect. But for both the models Model Bio-Sisal2 and Model GFRP3, overall the shear-diagonal tensile cracking pattern is much lesser than that of the Model ShearRC1. Also the ultimate failure load as obtained from the NLFEM of analysis is as presented in Table 1.

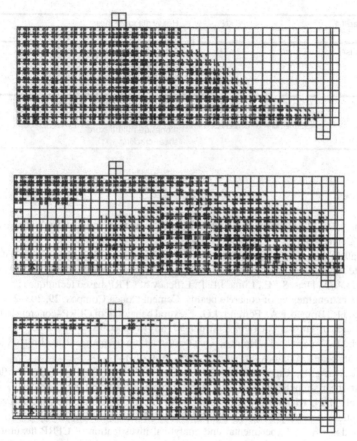

Fig. 6. Diagonal shear-tensile cracking and respective vector-plots of the FEM models (a) NLFEM generated crack-vector-plot of Model ShearRC1, (b) NLFEM generated crack-vector-plot of Model Bio-Sisal2, (c) NLFEM generated crack-vector-plot of Model GFRP3.

Table 1. Results of the restorative NLFEM of analysis

NLFEM of analysis and respective beam models	Thickness of FRP (mm) and ultimate tensile strength of FRP(MPa)	Maximum shear failure load at convergence criteria of load (KN)	Predominant failure pattern	Enhancement in shear load
Model ShearRC1	--	98 KN	Heavy diagonal shear tensile cracking	--
Model Bio-Sisal 2	3.98mm 234MPa	172 KN	Diagonal shear tensile cracking alongside mild flexure-shear cracking	0.75
Model GFRP3	1.4mm 678MPa	184 KN	Mild diagonal shear tensile cracking alongside mild flexure-shear cracking	0.88

References

1. Al-Amery, R., Al-Mahaidi, R.: Coupled flexural–shear retrofitting of RC beams using CFRP straps. Compos. Struct. **75**, 457–464 (2006)
2. Almusallam, T.H.: Load–deflection behavior of RC beams strengthened with GFRP sheets subjected to different environmental conditions. Cement Concr. Compos. **28**, 879–889 (2006)
3. Barros, J.A.O., Dias, S.J.E., Lima, J.L.T.: Efficacy of CFRP-based techniques for the flexural and shear strengthening of concrete beams. Cement Concr. Compos. **29**, 203–217 (2007)
4. Correia, J.R., Branco, F.A., Ferreira, J.G.: Flexural behaviour of GFRP–concrete hybrid beams with interconnection slip. Compos. Struct. **77**, 66–78 (2007)
5. Correia, J.R., Valarinho, L., Branco, F.A.: Post-cracking strength and ductility of glass–GFRP composite beams. Compos. Struct. **93**, 2299–2309 (2011)
6. Dong, J., Wang, Q., Guan, Z.: Structural behaviour of RC beams with external flexural and flexural–shear strengthening by FRP sheets. Compos. B **44**, 604–612 (2013)
7. Esfahani, M.R., Kianoush, M.R., Tajari, A.R.: Flexural behaviour of reinforced concrete beams strengthened by CFRP sheets. Eng. Struct. **29**, 2428–2444 (2007)
8. El-Ghandour, A.A.: Experimental and analytical investigation of CFRP flexural and shear strengthening efficiencies of RC beams. Constr. Build. Mater. **25**, 1419–1429 (2011)
9. Hashemi, S., Al-Mahaidi, R.: Flexural performance of CFRP textile-retrofitted RC beams using cement-based adhesives at high temperature. Constr. Build. Mater.er. **28**, 791–797 (2012)
10. Sundarraja, M.C., Rajamohan, S.: Strengthening of RC beams in shear using GFRP inclined strips – an experimental study. Constr. Build. Mater. **23**, 856–864 (2009)
11. Al-Rousan, R., Haddad, R.: NLFEA sulfate-damage reinforced concrete beams strengthened with FRP composites. Compos. Struct. **96**, 433–445 (2013)
12. Hawileh, R.A.: Nonlinear finite element modelling of RC beams strengthened with NSM FRP rods. Constr. Build. Mater. **27**, 461–471 (2012)
13. Hawileh, R.A., El-Maaddawy, T.A., Naser, M.Z.: Nonlinear finite element modelling of concrete deep beams with openings strengthened with externally-bonded composites. Mater. Des. **42**, 378–387 (2012)

14. Hawileh, R.A., Naser, M.Z., Abdalla, J.A.: Finite element simulation of reinforced concrete beams externally strengthened with short-length CFRP plates. Compos. B **45**, 1722–1730 (2013)
15. Jindal, A.: Finite element modelling of reinforced concrete exterior beam-column joint retrofitted with externally bonded fiber reinforced polymer (FRP). Master's Thesis, Department of civil engineering, Thapar University, Patiala, Punjab (2012)
16. Kachlakev, D., Miller, T.: Finite element modelling of reinforced concrete structures strengthened with FRP laminates Final Report SPR 316, Oregon department of transportation. For Oregon Department of Transportation Research Group200 Hawthorne SE, Suite B-240Salem, OR 97301-5192 and Federal Highway Administration, 400 Seventh Street, SW, Washington, DC 20590 (2001)
17. Luo, Y., Li, A., Kang, Z.: Parametric study of bonded steel–concrete composite beams by using finite element analysis. Eng. Struct. **34**, 40–51 (2012)
18. Özcan, D.M., Bayraktar, A., Sahin, A., Haktanir, T., Türker, T.: Experimental and finite element analysis on the steel fiber-reinforced concrete (SFRC) beams ultimate behavior. Constr. Build. Mater. **23**, 1064–1077 (2009)
19. Omran, H.Y., El-Hacha, R.: Nonlinear 3D finite element modelling of RC beams strengthened with prestressed NSM-CFRP strips. Constr. Build. Mater. **3**, 74–85 (2012)
20. Padmarajaiah, S.K., Ramaswamy, A.: A finite element assessment of flexural strength of prestressed concrete beams with fiber reinforcement. Cement Concr. Compos. **24**, 229–241 (2002)
21. Potisuk, T., Higgins, C.C., Miller, T.H., Yim, S.C.: Finite Element Analysis of Reinforced Concrete Beams with Corrosion Subjected to Shear. Hindawi Publishing Corporation. In: Advances in Civil Engineering, vol. 2011, Article ID 706803, 14 p. (2011)
22. Saifullah, I., Hossain, M.A., Uddin, S.M.K., Khan, M.R.A., Amin, M.A.: Nonlinear analysis of RC beam for different shear reinforcement patterns by finite element analysis. Int. J. Civ. Environ. Eng. **11**(01), 63–74 (2011)
23. Salem, G.G., Galishnikova, V.V., Elroba, S.M., Nikolai, I.V., Kharun, M.: Finite element analysis of self-healing concrete beams using bacteria. Materials (MDPI) **15**, 7506 (2022)
24. Santhakumar, R., Chandrasekaran, E.: Analysis of retrofitted reinforced concrete shear beams using carbon fiber composites. Electron. J. Struct. Eng. **4**, 66–74 (2004)
25. Sen,T.: Nonlinear computational crack analysis of flexural deficit carbon and glass FRP wrapped beams. In: Oscar Castillo, O., Bera, U.K., Jana, D.K. (eds.) Applied Mathematics and Computational Intelligence ICAMCI-2020, Springer Proceedings in Mathematics & Statistics. Springer Nature, Singapore, vol. 413, pp.161–170 (2023). https://doi.org/10.1007/978-981-19-8194-4_14
26. Willam, K.J., Warnke, E.P.: Constitutive model for triaxial behaviour of concrete. In: Seminar on concrete structures subjected to triaxial stresses. In: International Association of Bridge and Structural Engineering Conference, Bergamo, Italy, p. 174 (1974)
27. Wolanski, A.J.: Flexural behavior of reinforced and prestressed concrete beams using finite element analysis. Master of Science thesis, Faculty of the Graduate School, Marquette University, Milwaukee, Wisconsin (2004)
28. Barour, S., Zergua, A., Bouziadi, F., Kaloop, M.R., El-Demerdash, W.E.: Nonlinear numerical and analytical assessment of the shear strength of RC and SFRC beams externally strengthened with CFRP sheets, Hindawi. In: Advances in Civil Engineering, vol. 2022, Article ID 8741158, p. 17 (2022)
29. Pandimani: Computational modeling and simulations for predicting the nonlinear responses of reinforced concrete beams. Multidiscipline Model. Mater. Struct. **19**(4), 728–747 (2023)
30. Ravshanbek, M., Sobirjon, R.: Numerical modeling of combined reinforcement concrete beam. In: E3S Web of Conferences 401, 03007 (2023) CONMECHYDRO – 2023

31. Attiya, M.A., Abbas, H.N.: Non-linear analysis of reinforced concrete continuous deep beams with openings strengthened using CFRP sheets. Int. J. Sci. Technol. Res. **9**(01), 2350–2354 (2020)

32. Cruz, T.A., et al.: Comparative analysis of fiber reinforced polymers (FRP) on the behavior of reinforced concrete beam-column joint by simulation approach using Ansys workbench R22. Int. J. Progressive Res. Sci. Eng. **4**(06), 338–359 (2023)

33. IS-456: 2000; Indian Standard, Plain and reinforced concrete-code of practice (4th Revision)

Estimation of Mass Balance of Baspa Basin, Western Himalayas Using Remote Sensing Data

A. R. Deva Jefflin[1], M. Geetha Priya[1](✉), Dilsa Nasar[1], Sushil Kumar Singh[2], and Sandip Oza[2]

[1] CIIRC, Jyothy Institute of Technology, Bengaluru 560082, India
geetha.sri82@gmail.com

[2] Space Applications Centre, Indian Space Research Organization, Ahmedabad, India

Abstract. Glacial and snow meltwater holds immense importance as the primary origin of the Himalayan rivers, exerting a vital influence on hydrological processes and socio-economic endeavours within the mountainous communities. Mass balance is an essential parameter for evaluating glacier health and status. The Accumulation Area Ratio (AAR) method is employed in the current study to assess Baspa basin's specific mass balance for the Hydrological years 2013–2022. The estimated average ELA (equilibrium line altitude) and AAR of the basin has been estimated to be approximately 5265 ± 101 m.a.s.l. and 0.458 ± 0.13/year respectively. The study findings indicate that during the period of HY2021–2022, there was a significant reduction in ice mass, as evidenced by an accumulation area ratio of 0.23. The estimated cumulative specific mass balance of the basin is in the range of −4.306 ± 0.23 m.w.e to −0.86 ± 0.31 m.w.e using empirical equations. During the period from 2013 to 2022, the majority of the glaciers witnessed a decline in their mass, emphasizing the need to anticipate and adjust to the ongoing environmental changes affecting the world's water resources.

Keywords: Glaciers · Baspa Basin · Western Himalayas · AAR · Mass Balance

1 Introduction

Himalayan glaciers hold the distinction of being the third-largest ice mass on the Earth. The glaciers located in the Himalayan region are currently experiencing significant alterations in their ice mass due to shifting climatic conditions. This transformation not only affects the local hydrology but also has repercussions on the overall climate cycle of the planet [13]. Rivers emanating from the glaciers plays a predominant role in meeting the water requirements of inhabitants living in downstream areas by carrying the essential water supply derived from the glaciers and snow run-off [5]. Recent investigations reveal that the glaciers in Himalayan region have been experiencing an increased retreat since the onset of twenty-first century [14–16]. The more melting of glaciers is strongly correlated with climatic factors such as regional precipitation patterns and temperature variations. It is projected that the scenario of glacier mass will keep changing till 2070 [14, 17–19]. As a result, it is imperative to conduct a comprehensive evaluation of prospective water security in the Himalayan region. This assessment will enable

© The Author(s), under exclusive license to Springer Nature Switzerland AG 2024
K. K. Patel et al. (Eds.): icSoftComp 2023, CCIS 2030, pp. 205–217, 2024.
https://doi.org/10.1007/978-3-031-53731-8_17

the development of robust policies including strategies to effectively adapt for the challenges imposed by changing climate. By considering factors such as the accelerating retreat of Himalayan glaciers, as influenced by local precipitation patterns and temperature fluctuations, policymakers can get valuable insights into the future availability and sustainability of water resources. This scientific analysis will inform evidence-based decision-making processes, facilitating the development of adaptive measures to assure the long-term water availability and resilience of the Himalayan basins in the face of climate change impacts [14].

By monitoring the basin wide glacier mass balance, valuable information can be obtained regarding the overall status and condition of the glaciers, as well as the patterns and fluctuations they undergo. The glacier mass balance quantifies the volume of stored water or lost within the glacier during a hydrological year, reflecting the combined effects of inflow and outflow. It serves as an important metric for understanding the overall water storage dynamics within the glacier [20]. The assessment of mass balance of glaciers is a pivotal aspect in understanding the dynamics of glacier development and regression. Sustained negative mass balance, indicative of glacier decline, possesses the potential to induce both thinning and retreat of the glacier. Consequently, these processes can exacerbate reductions in the glacier's overall area and mass [4]. Due to increased melting, the glacier meltwater is projected to reach its maximum levels within a few decades, followed by a subsequent recession by the end of the century [21–23].

Mass balance estimation in the Himalayan region involves several methods namely the glaciological (field based), geodetic (altitude difference based) method, and AAR (empirical based) method. However, due to its unique orographic features, field measurements using glaciological methods are not commonly conducted for most glaciers in the region. As a result, the AAR and geodetic methods, which utilize satellite imagery, have been extensively employed to estimate mass balance on basin scales for Himalayan regions [2, 24]. The present study utilizes the AAR method to estimate the Baspa basin's specific mass balance for the period 2013 to 2022. A proper comprehension of the ELA is important for AAR computation. The ELA is an altitude at which the mass balance reaches equilibrium, indicating equilibrium between snow accumulation and ablation (melting) over the course of a hydrological year. The AAR technique depends on establishing a regression relationship between the AAR and specific mass balance [2, 25]. In the subsequent sections, study area, data sources, methodology followed by findings are discussed.

2   Study Area

The ongoing investigation focuses on the Baspa basin, situated in the Kinnaur district of Himachal Pradesh, India, is a sub-basin of the Sutluj river. The geographical coordinates of the basin ranges from 78°06′7.438″E, 31°31′45.433″N to 78°51′10.413″E, 31°10′19.671″N. The Fig. 1 displays the area of interest map for reference. As per RGI 6.0 (Randolph Glacier Inventory), the Baspa basin constitutes of 187 glaciers, with an area of 1106.4 km^2 (glaciated and non-glaciated). Among these glaciers, approximately 157 glaciers having area less than 1 km^2 (Fig. 2). The glaciated area in the Baspa basin accounts for about 176.34 km^2, which represents approximately 15.93% of the total area of the basin. The altitude of the Baspa basin ranges between 1753 and 6173 m above sea level (m.a.s.l.). The northern slopes of the Pir Panjal mountain range houses majority of the glaciers in the Baspa basin. [10]. The Baspa River originates from the Baspa Bamak and Arsomang glaciers, with Baspa Bamak glacier (largest) covering an expanse of 32.07 km². Other notable glaciers include Janapa Garang, Shaune Garang, Karu Garang, and Jorya Garang in the basin. Due to basin's high altitude profile, the runoff in the basin is primarily derived from snow and glacier meltwater [7]. Winter

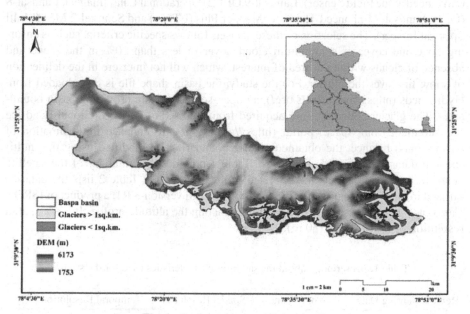

Fig. 1. Study area map of Baspa basin

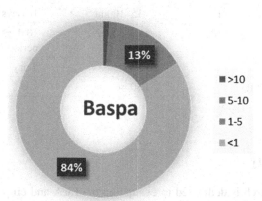

Fig. 2. Area wise glacier distribution in Baspa basin. Approximately 84% of the glaciers have an area smaller than 1 km², 13% of glaciers area is between 1 and 5 km², 2% of glaciers have an area between 5 to 10 km² and 1% of glaciers have an area more than 10 km².

precipitation in the Baspa basin is typically associated with westerly disturbances that bring in moisture from the Mediterranean region, leading to snowfall [8, 9].

3   Data Used

The mass balance estimation for the Baspa basin utilizes the visible and SWIR bands of multispectral datasets obtained from various satellites including IRS-P6 AWiFS (Advanced Wide Field Sensor), Landsat-9 OLI-2 (Operational Land Imager), Landsat-8 (OLI), Landsat-7 Enhanced Thematic Mapper Plus (ETM+) and Sentinel-2 MSI (Multi Spectral Imager). The selection of these datasets follows specific criteria, such as ensuring no cloud cover or a minimum cloud cover of less than 10% in the scene and absence of clouds within the area of interest, which will not interfere in the delineation of snow line over the glacier. For the study, the basin shape file is downloaded from Hydrosheds (https://hydrosheds.org/), an open-source website providing basin boundaries. The glacier boundaries are acquired from RGI 6.0 of GLIMS (Global Land Ice Measurement from Space) portal (https://glims.org/). To enhance the estimation of basin's mass balance, the obtained glacier boundaries are modified using the multi-spectral dataset and Google Earth Pro. Table 1 provides the summary of the satellite characteristics and datasets incorporated in the study, while Table 2 lists the datasets utilized in the present study. Additionally, CartoDEM version-3 R1, a product of ISRO's Cartosat-1 global mission, is utilized for extracting the altitude profile. It has a spatial resolution of 1 arc-second (30 m).

Table 1. Description about the satellites characteristics used for the study

Remote Sensing Data	No. of bands	Spatial Resolution	Temporal Resolution
Landsat-7	8	15–60 m	16 days
Landsat-8	11	15–100 m	16 days
Landsat-9	11	15–100 m	16 days
Sentinel-2	13	10–60 m	10 days (combined revisit period- 5 days)
IRS-P6 AWiFS	4	56 m	5 days
CartoDEM version 3 R1		30 m	

4   Methodology

The ongoing research is dedicated to evaluating the ELA and empirical mass balance of a Baspa basin. This assessment is achieved through the implementation of an approach that centers on the AAR concept, employing a comprehensive dataset captured by multispectral sensors spanning the timeframe from 2013 to 2022. The AAR-based methodology follows a systematic sequence, which begins with the creation of false color

Table 2. Details of all the multispectral dataset collected for the estimation of mass balance from 2013–2022 (NA represents the data gap due to unavailability due to cloud cover)

Year	June Path,Row/Tile id	June Satellite	June Date of Acquisition	July Path,Row/Tile id	July Satellite	July Date of Acquisition	August Path,Row/Tile id	August Satellite	August Date of Acquisition	September Path,Row/Tile id	September Satellite	September Date of Acquisition
2013	146,38	L-7	22	NA	NA	NA	146,38	L-7	25	146,38	L-8	18
2014	146,38	L-8	17	NA	NA	NA	146,38	L-8	20	146,38	L-8	21
2015	NA	NA	NA	NA	NA	NA	146,38	L-8	23	146,38	L-8	08
2016	T44RKV	S-2	04	146,38	L-8	08	146,38	L-8	25	146,38	L-8	10
2017	NA	NA	NA	146,38	L-8	27	NA	NA	NA	146,38	L-8	13
2018	T44RKV	S-2	24	NA	NA	NA	T44RKV	S-2	28	146,38	L-8	16
2019	146,38	L-8	15	146,38	L-8	01	NA	NA	NA	146,38	L-8	19
2020	96,47	IRS-6	27	146,38	L-8	03	98,47	IRS-6	24	146,38	L-8	21
2021	T44RKV	S-2	28	146,38	L-8	06	146,38	L-8	23	T44RKV	S-2	06
2022	T44RKV	S-2	13	T44RKV	S-2	18	146,38	L-8	18	146,38	L-8	11

composite (FCC) images. These FCC images are generated utilizing three specific bands of the multispectral data: Green, Red, and Short-Wave Infrared (SWIR). Incorporating the SWIR band within the FCC imaging process is particularly instrumental. Its inclusion serves a crucial purpose: discriminating between elements that share similar colors in the visual spectrum, namely snow and clouds. This differentiation is achieved because the reflectance characteristics of snow within the SWIR band differ markedly from those in the visible bands. Essentially, snow reflects significantly less SWIR light compared to the visible wavelengths. This disparity allows for a more accurate distinction between snow cover and cloud cover.

With the FCC images successfully obtained, the next step involves identifying a pivotal parameter known as the snow line. This snow line serves as a distinct demarcation point, acting as the boundary that separates the terrain covered by snow from the glaciated ice that lies exposed. This segmentation process is conducted for four specific months during each hydrological year: June to September. Crucially, the altitude data of the TSL (transient snow line) is deduced from the CartoDEM dataset, which provides altitude information. Among the various observed TSL values during a given hydrological year, the maximum averaged TSL is singled out as a representative measure and subsequently utilized as a stand-in for the ELA, as referred to in existing literature [24].

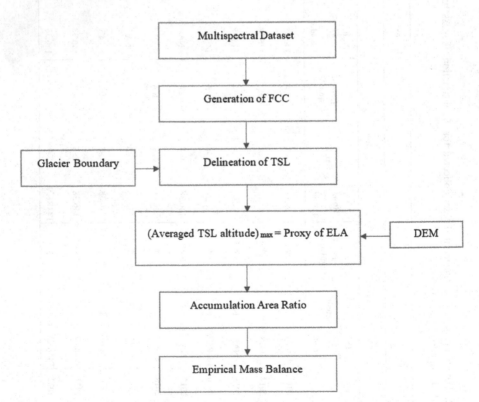

Fig. 3. Methodology utilized for mass balance estimation from 2013 to 2022

The estimated ELA, which is essentially the altitude at which the accumulation and ablation of ice balance out, serves as a key input in determining both the AAR and the mass balance. These parameters are extracted by applying a specific empirical relationship that has been established through prior research and observation. The entire process flow and sequence of operations is visually represented and elucidated in Fig. 3, which provides a graphical depiction of the outlined steps and their interconnections. Through the systematic execution of this methodological approach, the ongoing research endeavors to achieve empirical and comprehensive insights of the mass balance dynamics within the studied region over the specified time period.

5 Results

5.1 Transient Snow Line

The TSLs has been delineated for about 30 glaciers that are having an aerial extent greater than 1 km^2. in Baspa basin from 2013 to 2022. Totally 33 satellite images were collected for estimating Baspa basin's specific mass balance, and within these images, a total of 990 TSL need to be delineated. However, due to the occurrence of clouds and cloud shadows in the glacial region of the basin, a maximum of 890 snow lines were accurately delineated. The snow lines are drawn for each of the four ablation months within the hydrological year for every individual glacier. To visualize and provide a representative showcase of the delineated TSL altitude, Fig. 4 illustrates the TSL altitudes of the Baspa Bamak glacier, serving as a representative example. This visualization pertains to the years spanning from 2013 to 2022, encapsulating the progression of TSL altitudes over the specified period.

Over the course of the period from 2013 to 2022, a consistent pattern emerges in the altitude of the TSL. Notably, the highest averaged TSL altitude is consistently observed towards the end of August and early September, while the lowest averaged TSL altitude tends to occur in the months of June and late September. An interesting observation, however, arises from Fig. 4, where the displayed data indicates the highest averaged altitude in July. This anomaly can be attributed to data gaps that are present specifically in the hydrological year 2016–2017, which have influenced this particular visual representation.

In the broader context of the Baspa basin, the calculated altitude of the TSLs demonstrates a range spanning from 4846 ± 127 to 5248 ± 87 m.a.s.l. This range encapsulates the fluctuations in the snow line altitude that occur annually. The dynamics of this fluctuation are significantly influenced by both precipitation and temperature patterns, as established in previous research [1]. Moreover, the distinct characteristics of the terrain, including factors such as glacier slope and orientation, exert a noteworthy impact on the behavior exhibited by the glaciers.

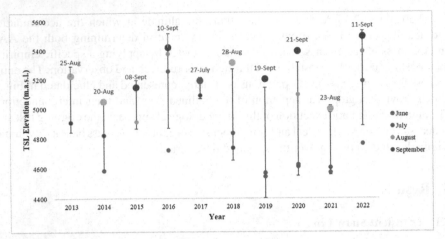

Fig. 4. TSL altitudes of Baspa Bamak glacier. The color dots represents the propagation of snow line in all four ablation months from 2013 to 2022.The maximum observed snow line is shown in big dot with date and the standard deviation is represented in black colored line

5.2 Equilibrium Line Altitude

An insightful visualization in Fig. 5 takes the form of a bar chart, effectively portraying the variability in Equilibrium Line Altitude (ELA) for selected ten benchmark (largest) glaciers oriented towards the North (six glaciers) and West (Four glaciers). Notably, the data reveals a consistent trend: North-facing glaciers generally tend to exhibit slightly higher ELA when contrasted with their West-facing counterparts. This discrepancy underscores the influence of glacier orientation on the ELA and adds a valuable layer of understanding to the interplay between glacial behavior and the surrounding topography.

The Fig. 6 presents a decade-long analysis of basin wide ELA values (based on 30 selected larger glaciers) across various hydrological years (HY) from 2013 to 2022. Notably, the ELA values exhibit a discernible fluctuation over this period, implying shifts in the equilibrium point where ice accumulation and ablation reach a balance. Although a linear trend isn't evident, the data demonstrates a cyclical pattern characterized by alternating higher and lower ELA values. For instance, the a ELA reaches its peak in 2022 at 5442 m.a.s.l., while hitting its lowest point in 2021 at 5151 m.a.s.l., representing a substantial variation of approximately 290 m between these two years. The year 2016 stands out with a notably elevated average ELA of 5363 m.a.s.l., potentially influenced by specific meteorological conditions. Hypsometry curve shown in Fig. 7 reveals that approximately 52.5% of the Baspa basin glaciated region is situated below the average ELA of 5265 m.a.s.l over the decade. These fluctuations suggest a complex interplay between temperature, precipitation, and glacial dynamics. This dataset provides valuable insights into the glaciers' response to shifting climate and underscores the importance of continued monitoring and analysis.

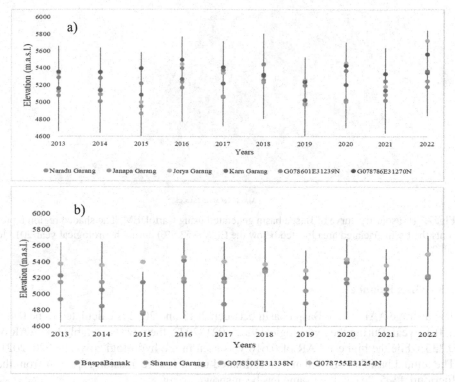

Fig. 5. Showing the variation of the ELA altitude based on the glacier orientation. The colored dots represents the distribution of ELA on the four ablation months in, a) north facing and b) west facing glaciers

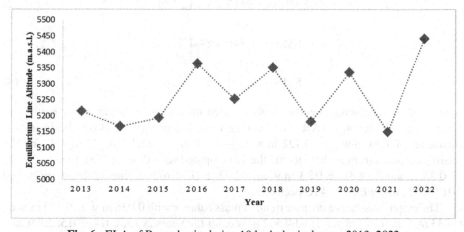

Fig. 6. ELA of Baspa basin during 10 hydrological years 2013–2022

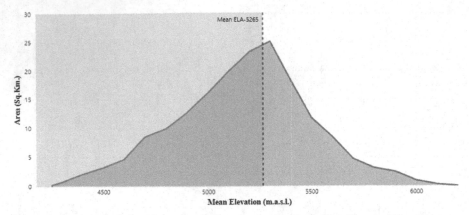

Fig. 7. Hypsometry curve of Baspa basin generated using CartoDEM. The shaded region represents the basin glaciated area located below the ELA (~52.5%) during hydrological year 2013 to 2022

5.3 Mass Balance

The average AAR of the Baspa basin between 2013 and 2022 is calculated to be 0.458 ± 0.13/year. During the hydrological year 2021–2022, there was a notably low AAR of 0.235 while the highest AAR of 0.618 was seen in the hydrological year 2020–2021. The annual basin's mass balance is estimated using four regression Eqs. 1–4 from the literature [3–5, 26] and the same has been shown in Fig. 8.

$$SMB = 2.4301 * Y - 1.20187 \tag{1}$$

$$SMB = 1.9433 * Y - 1.3149 \tag{2}$$

$$SMB = 1.746 * Y - 1.232 \tag{3}$$

$$SMB = 1.863 * Y - 1.228 \tag{4}$$

where, SMB represents specific mass balance measured in meter water equivalent (m.w.e.) and Y denotes AAR. The average mass balance of the Baspa basin is estimated as −0.086 m.w.e., −0.422 m.w.e., −0.431 m.w.e. and −0.372 m.w.e. and the basin's cumulative mass balance of the is computed as −0.86 ± 0.31 m.w.e., −4.22 ± 0.25 m.w.e., −4.31 ± 0.23 m.w.e., −3.73 ± 0.24 m.w.e. during the study period 2013–2022 using Eqs. 1–4 respectively.

The graph illustrates a positive trend of mass balance with 0.036 m.w.e., 0.175 m.w.e., 0.133 m.w.e., 0.215 m.w.e., and 0.302 m.w.e. for the years 2013, 2014, 2015, 2019, and 2021 respectively, obtained using Eq. 1. The positive mass balance is possibly due to the fact that Eq. 1 has been empirically derived from the field based data of Shaune Garang and Gor Garang glaciers situated in the Baspa basin [2].

During the hydrological year 2021–2022, substantial mass loss has been observed in the Baspa basin, indicated by negative mass balance values of −0.62 m.w.e., −0.85 m.w.e., −0.82 m.w.e., and −0.78 m.w.e. from all the four equations. The variations in SMB can be speculated to the distinct behavior of individual glaciers within the basin and constraints associated with regression equations [6]. It is observed that smaller glaciers with an areal extent smaller than 1 km², tend to exhibit negative mass balance. Additionally, even among glaciers located at similar altitudes, variations in mass balance can arise due to irregular topographical distribution of glacier area within the basin.

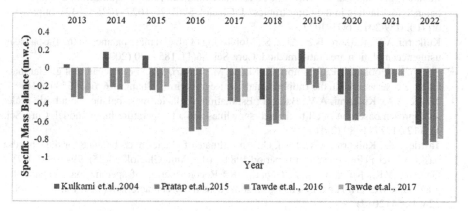

Fig. 8. Shows the SMB estimated using the four regression equations

6 Conclusion

A conventional AAR-based method employed to estimate of Baspa basin's mass balance, situated in the western Himalayas. This method utilizes multispectral images to identify the snow line altitude during each of the four ablation months within a HY to find the proxy ELA. The study's findings show an averaged ELA of 5265 ± 101 m.a.s.l. for 2013–2022, indicating that 52.5% of the basin glaciated area lies below it. This hints at a decline as the majority of the basin area is under the ELA, suggesting glacier mass loss and negative mass balance. According to the current study, the average yearly specific mass balance, estimated using four regression equations from 2013 to 2022, ranges between −0.08 and −0.43 m.w.e. This indicates a predominant negative mass balance in most Baspa basin glaciers, especially those at lower altitudes. The observed negative mass balance reflects the glaciers' vulnerable state, signifying ice and snow loss. This bears significant consequences, particularly for downstream power generation plants. The diminishing glaciers can adversely affect water availability and hydropower sustainability. Consequently, policymakers and government bodies must prioritize glacier preservation and management to ensure the well-being of the inhabitants and mitigate potential impacts of glacier retreat. This approach is vital to safeguarding water resources for both ecological and human needs in the region.

Acknowledgement. This research work is being supported by the Space Applications Centre-ISRO, Ahmadabad under the project "Cryosphere science and Applications Programme". The authors gratefully acknowledge the support rendered by PTICL and Dr. Krishna Venkatesh, Director, CIIRC, Jyothy Institute of Technology, Bengaluru.

References

1. Rabatel, A., et al.: Can the snowline be used as an indicator of the equilibrium line and mass balance for glaciers in the outer tropics? J. Glaciol. **58**(212), 1027–1036 (2012). https://doi.org/10.3189/2012JoG12J027
2. Kulkarni, A.V., Rathore, B.P., Alex, S.: Monitoring of glacial mass balance in the Baspa basin using accumulation area ratio method. Curr. Sci. **86**(1), 185–190 (2004)
3. Pratap, B., Dobhal, D.P., Bhambri, R., Mehta, M., Tewari, V.C.: Four decades of glacier mass balance observations in the Indian Himalaya. Reg. Environ. Change **16**, 643–653 (2016)
4. Tawde, S.A., Kulkarni, A.V., Bala, G.: Estimation of glacier mass balance on a basin scale: an approach based on satellite derived snowlines and a temperature index model. Curr. Sci. **111**(12), 1977–1989 (2016)
5. Tawde, S.A., Kulkarni, A.V., Bala, G.: An estimate of glacier mass balance for the Chandra basin, Western Himalaya, for the period 1984–2012. Ann. Glaciol. **58**(75), 99–109 (2017)
6. Gaddam, V.K., Kulkarni, A.V., Gupta, A.K.: Reconstruction of specific mass balance for glaciers in Western Himalaya using seasonal sensitivity characteristic(s). J. Earth Syst. Sci. **126**(4), 55 (2017)
7. Kulkarni, A.V., Alex, S.: Estimation of recent glacial variations in Baspa basin using remote sensing technique. J. Indian Soc. Remote Sens. **31**, 81–90 (2003). https://doi.org/10.1007/BF03030775
8. Mir, R.A., Jain, S.K., Saraf, A.K., Goswami, A.: Decline in snowfall in response to temperature in Satluj basin, Western Himalaya. J. Earth Syst. Sci. **124**(2), 365–382 (2015)
9. Dimri, A.P., Mohanty, U.C.: Location specific prediction of maximum and minimum temperature over the Western Himalayas. Meteorol. Appl. **14**(1), 79–93 (2007)
10. Raina, V.K., Srivastava, D.: Glacier Atlas of India. Geological Society of India, Bangalore (2008)
11. Gautam, C.K., Mukherjee, B.P.: Mass-balance vis-à-vis snout position of Tipra Bank Glacier District Chamoli, Uttar Pradesh. In: Proceedings of National Meet on Himalayan Glaciology, 5–6 June, pp. 141–148 (1989)
12. Singh, R.K., Sangewar, C.V.: Mass balance variation and its impact on glacier flow movement at Shaune Garang glacier, Kinnaur, Himachal Pradesh. In: Proceedings of National Meet on Himalayan Glaciology, 5–6 June 1989, pp. 149–152 (1989)
13. Bolch, T., Pieczonka, T., Benn, D.I.: Multi-decadal mass loss of glaciers in the Everest area (Nepal Himalaya) derived from stereo imagery. Cryosphere **5**(2), 349–358 (2011). https://doi.org/10.5194/tc-5-349-2011
14. Barnett, T.P., Adam, J.C., Lettenmaier, D.P.: Potential impacts of a warming climate on water in snow-dominated regions. Nature **438**, 303–309 (2005)
15. Azam, M.F., Wagnon, P., Berthier, E., Vincent, C., Fujita, K., Kargel, J.S.: Review of the status and mass changes of Himalayan-Karakoram glaciers. J. Glaciol. **64**(243), 61–74 (2018)
16. Gaddam, V.K., Kulkarni, A.V., Gupta, A.K.: Assessment of the Baspa basin glaciers' mass budget using different remote sensing methods and modeling. Geocarto Int. **35**(3), 296316 (2018)

17. Laurent, L., Buoncristiani, J.-F., Pohl, B., Zekollari, H., Farinotti, D., Huss, M., et al.: The impact of climate change and glacier mass loss on the hydrology in the Mont-Blanc massif. Sci. Rep. **10**(1), 10420 (2020). https://doi.org/10.1038/s41598-020-67379-7

18. Aesawy, A.M., Hasanean, H.M.: Annual and seasonal climatic analysis of surface air temperature variations at six southern Mediterranean stations. Theoret. Appl. Climatol. **61**(1), 55–68 (1998). https://doi.org/10.1007/s007040050051

19. Bhutiyani, M.R., Kale, V.S., Pawar, N.J.: Climate change and the precipitation variations in the north western Himalaya: 1866–2006. Int. J. Climatol. **30**(4), 535–548 (2010). https://doi.org/10.1002/joc.1920

20. Paterson, W.S.B.: Physics of Glaciers, 3rd edn. Butterworth-Heinemann, Oxford (2000)

21. Huss, M., Farinotti, D., Bauder, A., Funk, M.: Modelling runoff from highly glacierized alpine drainage basins in a changing climate. Hydrol. Process. **22**(19), 3888–3902 (2008). https://doi.org/10.1002/hyp.7055

22. Engelhardt, M., Schuler, T.V., Andreassen, L.M.: Contribution of snow and glacier melt to discharge for highly glacierized catchments in Norway. Hydrol. Earth Syst. Sci. **18**(2), 511–523 (2014). https://doi.org/10.5194/hess-18-511-2014

23. Raman, A., Kulkarni, A.V., Prasad, V.: Glacier mass balance estimation in Garhwal Himalaya using improved accumulation area ratio method. Environ. Monit. Assess. **194**, 583 (2022). https://doi.org/10.1007/s10661-022-10261-y

24. Gardelle, J., Berthier, E., Arnaud, Y., Kaab, A.: Region-wide glacier mass balances over the Pamir-Karakoram-Himalaya during 1999–2011. Cryosphere **7**(6), 1885–1886 (2013). https://doi.org/10.5194/tc-7-1263-2013

25. Kulkarni, A.V.: Mass balance of Himalayan Glaciers using AAR and ELA methods. J. Glaciol. **38**(128), 101–104 (1992). https://doi.org/10.3189/s0022143000009631

26. Nagajothi, V., Priya, M.G., Sharma, P., Bahuguna, I.M.: Mass balance of glaciers in Bhaga basin, Western Himalaya: a geospatial and temperature-weighted AAR based model approach. Curr. Sci. **119**(12), 00113891 (2020)

System and Applications

OntoOpinionMiner: An Opinion Mining Algorithm for Drug Reviews

Rashi Srivastava[1]([✉]) [iD] and Gerard Deepak[2]

[1] Central University of Karnataka, Kalaburagi, Karnataka, India
rashisrivastava2001@gmail.com
[2] Manipal Institute of Technology, Manipal, Karnataka, India

Abstract. The internet is now swamped with user data such as thoughts, comments, reviews, etc. due to the rise of social media in the modern day. This enables the creators to have a deeper understanding of the benefits and drawbacks of their creation totally based on the opinions of end users. Opinion mining has been one of the most studied fields along the lines of data mining and natural language processing in recent years. However, the methods to extract the important parts of the healthcare sector are still lacking in an abundance of literature. Patients constantly look for reviews of a specific drug from other users. In order to estimate the drug satisfaction rate among experienced patients, the proposed approach includes a novel three-fold feature selection and classification architecture. For improved outcomes, the Random-Forest Classifier and Long Short-Term Memory (LSTM) have been combined in the construction of this classifier. The outcomes obtained support the claim that this strategy is superior than traditional individual strategies. The proposed methodology has an average precision of 95.83.

Keywords: Information Scent · Lesk algorithm · LSTM · Opinion Mining · Random-Forest Classifier · SimantoSim

1 Introduction

The interaction between humans and computers taking into account the natural language is dealt with by a distinct fragment of Artificial Intelligence known as Natural Language Processing (NLP). It enables computers to comprehend the language of humans through a set of computational methods which are used to analyze natural language and reconstruct them into computer understandable terms. Opinion mining traces down its root to NLP which is a process used to detect and analyze the textual information, opinionated outlook, and sentiments. It stands at the crossroads of computational semantics and retrieval of information. The concepts and frameworks of opinion mining are collated from these fields. Opinion mining can be broadly defined as a technique for text analysis using NLP and computational techniques to draw out opinions, emotions, perceptions, and sentiments hidden within the texts. The emergence of opinion

K. K. Patel et al. (Eds.): icSoftComp 2023, CCIS 2030, pp. 221–234, 2024.
https://doi.org/10.1007/978-3-031-53731-8_18

mining has made it utterly convenient to assess the performance and potency of a particular product or business through the comments and reviews streamed on the web. Researchers have been extensively using it in various domains to track down the acceptance of products, materials, services, and many other things in the market.

The health care domain has been a little unexplored in terms of opinion mining. It is really important to pass the drugs through surveillance to check with their safety once released for the public and so it becomes crucial to understand how a particular drug affects the patients in real-time.

Previous works in this domain have pivoted mostly around drug reviews on the web. An enormous amount of which makes it difficult to root out the gist of consumer discourse. Hence, to acquire the idea on an epidemiological scale on how the drug is being utilized by common people and its efficiency, researchers have been employing opinion mining for the same. In the framework put forth, the investigation has been done on drug satisfaction among patients using opinion mining. The prime objective of this study is to come up with an efficient architecture which will inspect the patients' comments on drug satisfaction quotient and classify it into negative and positive review which can be further used by other patients before consuming the drug. A novel approach has been presented consolidating the Random-Forest classifier and LSTM to obtain a final classification. Feature Extraction is done on a single dataset in a three-fold fashion for better results. The final results show that this framework is more efficient than other existing strategies.

Motivation: Taking into consideration the existing work done by other researchers, it can be seen that opinion mining has been predominantly used for the general domains, such as products, materials, travel, housing, and many others. But the health care domain has been overlooked, while the application of opinion mining can significantly help in pharmacovigilance to manufacturers as well as consumers. Henceforth, a new strategy is proposed taking into regard the drug satisfaction level among patients to trace down the effectiveness or adverse reactions to a drug-using social repository.

Contribution: A novel methodology has been proposed for drug satisfaction level opinion mining wherein a combination of LSTM and Random Forest Classifier is used for final classification. Another innovative contribution includes the incorporation of a three-fold feature extraction method using term frequency, ontology alignment through Lesk similarity, and finally feature weighting based on Information Scent. Integration of all these novel concepts has helped in amplifying the performance of this work put forward.

Organization: The following describes how the paper is organised: Sect. 2 covers the literary survey on work done in the field of interest. The proposed ethodology/technical approach has been outlined in Sect. 3. Implementation of this work has been given in Sect. 4. Performance analysis is discussed in Sect. 5. The conclusion has been formulated in Sect. 6.

2 Literature Survey and Related Works

Research work taken up on the lines of opinion mining for the health care domain has been sparse. Yet, a few scholars have tried to efficiently incorporate the application of opinion mining in this domain to generate an idea on drug reviews. P. Padmavathy et al. [1] proposed a hybrid two-pass classifier to draw out patient's opinions on drug-level satisfaction. A study on Aspect-Based Opinion Mining was put forth by Diana Cavalcanti et al. [2], where a supervised classification system was utilized to classify opinion based on aspect type. This experiment was carried on the dataset for three different diseases. Another idea was brought into the light to analyze drug satisfaction among patients using Supervised learning through the works of Vinodhini Gopalakrishnan et al. [3]. Opinion Mining via Neural Networks was applied on two different datasets. An approach for a aspect mining based on probabilities for drug reviews was introduced by Victor C. Cheng et al. [4]. To trace out drug aspects clustering algorithm is encompassed on four different drugs. In the works of Muhammad Zubair Asghar et al. [5], a study was performed for subjective lexicon construction to carry out opinion mining on drug reviews. SentiWordNet was used to expand terms and assign polarity scores based on which a final lexicon was triangulated. G. Leena Giri, Gerard Deepak et al. [6] presented a fresh approach on Ontology recommendation through their work. For the computation of semantic similarity, SemantoSim presented a strategic algorithm. Vrushali Moon et al. [7] developed a novel strategy of extracting drug reviews by specific aspects, for instance, age, gender, etc., and thereafter summarizing them into meaningful opinions. Another state-of-the-art opinion mining framework was brought forward by J. Satish Babu et al. [8], wherein the SVM algorithm has been implemented for drug review's opinion mining. Thu Dinh et al. [9] proposed an approach of a transfer learning algorithm for exploring drug reviews online utilizing text analytics.

3 Proposed Methodology

A new strategy of implementing opinion mining for drug satisfaction levels in clients is proposed through this work. The study is implemented on the drug reviews data repository as an input for feature selection and classification. The implementation of three-fold feature extraction is done by manipulating Tf-Idf and information scent followed by opinion alignment under Sunflower Optimization and finally realizing the feature weights based on SemantoSim value. The performance of this hybrid classifier constructed with a combination of LSTM and Random-Forest is strengthened by opting for the best weight after merging the two results. Ontology alignment is incorporated using Lesk Similarity and SemantoSim on two different ontologies, namely Drug Efficacy Ontology and Drug Side-effect ontology.

On reception of the Drug Review dataset from UCI ML Drug Review repository, which contains reviews for about 3436 unique drugs divided into 23 different classes, this dataset undergoes qualification based on NLP preprocessing techniques. Significance has been given to 5 features of the dataset, i.e., Medical Condition, User Review, Rating, Useful Count and Side-effects. The first phase of pre-processing is carried off by tokenizing each of the textual data present in the corpus to smaller units called tokens followed by lemmatization, wherein, differently inflected forms of a term are analyzed as a single item using the morphological analysis and vocabulary of words. After these two steps, stop words are removed from the given text corpus. These three crucial phases of pre-processing make the data suitable for further operations and model implementation. The final stage of pre-processing is realized through Named Entity Recognition (NER) which helps in identifying the prime elements in a text. While dealing with large data, extraction of main entities in a corpus helps in sorting unstructured data and detecting important information. The strategy of pre-processing is carried out in such a way that no data is lost and a well-structured form of the dataset is obtained which can be used for further course of action.

To classify the documents and draw out an inference employing opinion mining, we first need to construct a feature vector to be fed as input data. To achieve this, a novel three-fold feature extraction method is taken into consideration. With every advancing step, a detailed feature mapping is acquired. The rudimentary phase starts with information retrieval using Term Frequency and Inverse Document Frequency (Tf-Idf) to outline the significance of a specific phrase or word to the given textual data. The combined Tf-Idf algorithm weighs the provided term's weight against the number of times a word appears in the document, which consequently determines the amount of information a particular word lays out. Subsequently, information scent is incorporated to obtain a feature set built on the reliable cues taken from the terms in the text corpus which leads to the formation of a relevant set of words. This takes it closer to mining accurate opinion on a particular drug. Usage of information scent ensures that the correct path is being taken towards the goal. Computation of Tf-Idf score on a logarithmic scale is depicted in Eq. 1.

$$W_{a,b} = tf_{a,b} \times log\frac{N}{df_a} \tag{1}$$

where, $tf_{a,b}$ represents the frequency of a in b, df_a illustrates the total number of documents having a and N refers to the count of documents present in the corpus for the term a in a document b.

Lesk similarity algorithm is maneuvered in the contiguous juncture to encompass Ontology Alignment. Two different ontologies o and o' are used for this purpose; namely Drug Efficacy Ontology and Drug Side-effect Ontology. Given the two ontologies, certain input parameters are reviewed based on which correspondence is traced out between the data from the feature set secured in the first phase of feature extraction with the entities from o and o'. Following this, the shortest distance is found between two neighboring words to remove ambiguity and trace out the resonance by further processing under Lesk Algorithm. Following a semantic mapping of the subsequent words with the provided ontology topics, a coherence is established. In order to closely imitate the process of ontology alignment, SemantoSim, a tool to define semantic similarity along the lines of PMI score, is used to implement another phase of semantic matching on tokenized words. SemantoSim evaluates the relatedness of a term m in connection to a term n based on their probability score in accordance with Eq. 2, where $p(m, n)$ is the likelihood of m occurring with n and $p(m)$ and $p(n)$ are their probabilities.

$$SemantoSim(m, n) = \frac{pmi(m, n) + p(m, n)log[p(m, n)]}{p(m) \cdot p(n) + log[p(n, m)]} \qquad (2)$$

To make our approach and this phase of ontology alignment more robust, semantic matching and term relatedness score are assigned under the influence of the Sunflower Optimization Technique (SFO). It is a meta-heuristic algorithm invigorated by the movement of sunflower and considering pollination between adjacent sunflowers. To put it in technical terms and relevance with this work, it is an iterative optimization algorithm, where a parameter is initialized i.e., a term is selected randomly and the best value for this parameter is foraged within nearby terms. The best objective for each adjacent term is randomly selected, as a result of which; with each iteration the parameter is updated giving rise to a new parameter i.e., a new term is fetched for words in the near locality to frame better opinion on drug review. These final parameters are optimized to achieve higher accuracy with less time complexity and the model is made more sound towards statistical noise.

After the completion of this phase, we have another set of features aligned in vectors representing opinion for their root data. The succeeding feature extraction phase calls for feature weighting based on Information Scent and SemantoSim values. For each term in the feature vector, an associated ontology concept exists whose weight is measured now. Here, the SemantoSim value is accompanied by information scent for each term which aids in pursuing the right path towards forming an opinion for each drug review in the corpus. Weight for each of these features is estimated and the existing vector is worked out with term filtration in the following step. At the end of this phase, a feature vector model is formed with multiple scores for each term. Feature matrix can be extracted from this model iteratively to be processed under the final classification and opinion mining phase.

With the completion of feature vector formation, the final classification is realized. Here, opinions are secured for corresponding drugs; stemmed from their reviews delineated in the form of feature vectors, which are subsequently used to deduce the data and camp them under the umbrella of negative and positive reviews obtained from the patients in line with their user experience. For this purpose, a hybrid classifier has been devised administering Random Forest on top of LSTM. The performance of LSTM is highly reliable when the data consists of more information, while Random Forest learns the information representing the complete dataset.

LSTM: Processing sequences and time-series data is a good fit for the recurrent neural network (RNN) architecture known as long short term memory (LSTM). It is intended to get over the drawback of conventional RNNs, which is the vanishing gradient issue. LSTM networks are especially helpful for applications like speech recognition and natural language processing because of its memory mechanism, which enables them to record long-range dependencies in data.

LSTM is made up of three gates which is represented by Eqs. (3), (4) and (5).

$$x_t = \sigma(wt_x[out_{t-1}, ts_t] + b_x) \tag{3}$$

$$y_t = \sigma(wt_y[out_{t-1}, ts_t] + b_y) \tag{4}$$

$$z_t = \sigma(wt_z[out_{t-1}, ts_t] + b_z) \tag{5}$$

Input, forget, and output gates are represented here by $x_t, y_t, and\ z_t$, respectively. σ depicts sigmoid function, wt stands for respective gate neurons, out_{t-1} is the output for previous LSTM block and b represents respective biases.

Random Forest: Its foundation is decision trees, and in order to produce predictions that are more reliable and accurate, it aggregates the predictions of several decision trees. Random feature selection at each decision tree node and random training data selection through bootstrapping-the process of generating various subsets of the data-are the two sources of randomization that contribute to Random Forest's "random" quality. Random Forest enhances generalisation performance and lessens overfitting by aggregating the output of many decision trees.

Hence, the LSTM model constituting multiple layers is trained first, after which the features are extracted from its last cell and are concatenated with the primary input data to be given as training data for Random Forest Classifier. To obtain enhanced performance from this hybrid classifier, feature weights are selected optimally utilizing sunflower optimization in previous steps. Eventually, the testing process is carried out when the reviews are categorized into positive or negative and the state-of-the-art performance is evaluated based on certain metrics.

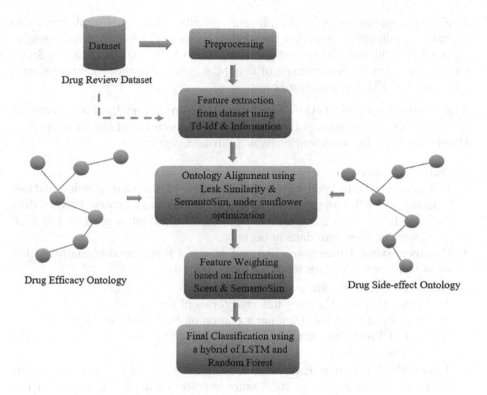

Fig. 1. System Architecture of OntoOpinionMiner

Figure 1 shows the OntoOpinionMiner system architecture in its entirety. The system begins with the input of data, which proceeds through various processing steps. Following the completion of pre-processing, the three-fold feature extraction phase begins with the use of various feature extraction strategies, and at the end, a structured feature vector is created with improved data. When the data reaches this stage, all the superfluous information has been removed. The final categorization is done here and the data is now qualified enough to be delivered for training.

4 Implementation

The proposed OntoOpinionMiner is put forth in accordance with the given algorithm. By supplying the raw data to the pre-processing stage, it is first made consistent and turned into a comprehensible format. Textual data is tokenized, lemmatized, and stop words are eliminated after that. Then, NER is incorporated to pick out the primary elements from the text corpus. Once the dataset is cleaned according to model demands, the Tf-Idf score is then calculated for each of these tokenized terms, taking into consideration information scent value. This vectorized data is subsequently directed for ontology alignment using Lesk

Similarity measure and SimantoSim score calculation. SFO is implemented to optimize the algorithm and refine the model accuracy. Thereafter, feature weighting is carried through, contingent on SemantoSim value and Information Scent. Finally, classification is performed utilizing the hybrid classifier constructed with a fusion of LSTM and Random Forest.

Algorithm Implementation for OntoOpinionMiner: In this section, we provide a detailed account of the practical implementation of the OntoOpinion-Miner algorithm for sentiment analysis in drug reviews:

1. **Data Preparation:**
   The process commences with the acquisition of the Drug Review dataset from the UCI ML Drug Review repository, containing a diverse array of drug reviews, divided into 23 distinct categories. The dataset is subjected to NLP preprocessing to ensure data uniformity.
2. **Preprocessing Phase:**para The first phase of preprocessing comprises key steps to prepare the data for subsequent analysis:
   (a) Tokenization: Dividing text into tokens.
   (b) Lemmatization: Unifying inflected forms of terms.
   (c) Stop Word Removal: Eliminating common but uninformative words.
   (d) Named Entity Recognition (NER): Identifying and extracting essential elements.
3. **Three-Fold Feature Extraction:** A novel three-fold feature extraction method is employed to create feature vectors, each playing a pivotal role in sentiment analysis:
   (a) Tf-Idf Feature Extraction: Computing Term Frequency-Inverse Document Frequency (Tf-Idf) scores for each term to gauge its significance.
   (b) Information Scent: Extracting reliable cues from terms in the text corpus, refining the feature set.
4. **Ontology Alignment:**
   Feature vectors are aligned with Drug Efficacy Ontology and Drug Side-effect Ontology, streamlining the connection between terms and ontology concepts:
   (a) Lesk Similarity: Minimizing ambiguity by determining the shortest distance between words.
   (b) SemantoSim: Enhancing semantic matching of tokenized words based on Probability of Mutual Information (PMI) score.
5. **Optimization with Sunflower Algorithm (SFO):** The iterative Sunflower Optimization Algorithm (SFO) fine-tunes semantic matching parameters, boosting model accuracy and reducing time complexity.
6. **Feature Weighting:**
   Term weights are assigned within the feature set, incorporating SemantoSim values and Information Scent. These weights provide accurate term contributions to the final classification.
7. **Sentiment Classification:** The concluding phase encompasses sentiment classification, categorizing drug reviews into positive or negative sentiment categories.

Algorithm 1. Drug Review Opinion Analysis

1: **Input:** A Drug Review (DR)
2: **Output:** Positive or Negative opinion regarding drug usage
3: **procedure** OPINIONANALYSIS(DR)
4:　　Initialize pre-processing techniques on DR
5:　　**for** each term in DR **do**
6:　　　　Calculate Tf-Idf score
7:　　　　Incorporate Information Scent for each term
8:　　　　Give out a preliminary feature vector (wa, b scores)
9:　　　　Structure a set of extracted words by conjoining obtained words into WF
10:　　　　Compose sentences containing these extracted words into SF
11:　　**end for**
12:　　**for** each word in SF **do**
13:　　　　Compute semantic similarity with ontology concepts from o and o'
14:　　　　Generate a result set of words WO based on each ontology alignment
15:　　　　Formulate sentences SO with these words
16:　　　　Estimate the shortest distance using Lesk Algorithm to draw out the relevance between adjacent words in each sentence and make it unambiguous
17:　　　　Compute SemantoSim probability score of terms in a set of two words depicted by m and n
18:　　**end for**
19:　　Optimize the semantic matching process using Sunflower Optimization Algorithm:
20:　　**for** each iteration **do**
21:　　　　Initialize value for parameters for sunflower, SF
22:　　　　Set $i = 0$
23:　　　　Generate SF based on initial parameters
24:　　　　Evaluate the population SF$(i) \in [lb, ub]$, $I = 1, 2, 3, \ldots, n$
25:　　　　Calculate the objective flower (SF(i))
26:　　　　Adjust the sunflower for best solution
27:　　　　Obtain new SF
28:　　　　Update the initial SF
29:　　**end for**
30:　　Compare the SFs and select the one with the best values
31:　　Formulate a new feature set F_F with best values
32:　　**for** each term in F_F **do**
33:　　　　Perform feature weighting based on SemantoSim scores
34:　　　　Follow information scent
35:　　　　Generate final feature vector
36:　　**end for**
37:　　Classify the final feature vector using a hybrid classifier of LSTM and Random-Forest
38: **end procedure**

5   Results and Performance Evaluation

The execution of the proposed OntoOpinionMiner was realised on the Drug Review dataset. The experiment was executed based on the algorithm put forth on 215,063 instances with 6 different attributes. To evaluate the performance of OntoOpinionMiner certain evaluation metrics have been considered; namely, Precision, Recall, Accuracy, F-measure, and FNR score. According to Eq. 6, a measure of precision is one that calculates the proportion of positive instances that are actually true. Recall provides the percentage of all appropriate instances that have been accurately extracted, as shown in Eq. 7. Equation 8's definition of accuracy is the ratio of correctly predicted outcomes to the total number of samples used as input. Equation 9's equation for the F-measure measures a test's accuracy, while Eq. 10's equation for the FNR score calculates the rate of false-negative values (Table 1).

$$Precision = \frac{TP}{TP + FP} \tag{6}$$

$$Recall = \frac{TP}{TP + FN} \tag{7}$$

$$Accuracy = \frac{TP + TN}{TP + TN + FP + FN} \tag{8}$$

$$F - Measure = \frac{2 \cdot Precision \cdot Recall}{Precision + Recall} \tag{9}$$

$$FNR = 1 - Recall \tag{10}$$

Table 1. Comparative analysis of performance of the proposed OntoOpinionMiner with other approaches

Search Technique	Precision %	Recall %	Accuracy %	F-Measure	FNR
BP1	81.42	76.69	79.97	75.92	0.23
BP2	84.63	80.12	82.77	78.08	0.20
BP3	82.27	75.21	78.28	79.04	0.25
OOM	95.83	90.71	92.72	93.75	0.092
OOM Mod	85.21	80.79	83.38	82.56	0.19
OOM F	86.23	81.78	84.98	82.98	0.18

where, BP1 → Two-Pass Opinion Mining [1], BP2 → Subject Lexicon Construction [5], BP3 → Probabilistic Aspect Mining [4], OOM → Proposed OntoOpinionMiner, OOM Mode → Eliminating Ontologies from proposed

OntoOpinionMiner, OOM F → Eliminating Feature Weighting from proposed OntoOpinionMiner.

The performance of this OntoOpinionMiner methodology with other baseline procedures has been compared using average values of these parameters. OntoOpinionMiner's comprehensive computational analysis is provided in Table 2, along with a comparison of the results with other similar methodologies. As shown in Table 2, the suggested framework on average achieves 95.83% average precision, 90.71% average recall, 92.72% average accuracy, 93.75% average F-Measure score, and 0.092 average FNR rate. The results produced by OntoOpinionMiner for each of the evaluation metrics considered are highly significant and outperform the current baseline approaches.

The proposed OntoOpinionMiner performs admirably owing to the three-fold feature extraction strategy. Three different feature extraction strategies have been carried out to build the final feature vector incorporating various reliable algorithms and techniques. Implementation of Tf-Idf score calculation in coherence with information scent filters out most of the statistical noise. Implementation of ontology alignment taking into account two different ontologies, Drug Efficacy ontology, and Drug Side-Effect Ontology further helps in boosting the accuracy by using the proposed framework. Ontology Alignment is the process wherein semantic similarity is measured for a particular term with entities from two different ontologies which helps in demarcating the efficacy or side-effect of a particular drug profusely in this case, which enhances the overall model performance. Furthermore, the implementation of SFO, a metaheuristic optimization algorithm that employs root velocity terms and pollination terms, makes the model sounder and more robust. Since ontology alignment is carried out under the influence of the Lesk Similarity algorithm and estimation of SemantoSim value; this again contributes towards the proposed model's better results.

Another reason for the high performance percentage ensured by the presented algorithm is because of feature weighting measures which help in the removal of misleading data and also reduces the computational cost of modeling. Finally, the incorporation of the hybrid classifier approach accelerates the model performance even more.

It has been found that the results of proposed OntoOpinionMiner were more precise than those of any framework used. By putting the above-mentioned strategies into practise, this high performance was achieved. Despite the fact of baseline strategy 1 having two-pass classifier technique that aids in the model's achievement of a high accuracy rate, over the long term, it renders the model much more unstable. Furthermore, a combination of SVM with Neural networks does not produce much remarkable results. Another novel approach of employing Subjective Lexicon Construction for drug review opinion mining doesn't help much in achieving higher accuracy. The incorporation of SentiWordNet alone results in feature selection imbalance which eventually affects the model performance. In baseline paper 3, another new approach has been taken up of probabilistic aspect mining; though only relevant aspects are taken into consideration for sentiments are extracted, but due to proper data segmentation issue the overall efficiency of the model drops down.

For comparative analysis, two more approaches were taken into consideration, where ontologies were removed from OntoOpinionMiner and feature weighting was eliminated from the proposed methodology respectively. Removing ontologies from the proposed approach drastically affects the model accuracy as the incorporation of ontologies enhances the background processing of a model. It refines the process of feature extraction and filters out the irrelevant entities, which eventually qualifies the input data as being sounder and more misleading. Elimination of the feature weighting strategy results in redundant data being left along with useful data which affects the data consistency and further makes the data statistically unstable. Furthermore, all baseline models only provide predictions for a smaller number of data points, but a sizable amount of data has been used to conduct experiments on the proposed framework. Additionally, the new approach of combining each of the previously mentioned strategies together helps to improve the performance and accuracy of the model. Each algorithm compliments the others, which helps make the proposed OntoOpinionMiner capable of outperforming all baseline methods currently in use.

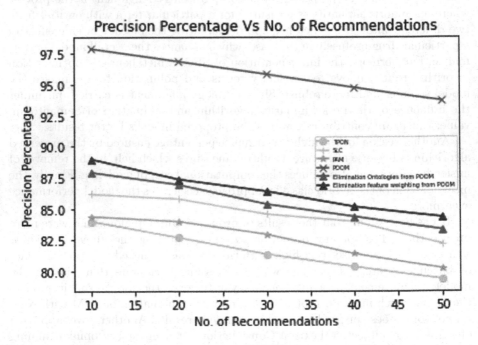

Fig. 2. Precision percentage Vs. No. of recommendations

Figure 2 shows a study of the proposed OntoOpinionMiner's precision percentage in relation to the number of recommendations in comparison to other baseline frameworks currently in use. It is evident that the proposed system performs in a very substantial and outstanding manner. By reaching a precision percentage of 97.5% against the number of recommendations, the suggested methodology exceeds all currently used here-outlined strategies. The integration of this hybrid LSTM-RandomForest classifier system has greatly raised the yields of the suggested framework. After considering all the data and corresponding model performances, it may be concluded that the methodology suggested is more effective than other current frameworks. As a result, it can be stated that OntoOpinionMiner is the best-in-class model for opinion mining on drug evaluations and determining a patient's condition for a certain prescription based on his or her experience.

Table 2. Sentimental Labels vs No. of Reviews

Label	Number of Reviews
Positive	101035
Negative	53430
Neutral	6953

6   Conclusion

An innovative strategy has been put forth via this paper for deriving review on drugs utilizing opinion mining technique based on a patient's experience. The proposed OntoOpinionMiner puts before a classification algorithm to separate negative and positive reviews on a particular drug incorporating various novel strategies and methodologies. A hybrid of RandomForest built on top of LSTM has been used as a prime classifier to get the final review prediction. In order to create a calculated and appropriate pipeline for the efficient classification of all the textual data available in the corpus, this methodology combines multiple algorithms. The three-fold feature extraction method is the highlight of the whole algorithm as it helps to refine and stabilize the data in a way such that the overall performance of the model is elevated. Usage of SFO as an optimization algorithm makes the overall model robust and removal all the inconsistent data to avoid final stage chaos and skepticism. The model is clearly the best in its class based on the values obtained for evaluation measures. The proposed methodology has an average precision of 95.83% and an average recall value of 90.71%. Additionally, 92.72% of the eight emotions considered were correctly predicted on average. The proposed method secures a FNR rate of 0.092 and an average F-Measure value of 93.75%. All of these estimates contribute to a rough conclusion that OntoOpinionMiner is the top model in its class for mining patient comments for medication review opinions.

Future work on this system could likely concentrate on boosting its overall relevance. To get even better results for assessment metrics, it is also possible to apply other strategies for estimating semantic similarity or co-occurrence. To get better results, attempt to combine different neural networks with classifiers from machine learning can turn fruitful.

References

1. Padmavathy, P., Pakkir Mohideen, S., Gulzar, Z.: A novel architecture for a two-pass opinion mining classifier. In: Chillarige, R.R., Distefano, S., Rawat, S.S. (eds.) ICACII 2019. LNNS, vol. 119, pp. 27–35. Springer, Singapore (2020). https://doi.org/10.1007/978-981-15-3338-9_4
2. Cavalcanti, D., Prudêncio, R.: Aspect-based opinion mining in drug reviews. In: Oliveira, E., Gama, J., Vale, Z., Lopes Cardoso, H. (eds.) EPIA 2017. LNCS (LNAI), vol. 10423, pp. 815–827. Springer, Cham (2017). https://doi.org/10.1007/978-3-319-65340-2_66
3. Gopalakrishnan, V., Ramaswamy, C.: Patient opinion mining to analyze drugs satisfaction using supervised learning. JART 15(4), 311–319 (2017). https://doi.org/10.1016/j.jart.2017.02.005
4. Cheng, V.C., Leung, C.H.C., Liu, J., Milani, A.: Probabilistic aspect mining model for drug reviews. IEEE Trans. Knowl. Data Eng. 26(8), 2002–2013 (2014). https://doi.org/10.1109/TKDE.2013.175
5. Asghar, M., Khan, A., Ahmad, S., Ahmad, B.: Subjectivity lexicon construction for mining drug reviews. Sci. Int. 26, 145–149 (2013)
6. Leena Giri, G., Deepak, G., Manjula, S.H., Venugopal, K.R.: OntoYield: a semantic approach for context-based ontology recommendation based on structure preservation. In: Chaki, N., Cortesi, A., Devarakonda, N. (eds.) Proceedings of International Conference on Computational Intelligence and Data Engineering. LNDECT, vol. 9, pp. 265–275. Springer, Singapore (2018). https://doi.org/10.1007/978-981-10-6319-0_22
7. Liu, J., Kaiser, C., Bodendorf, F., Zhuang, C.: Extraction of drug reviews by specific aspects for sentimental analysis (2016)
8. Satish Babu, J., Rao, C.S.K., Banerjee, D., Sagar Imambi, S., Krishna Mohan, G.: Opinion mining for drug reviews. IJITEE 8(7), 1314–1318 (2019)
9. Dinh, T., Chakraborty, G., Mcgaugh, M.: Exploring online drug reviews using text analytics, sentiment analysis, and data mining models (2020)
10. Ngai, J., Kalter, M., Byrd, J.B., Racz, R., He, Y.: Ontology-based classification and analysis of adverse events associated with the usage of chloroquine and hydroxychloroquine. Front. Pharmacol. 13, 812338 (2022). https://doi.org/10.3389/fphar.2022.812338
11. Dara, S., et al.: Machine learning in drug discovery: a review. Artif. Intell. Rev. 55(3), 1947–1999 (2022). https://doi.org/10.1007/s10462-021-10058-4
12. Reboredo, P.C., Blanco, J.L.: A review on machine learning approaches and trends in drug discovery. Comput. Struct. Biotechnol. J. 19, 4538–4558 (2021). https://doi.org/10.1016/j.csbj.2021.08.011
13. Singh, R., Singh, R.: Applications of sentiment analysis and machine learning techniques in disease outbreak prediction - a review. Mater. Today Proc. 81, Part 2, 1006–1011 (2023). https://doi.org/10.1016/j.matpr.2021.04.356
14. Ghosh, A., Umer, S., Khan, M.K., Rout, R.K., Dhara, B.C.: Smart sentiment analysis system for pain detection using cutting edge techniques in a smart healthcare framework. Cluster Comput. 26, 119–135 (2023). https://doi.org/10.1007/s10586-022-03552-z

MTL-rtFND: Multimodal Transfer Learning for Real-Time Fake News Detection on Social Media

Sudha Patel[1] and Shivangi Surati[2]([✉])

[1] Kadi Sarva Vishwavidyalaya, Gandhinagar, Gujarat, India
[2] Department of Computer Science and Engineering, School of Technology, Pandit Deendayal Energy University, Gandhinagar, Gujarat, India
shivangi.surati@gmail.com

Abstract. Social media platforms have become crucial channels for the rapid dissemination of news in various formats, including text, images, audio, and video. Ensuring the authenticity of news content at primary stage is crucial to prevent the spread of false information. In order to gather semantic and contextual data for the identification of false news, current state-of-the-art majorly concentrated on text-based techniques, leveraging pre-trained word embedding and language models. However, these approaches suffer from the limitations viz. inefficiency in extracting context-based features, reliance on pre-trained models trained on more compact corpora, and static-masking utilization. In order to get over these issues, a new Framework for Transfer Learning based on Content for real-time False or Fake News Detection (MTL-rtFND) that integrates multimodal transfer learning methodologies has been proposed in this paper. It consists of a Data preprocessing block, Multimodal Feature Extraction Block (MFEB) and a Classification Block (CB). In the MFEB, a multimodal pre-trained mode, such as a fusion of visual and textual representations is leveraged to efficiently capture context-based features. This pre-trained model is trained on larger-scale multimodal datasets, enabling the extraction of richer contextual information. The resulting multimodal feature vectors from the MFEB are fed into the CB that employs a real time classification of news articles as fake or legitimate using deep neural network. The proposed MTL-rtFND model evaluation for real world datasets, demonstrating its effectiveness for real-time detection of fake news. The experimental results show significant improvements compared to state-of-the-art methods, obtaining an average gain in accuracy of 6.34% across multiple text embedding techniques. The findings demonstrate the capacity of multimodal transfer learning to improve fake news detection in real-time scenarios, thus contributing to mitigate the undesirable impact of misinformation on social media platforms.

Keywords: Large Language Models · Transfer Learning · Real Time Fake News Detection

K. K. Patel et al. (Eds.): icSoftComp 2023, CCIS 2030, pp. 235–247, 2024.
https://doi.org/10.1007/978-3-031-53731-8_19

1 Introduction

Online social media like Twitter, Sina Weibo, Facebook, and Reddit are now often used by users to share and quickly distribute news in a variety of media, including text, photographs, and videos. Unfortunately, this freedom of information sharing has produced the widespread multitude of misinformation and fake news, with significant negative consequences for the economy, politics, and social security [1]. Consequently, the need for accurately identifying fake and real news at an early stage is very critical. To address this challenge, several approaches have been developed for false news detection on social media platforms. These approaches can broadly be categorized as content-based and social context-based methods. Content-based methods primarily consider analyzing the textual content, including the title, text body, and associated images or videos to extract essential features for identifying fake news. In contrast, social context-based methods leverage user profiles, user responses, post activity, and propagation structures to infer the authenticity of news articles [2].

Existing text-based false news detection methods involve Machine Learning (ML) and Deep Learning (DL) algorithms [3]. Machine learning based approaches require extensive feature engineering and are often time-consuming, while DL-based architectures extract features automatically using pre-trained word embedding techniques. However, these approaches still face limitations in capturing long-term dependencies between words and generating improved textual representations of the sentences [4].

To overcome these limitations, the primary contributions of this research paper are as follows:

- A novel MTL-rtFND (Multimodal Transfer Learning for real-time Fake News Detection) model that combines multimodal feature extraction with a basic feed-forward neural network for classification is proposed and described in detail. A transfer learning and multimodal methods for real-time false news identification are utilized in the proposed strategy.
- Significant improvements in fake news detection are achieved using semantic and contextual features effectively captured from textual content.
- The experimental assessments are performed on benchmark datasets [5–7] to ascertain if the suggested strategy is superior to cutting-edge methods.

The rest of the paper is arranged as follows. A review of previous works related to fake news identification is provided in Sect. 2 highlighting the research gaps. The detailed architecture and different blocks of the proposed MTL-rtFND model are described in Sect. 3. Next, the experimental backgrounds, comparison models and evaluation metrics are discussed in Sect. 4. The empirical outcomes of the MTL-rtFND model are shown in Sect. 5, along with a comparison to baselines. Finally, in Sect. 6, the study is concluded, and future research guidelines are discussed.

2 Related Work

A brief summary of existing research about the detection of false news is presented in this section. The prior work in this domain has majorly focused on utilizing the linguistic elements in news stories for spotting fake news [8]. Within this context, two main types of approaches have emerged: machine learning methods and deep learning methods. Methods based on machine learning depend on text features and involve extensive feature engineering, while deep learning based methods leverage deep learning architecture to automatically extract relevant features from the text [4].

2.1 Machine Learning Based Fake News Detection Methods

Investigators have used ML techniques along with conventional text encoding for detecting false information in social media news articles [9]. They used n-gram and TF-IDF techniques with ISOT dataset, followed by traditional Machine learning approaches for fake news recognition [9]. A machine learning method for detecting false news included 23 supervised classifiers and the feature extraction method TF-IDF [9]. Another language-independent ML model extracted features from five domain datasets in three languages using Bag-of-Words. For effective false news categorization, it merged linguistic-based characteristics from news articles into a single feature vector [10–14]. However, while neglecting the contextual and semantic value of words depending on human feature engineering, which may take time, these machine learning-based textual content models do provide improved predictions [10–14].

2.2 Deep Learning Based Fake News Detection Methods

In a number of sectors, models based on deep learning have grown in popularity recently, such as NLP tasks, computer visualization, autonomous systems and voice analysis. Automated identification of features has the advantage of eliminating the requirement for manual engineering [15].

A hybrid technique called CSI that combines RNNs and LSTM models to detect fake news using temporal patterns is proposed in [16]. For the categorization of false news, another prediction model [17–20] makes use of linguistic cues, news word embedding (like GloVe), LSTM, and CNNs. CNN models have been explored to extract important features from news articles in [18], However, LSTM has been used to identify bogus news based on Facebook user characteristics [19].

A DL framework called BerConvoNet that combines CNN and BERT to extract several attributes and incorporate contextual news is presented in [21]. To extract informative characteristics, a detailed model for representing contextual text employs models for pre-trained languages based on transformers (like Funnel transformer, RoB-ERTa, BERT, and GPT-2) [21]. Multimodal fake news detection models that merge BERT's textual features with visual features show improved accuracy compared to traditional methods [22–26] (Table 1).

2.3 Open Challenges

Despite improvements in false news identification, there are still a number of challenges.

- **Multimodal Integration:** The integration of multiple modalities, like written content, visuals, and videos, poses a significant challenge in multimodal fake news detection [27].

Table 1. Comparison between ML and DL models

ML Based Model	DL Based Model
It takes feature engineering to identify bogus news	Features are automatically extracted for false or fake news classification
BoW and TF-IDF are two examples of traditional ML-based text encoding algorithms that fail to capture the semantic meaning of words	For the purpose of identifying false news, trained word-embedding methods like word2vec, GloVe, and transformers record contextual data that are semantically linked
Detection accuracy is typically lower than in DL-based models	Ordinarily, detection accuracy is greater than in ML-based models

- **Real-time Detection**: In order to slow the distribution of fake information quickly on social media sites, real-time fake news identification is essential [28].
- **Transfer Learning in Multimodal Context:** Transfer learning techniques have shown promising results in various domains, including image and text analysis. However, applying transfer learning effectively in the multimodal context for fake news detection remains a challenge [29].
- **Limited Labeled Data:** Obtaining a large-scale labelled dataset specifically for multimodal fake news identification is exciting due to the complication and subjectivity of labelling news articles [30].
- **Contextual Understanding:** Understanding the complex contextual relationships and subtleties in fake news detection is a challenging task. Incorporating contextual information, such as temporal, social, and cultural context, along with multimodal features, can expand the accuracy and robustness of fake news detection models [31].

Addressing these challenges is crucial to enhance the effectiveness and efficiency of multimodal transfer learning for real-time fake news detection. Overcoming these challenges will contribute to the development of more reliable systems that can identify and combat the spread of false news in situations that are current.

3 Proposed Architecture

The proposed architecture of MTL-rtFND (Multimodal Transfer Learning for real-time Fake News Detection) includes several blocks and modules to enhance its effectiveness. These components include-

- Data Preprocessing Block for normalizing and transforming input data, Text Encoding, Image Encoding, and Video Encoding Modules to capture relevant features from different modalities
- Multimodal Fusion Block to integrate the encoded features. Additionally, a Transfer Learning Module can leverage pre-trained models for improved performance
- Real-time Processing Module includes subparts such as Incoming Data Stream Handling, Parallel Processing, Feature Extraction and Encoding, Multimodal Fusion and Integration, Incremental Learning and Model Updating, Decision Making and Output Generation, and Low-Latency Response and Output Delivery are vital for timely

processing of incoming news articles. A Classification Block performs the actual fake news detection.

Within Overall, the proposed architecture combines multimodal transfer learning techniques with real-time processing capabilities to effectively identify fake news. As given in Fig. 1 the suggested model's architecture is divided into blocks as follows:

3.1 Data Pre-processing Block

The Data Preprocessing Block constructs a thorough representation of the news items by assembling textual, visual, and temporal information and the integration of three modules. This multimodal representation is crucial for the subsequent stages of the architecture, such as multimodal fusion and transfer learning, enabling the model to leverage diverse and complementary information from different modalities for accurate real-time fake news detection.

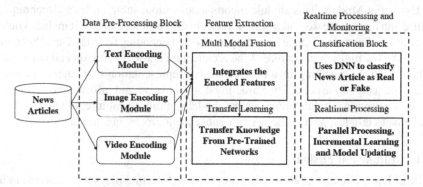

Fig. 1. Proposed architecture of MTL-rtFND

3.1.1 Text Encoding

Text Encoding module plays a crucial role in capturing semantic and contextual information from the textual content of news articles. This module utilizes advanced techniques such as word embedding (viz. Word2Vec) or transformer-based models for learning languages (viz. BERT) to generate high-dimensional vector representations of words. By leveraging these techniques, the Text Encoding module captures the intricate relationships and meanings of words, enabling the model to understand the nuances and subtleties present in the text.

3.1.2 Image Encoding

The Image Encoding module focuses on encoding visual features extracted from images associated with news articles. To extract pertinent visual patterns and characteristics, it makes use of potent deep learning techniques based pre trained models viz. VGG or

ResNet or Convolutional Neural Networks (CNNs). These visual representations enable the model to understand the visual context of news articles and incorporate important visual cues into the fake news detection process.

3.1.3 Video Encoding

The video Encoding module handles the encoding of video data, extracting informative features from the temporal dynamics and motion patterns within videos. Techniques viz. 3D CNNs or spatiotemporal feature extraction methods are employed to capture the temporal evolution and visual content of videos. By encoding video data, the module enables the architecture to effectively analyze and comprehend the visual aspects of news articles presented in video format.

3.2 Feature Extraction Module

The required characteristics must be extracted from the pre-processed data by the Feature Extraction Module. This module encompasses various submodules and techniques to capture and represent essential information from different modalities. It includes outputs from the Text Encoding, Image Encoding, and Video Encoding from the Data Preprocessing Block. The multimodal nature of the architecture allows the fusion and combination of these features, resulting in a comprehensive representation that integrates textual, visual, and temporal aspects. By extracting discriminative and informative features, the Feature Extraction Module enhances the model's ability to capture key characteristics of fake news articles, paving the way for subsequent processing and analysis.

3.2.1 Multimodal Fusion

The Multimodal Fusion module in the proposed architecture plays a crucial role in integrating the encoded features from different modalities. It combines the textual, visual, and temporal representations obtained from the Feature Extraction Module, leveraging various fusion techniques. These methods are – (i) early fusion, in which the characteristics from each mode are concatenated or joined early, or (ii) late fusion, in which the features are mixed later, as during categorization. Attention mechanisms or ensemble methods can also be utilized to effectively weigh and combine the multimodal features. By fusing the information from multiple modalities, the Multimodal Fusion module enhances the model's capability to capture complementary and synergistic patterns, leading to a more robust and comprehensive representation for fake news detection.

3.2.2 Transfer Learning

The Transfer Learning module in the proposed architecture exploits the power of transfer learning techniques to expand the effectiveness of the model for detecting false news. It uses embedding or models that have already been trained (viz. BERT, RoBERTa) or other models based on transformer, to transfer knowledge from large-scale pretraining tasks to the specific task of fake news detection. By utilizing these pre-trained models, the Transfer Learning module captures rich contextual information, semantic representations, and language patterns that are crucial for accurately distinguishing between fake and genuine news articles. To adjust the pre-trained models to the unique requirements of the false news detection task, fine-tuning techniques can be used. Transfer learning is incorporated into the module to improve the model's capacity for generalization and successful false news detection, even in the presence of scant labelled data and intricate language variations.

3.3 Realtime Processing and Monitoring

The Realtime Processing and Monitoring section focuses on handling real-time data and ensuring timely and efficient processing. This section comprises two key components: the Classification Block and the Realtime Processing module.

3.3.1 Classification Block

The Classification Block is a vital component of the real-time processing system. It receives the multimodal feature representation extracted from the previous stages and performs the final classification of news articles into fake or genuine. The classification decision is often taken by this block using machine learning or deep learning methods, such as Recurrent Neural Networks (RNNs) and feed-forward neural networks, or Support Vector Machines (SVMs). Activation functions and loss functions are applied to obtain the desired output that can be binary labels, confidence scores, or probability distributions. The Classification Block plays a crucial role in providing accurate and timely predictions for real-time fake news detection.

3.3.2 Realtime Processing

The Realtime Processing module within the proposed architecture focuses on enabling efficient and low-latency processing of incoming news articles. This module incorporates various subcomponents to handle the real-time data stream effectively. It includes the management of the incoming data stream, parallel processing techniques to ensure efficient execution of tasks, and feature extraction and encoding of new articles in real time. Additionally, the Realtime Processing module may involve techniques such as incremental learning and model updating to adapt the system to new data and evolving trends. It also encompasses decision-making processes, where the processed features are utilized to determine the authenticity of news articles in real time. Low-latency response and output delivery mechanisms are implemented to provide timely results. This module enables the proposed architecture to handle the continuous flow of news articles and ensure real-time detection of fake news.

4 Implementation

Information on the benchmark dataset, experimental setup, assessment metrics, and baseline model used to measure the performance of the suggested model is provided in this section.

4.1 Dataset

Shu introduced the FakeNewsNet repository, a benchmark false news dataset from the real world in 2018 [32]. The dataset comprises two sets of news editorials focusing on politics and entertainment, labelled Gossipcop [34] and PolitiFact [33]. For this research, only the textual content of the news articles was utilized. Politifact and Gossipcop were used to gather ground-truth labels [33, 34]. A review of the benchmark dataset on misinformation, with numbers in brackets reflecting the total number of news items after data preprocessing is given in Table 2.

Table 2. Benchmark dataset details

Dataset	Politifact	Gossipcop
Actual (Real) News	624 (499)	16817 (15223)
False (Fake) News	432 (376)	5323 (4784)

4.2 Experimental Setup

Standard gear, including a single-core Xeon processor with a Tesla K80 GPU with 2496 CUDA cores and 12 GB GDDR5 VRAM, 12.6 GB RAM, and a 33 GB disc capacity, were used for the experiments on Google Colab. The cleaned datasets were split 80 for training and 20 for testing. Python was used for implementation, with TensorFlow for deep learning and Scikit-learn for conventional ML. NLTK (Natural Language Toolkit) handled text preprocessing. News text was transformed using count vectorizer and TF-IDF for ML models, while DL models used pre-trained word embedding like Glove, Word2vec to detect false news. The RoBERTa Tensor-Flow implementation is used in the proposed MTL-rtFND model. During training using binary cross-entropy loss, RoBERTa's default hyperparameter values were applied. These values were epsilon (1e−8 in Adam optimizer), dropout (0.1), learning rate (1e−5, 2e−5), and batch size (32). To identify if news items are fraudulent or legitimate, the classification layer consists of two FFN (Feed Forward Neural) layers with sizes of 128 with ReLU and 32 with softmax activation functions.

4.3 Performance Metrics

The proposed MTL-rtFND model's performance for fake news identification was quantitatively analyzed using traditional evaluation metrics viz. F1-score, recall, precision, and classification accuracy. Accuracy is the ratio of correctly predicted news articles to the total tested articles (Eq. 1). According to Eq. 2, precision is defined as the ratio of accurately predicted fake news to all predicted fake news. Recall indicates the percentage of correctly predicted fake news among the actual fake news (Eq. 3). The harmonic mean of recall and precision is determined by the F1-score (Eq. 4). The MTL-rtFND model was compared with traditional machine learning and deep learning-based models and with the state-of-the-art methods (SoTA) for the detection of fake or false news based on content.

$$Accuracy = (|TP| + |TN|)/(|TP| + |TP| + |FP| + |FN|) \qquad (1)$$

$$Precision = |TP|/(|TP| + |FP|) \qquad (2)$$

$$Recall = |TP|/(|TP| + |FN|) \qquad (3)$$

$$F1 \text{ - } score = 2 * Precision * Recall/(Precision + Recall) \qquad (4)$$

5 Result Analysis

The effectiveness of the MTL-rtFND model is assessed using two datasets that contain genuine examples of false news. The findings of the experiment are compared with traditional machine learning, deep learning and state-of-the-art methods as given in [1] across three sets of experiments, focusing on classification accuracy and F1-score as important evaluation metrics.

5.1 Contrasting Traditional Text Embedding and Machine Learning Models

The "MTL-rtFND" model for real-time fake or false news detection (FND) on social network is experimented in this study. Using TF-IDF and Bag-of-Words (BoW) textual encodings, the effectiveness of the proposed model is evaluated and compared with that of conventional machine learning models. The evaluation findings are presented in Table 3, while graphical representation is shown in Fig. 2.

While adopting BoW textual encoding, the suggested MTL-rtFND model achieves remarkable accuracies of 93.47% using Politifact dataset and 92.38% with Gossipcop dataset. On Politifact, these scores exceed the base classifiers (KNN, NB, DT, SVM) by 10.26% to 26.26%, while on Gossipcop, they outperform them by 10.66% to 14.97%. Furthermore, the presented MTL-rtFND model considerably raises the F1 score on both datasets as shown in Fig. 2. These findings show that on the Politifact dataset, the proposed MTL-rtFND model beats the other ML-based classifiers in all assessment measures. The model's higher performance can be due to its capacity to make use of semantic and contextual data, which traditional ML-based models may not fully take into account.

Further, the proposed MTL-rtFND model applied to the Gossipcop dataset outperforms BoW and machine learning based models in terms of F1-score, recall, and accuracy with the exception of F1-score and precision for the Random Forest model. But it's important to observe that the RF model exhibits poor performance in predicting real news (TN) on the Gossipcop dataset, leading to a 5.34% lower accuracy score compared to the proposed MTL-rtFND model.

The primary attention of our classification method is to accurately predict real update and fake news both, not just one. By considering this dual objective, the proposed MTL-rtFND model demonstrates its effectiveness in addressing the challenge of real-time fake or false news detection on social network, outperforming traditional machine learning based models and BoW encoding in terms of each assessment metrics.

Table 3. Comparison of the proposed model's results with that of existing ML classifiers

Base Classifier	Text Embedding	Politifact				Gossipcop			
		Accuracy	Precision	Recall	F1-score	Accuracy	Precision	Recall	F1-score
Support Vector Machine	Bag of Words	0.81	0.84	0.81	0.82	0.80	0.85	0.87	0.86
	TF-IDF	0.83	0.87	0.83	0.85	0.85	0.94	0.87	0.90
Random forest	Bag of Words	0.84	0.85	0.85	0.85	0.85	0.96	0.85	0.91
	TF-IDF	0.84	0.89	0.84	0.86	0.85	0.96	0.85	0.91
Decision Tree	Bag of Words	0.72	0.85	0.70	0.76	0.79	0.86	0.86	0.86
	TF-IDF	0.75	0.78	0.76	0.77	0.78	0.86	0.85	0.86
K-Nearest Neighbors	Bag of Words	0.67	0.67	0.69	0.68	0.77	0.86	0.84	0.85
	TF-IDF	0.75	0.86	0.68	0.76	0.80	0.96	0.81	0.88
Naïve Bayes	Bag of Words	0.82	0.76	0.89	0.82	0.81	0.82	0.92	0.86
	TF-IDF	0.83	0.86	0.78	0.82	0.84	**0.97**	0.84	0.90
MTL-rtFND	Bag of Words	0.85	0.90	0.86	0.88	0.86	0.92	0.86	0.92
	TF-IDF	**0.91**	**0.92**	**0.90**	**0.89**	**0.91**	0.93	**0.92**	**0.93**

Due to its capacity to include rich data from several modalities, multimodal embedding algorithms outperform conventional TF-IDF and BOW techniques for the identification of bogus news. By combining textual, visual, and sometimes audio data, they achieve a deeper contextual understanding, making them more adept at identifying subtle cues and inconsistencies indicative of fake news. Their robustness to noise and cross-modal learning enables effective handling of data imbalance, even with limited fake news examples. Moreover, multimodal embedding can easily adapt to new modalities, ensuring the model remains up-to-date and future-proof. The success of these methods relies on the quality and diversity of training data, appropriate architecture selection, and fine-tuning with a relevant objective function. Thus, the proposed method outperforms the classical ML approaches.

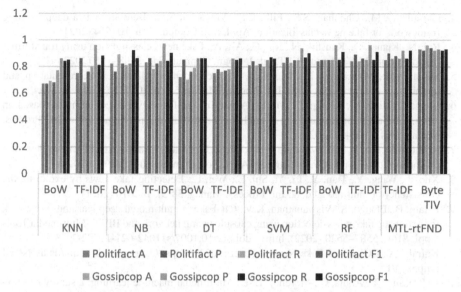

Fig. 2. Result comparison of Proposed Model with available ML classifiers

6 Conclusion and Future Work

The novel framework MTL-rtFND for real-time false news detection on social networks is introduced and presented in this research work. The majority of prior strategies employed text based techniques, however, MTL-rtFND uses multimodal transfer learning techniques to address the limitations of these approaches. By incorporating a Multimodal Feature Extraction Block (MFEB) and a Classification Block (CB), MTL-rtFND efficiently captures context-based features using a pre-trained multimodal model trained on larger-scale datasets. Currently, only MTL-rtFND with multimodal embedding techniques has been experimented and results are compared with the existing methods. The proposed multimodal embedding methods have proven to be outperformed by 5.34% while only considering text data currently, with the comparison of the state-of-the-art methods.

In the future, the approach can be implemented on real-world datasets and multimodality. By further exploring these areas, the work can be continued to advance the field of real-time fake news detection, mitigating the adverse impacts of misinformation on social media networks and fostering a more informed society.

References

1. Palani, B., Elango, S.: CTrL-FND: content-based transfer learning approach for fake news detection on social media. Int. J. Syst. Assur. Eng. Manag. **14**(903–918), 2016 (2023). https://doi.org/10.1007/s13198-023-01891-7
2. Chaturvedi, B., Dubey, S.: An efficient approach based on machine learning and performance improvement for fake news detection. Int. J. Emerg. Technol. Innov. Res. **8**, 331–336 (2021)

3. Choudhary, M., Chouhan, S.S., Pilli, E.S., Vipparthi, S.K.: Berconvonet: a deep learning framework for fake news classification. Appl. Soft Comput. **110**, 107614 (2021)
4. Rai, N., Kumar, D., Kaushik, N., Raj, C., Ali, A.: Fake news classification using transformer based enhanced LSTM and BERT. Int. J. Cogn. Comput. Eng. **3**, 98–105 (2022)
5. Samadi, M., Mousavian, M., Momtazi, S.: Deep contextualized text representation and learning for fake news detection. Inf. Process. Manag. **58**(6), 102723 (2021)
6. Song, C., Ning, N., Zhang, Y., Wu, B.: A multimodal fake news detection model based on crossmodal attention residual and multichannel convolutional neural networks. Inf. Process. Manag. **58**(1), 102437 (2021)
7. Sahoo, S.R., Gupta, B.B.: Multiple features-based approach for automatic fake news detection on social networks using deep learning. Appl. Soft Comput. **100**, 106983 (2021)
8. Xue, J., Wang, Y., Tian, Y., Li, Y., Shi, L., Wei, L.: Detecting fake news by exploring the consistency of multimodal data. Inf. Process. Manag. **58**(5), 102610 (2021)
9. Palani, B., Elango, S., Viswanathan, K.V.: CB-Fake: a multimodal deep learning framework for automatic fake news detection using capsule neural network and BERT. Multimedia Tools Appl. **81**(4), 5587–5620 (2022). https://doi.org/10.1007/s11042-021-11782-3
10. Koirala, A.: COVID-19 Fake News Dataset (2021). https://data.mendeley.com/datasets/zwf dmp5syg/1
11. Baltrušaitis, T., Ahuja, C., Morency, L.-P.: Multimodal machine learning: a survey and taxonomy. IEEE Trans. Pattern Anal. Mach. Intell.EEE Trans. Pattern Anal. Mach. Intell. **41**(2), 423–443 (2018)
12. Cui, L., Wang, S., Lee, D.: SAME: sentiment-aware multi-modal embedding for detecting fake news. In: Proceedings of the 2019 IEEE/ACM International Conference on Advances in Social Networks Analysis and Mining, pp. 41–48 (2019)
13. Jiang, T., Zeng, J., Zhou, K., Huang, P., Yang, T.: Lifelong disk failure prediction via GAN-based anomaly detection. In: 37th IEEE International Conference on Computer Design, ICCD 2019, Abu Dhabi, United Arab Emirates, 17–20 November 2019, pp. 199–207. IEEE (2019)
14. Ma, X., Zeng, J., Peng, L., Fortino, G., Zhang, Y.: Modeling multi-aspects within one opinionated sentence simultaneously for aspect-level sentiment analysis. Fut. Gener. Comput. Syst. **93**, 304–311 (2019)
15. Guo, Y.: A mutual attention-based multimodal fusion for fake news detection on the social network. Appl. Intell. **53**, 15311–15320 (2023)
16. Wang, B., Feng, Y., Xiong, X., et al.: Multi-modal transformer using two-level visual features for fake news detection. Appl. Intell. **53**, 10429–10443 (2023). https://doi.org/10.1007/s10 489-022-04055-5
17. Hua, J., Cui, X., Li, X., Tang, K., Zhu, P.: Multimodal fake news detection through data augmentation-based contrastive learning. Appl. Soft Comput. **136**, 110125 (2023). https://doi.org/10.1016/j.asoc.2023.110125
18. Praseed, A., Rodrigues, J., Thilagam, P.S.: Hindi fake news detection using transformer ensembles. Eng. Appl. Artif. Intell. **119**, 105731 (2023). https://doi.org/10.1016/j.engappai.2022.105731
19. Rastogi, S., Bansal, D.: A review on fake news detection 3T's: typology, time of detection, taxonomies. Int. J. Inf. Secure. **22**, 177–212 (2023). https://doi.org/10.1007/s10207-022-006 25-3
20. Jing, J., Wu, H., Sun, J., Fang, X., Zhang, H.: Multimodal fake news detection via progressive fusion networks. Inf. Process. Manage. **60**(1), 103120 (2023). https://doi.org/10.1016/j.ipm.2022.103120
21. Farooq, M.S., Naseem, A., Rustam, F., Ashraf, I.: Fake news detection in Urdu language using machine learning. PeerJ Comput. Sci. **9**, e1353 (2023). https://doi.org/10.7717/peerj-cs.1353

22. Jarrahi, A., Safari, L.: Evaluating the effectiveness of publishers' features in fake news detection on social media. Multimed Tools Appl. **82**, 2913–2939 (2023). https://doi.org/10.1007/s11042-022-12668-8

23. Ramkissoon, A.N., Goodridge, W.: Enhancing the predictive performance of credibility-based fake news detection using ensemble learning. Rev. Socionetw. Strat. **16**, 259–289 (2022). https://doi.org/10.1007/s12626-022-00127-7

24. Ali, A.M., Ghaleb, F.A., Al-Rimy, B.A.S., Alsolami, F.J., Khan, A.I.: Deep ensemble fake news detection model using sequential deep learning technique. Sensors **22**, 6970 (2022). https://doi.org/10.3390/s22186970

25. Huang, Y.-F., Chen, P.-H.: Fake news detection using an ensemble learning model based on Self-Adaptive Harmony Search algorithms. Exp. Syst. Appl. **159**, 113584 (2020)

26. Ansar, W., Goswami, S.: Combating the menace: a survey on characterization and detection of fake news from a data science perspective. Int. J. Inf. Manag. Data Insights **1**, 100052 (2021)

27. Nistor, A., Zadobrischi, E.: The influence of fake news on social media: analysis and verification of web content during the COVID-19 pandemic by advanced machine learning methods and natural language processing. Sustainability **14**, 10466 (2022)

28. Now, N.X., Chua, H.N.: Detecting fake news with tweets' properties. In: Proceedings of the 2019 IEEE Conference on Application, Information and Network Security (AINS), Pulau Pinang, Malaysia, 19–21 November (2019)

29. Choraś, M., et al.: Advanced Machine Learning techniques for fake news (online disinformation) detection: a systematic mapping study. Appl. Soft Comput. **101**, 107050 (2021)

30. Kumari, R., Ekbal, A.: AMFB: attention-based multimodal factorized bilinear pooling for multimodal fake news detection. Exp. Syst. Appl. **184**, 115412 (2021)

31. Trueman, T.E., Kumar, A., Narayanasamy, P., Vidya, J.: Attention-based C-BiLSTM for fake news detection. Appl. Soft Comput. **110**, 107600 (2021)

32. Shu, K., Mahudeswaran, D., Wang, S., Lee, D., Liu, H.: FakenewsNet: a data repository with news content, social context and dynamic information for studying fake news on social media. CoRR abs/1809.01286 (2018)

33. Misra, R.: Politifact (2021). https://www.politifact.com/

34. Lewittes, M.: Celebrity News Pop Culture (2009). https://www.gossipcop.com/

Classification of Exaggerated News Headlines

Mapitsi Roseline Rangata[✉][iD] and Tshephisho Joseph Sefara[iD]

Council for Scientific and Industrial Research, Pretoria, South Africa
{mrangata,tsefara}@csir.co.za

Abstract. The amount of data online is increasing as companies generate news articles daily. These news articles contain headlines that have a level of exaggeration aimed to win the readers. In addition, these companies are competing against one another; hence creating appealing and exaggerated news headlines is one of the options to win the readers. Some of the exaggerated headlines contain some level of misleading information. Hence, this paper aims to apply machine learning methods and natural language processing to detect and identify exaggerated news headlines in South African context. Machine learning models such as logistic regression, decision trees, support vector machines, and XGBoost are trained on data that contain labelled news headlines as binary classification. The models produced good results, with XGboost and SVM obtaining 70% in terms of accuracy. Furthermore, the F measure was used to evaluate the models and decision trees obtained 56% followed by SVM with 53%. The classification of exaggerated news headlines is a difficult task. Therefore, we oversampled the data to obtain balanced labels. The performance of the models was increased. SVM obtained 84% followed by logistic regression, XGBoost, and decision trees with accuracy of 78%, 72% and 71%, respectively.

Keywords: Classification · News headlines · Machine learning · Natural language processing · Exaggerated News

1 Introduction

Text classification is a method or technique that automatically assigns labels or categories to given text or documents by utilising machine learning and statistical methods. Text classification is one of the important parts of natural language processing (NLP). It contains subtasks such as text-based language identification, topic classification, authorship classification, document classification, news classification, and more. The aim of text classification is to categorise given texts or documents by labels. This plays an important role in systems such as information retrieval and organisation [8].

The classification of news is becoming an urgent research area. Since 2016 US elections, the classification of news as real or fake gained popular awareness. Fake news detection can be done using machine learning models and NLP techniques

K. K. Patel et al. (Eds.): icSoftComp 2023, CCIS 2030, pp. 248–260, 2024.
https://doi.org/10.1007/978-3-031-53731-8_20

[4]. Zhang et al. [23] discussed a comprehensive overview of fake news online, the characterisation of the impact of fake news, and the methods used to detect fake news. These methods involve practical-based approaches such as online fact checking and research-based approaches, which involve the use of machine learning.

The media is one of the most significant sources for providing information about what is happening in the world. However, with the large number of users of social networks or the Internet, the media tend to oversell or mislead readers with exaggerated news titles. This poses the credibility of the news and makes it difficult for readers to identify non-exaggerated news. Researchers have explored the classification of misleading or fake news with supervised and unsupervised machine learning algorithms [22]. Machine learning algorithms are the most commonly and widely used by researchers to predict or classify a given set of data. The aim of this paper is to classify news headlines from the South African news data as exaggerated or non-exaggerated using machine learning models. The data set is acquired from the study by Sefara et al. [18], where the authors proposed a method to label exaggerated news headlines.

The contributions of the paper are as follows:

- We transform the data using Term Frequency-Inverse Document Frequency (TFIDF) to generate features.
- We train machine learning models on the data.
- We propose the baseline classification results of the models on the proposed data.
- We publish the code and results on Github[1] to allow future benchmarking.

This paper is organised as follows. The background is discussed in the next section. Section 3 explains the proposed architecture. Section 4 explains data collection and data engineering. The methods used to build machine learning models are discussed in Sect. 5. The evaluation of the machine learning models is discussed in Sect. 6. The findings and analysis of the results are discussed in Sect. 7, while Sect. 8 concludes the paper with future work.

2   Background

There is little or limited research that asserts the study of exaggerated news; however, most studies have explored research on fake news detection and classification. Researchers such as Snell et al. [20] took the liberty to manually classify news articles as fake or real. They began to identify news articles on different topics. Then manually evaluate each article to verify whether the article is real or fake by looking at several aspects such as capitalisation in headlines, comparison of an article from a different source to a trusted source, and others. The final step in their method was for each article to be validated by a different team member. Jehad et al. [5] classified news articles as fake or real using machine

---

[1] https://github.com/JosephSefara/exaggerated-news-titles.

learning models; Random Forest and Decision Tree, and incorporated TF-IDF as a feature extractor to improve model performance. Decision Tree appeared to perform better with feature extractor than without feature extractor.

The application of deep learning to fake news has sparked considerable interest; therefore, authors such as Mehta et al. [9] have proposed a bidirectional encoder representation (BERT) method for the classification of fake news in two separate datasets, which has demonstrated a significant improvement in performance compared to supervised machine learning classifiers such as Support Vector Machine (SVM) and others. Furthermore, Aggarwal et al. [1] proposed the BERT model to classify political news articles as fake or real. They also compared BERT to XGBoost and LSTM for performance evaluation in terms of accuracy, and BERT was the top performer with higher accuracy. Furthermore, a method for combining two word embedding models and their related entities with deep learning-based MLP method was proposed by [21] with their results suggesting that the count vectorizer and Glove together with MLP perform better than the Random Forest classifier when used with TF-IDF as a feature extractor. Jehad et al. [6] proposed a method based on TF-IDF and Multi-Layer Perceptron (MLP) to classify news as real or fake. Their results showed higher precision compared to related work covered in their study. Samadi et al. [16] have explored the use of different features such as topic extraction or categories, sentiment in text, and the use of various entities. These features were fed into a neural network classifier with semantic-based features to classify news articles as fake or real. In their analysis, the use of these features has been shown to improve the performance of the neural network classifier in terms of accuracy. Lai et al. [7] made a comparison between several machine learning models and neural network models for classification in twitter news data as fake or real. In their findings, the overall neural networks models seemed to perform better than the machine learning models. In a nutshell, there is more literature on fake news classification, with most applications based on deep learning approaches, while the application on classification of exaggerated news titles is limited. Hence, in this paper, we conduct research for the classification of exaggerated news titles in the South African context.

3   Proposed Architecture

The high-level architecture of the proposed method is illustrated in Fig. 1. For training the models, the news headline data is preprocessed to eliminate features that are not important for model training. This includes the removal of numbers, the removal of special characters, the removal of links, and the removal of stopwords. Stopwords are functional words that frequently occur within the data. Removing these words helps the model distinguish and learn the importance of features for each predicted label. Features are extracted in a form of n-gram of sizes between 1 and 3. N-grams are sequences of N words or tokens in a document that may overlap. At this stage, a vocabulary is being created. A vocabulary contains a list of features. This list can be later used to identify the importance of features. The data are divided into 80% for training the machine learning models and 20% for testing the models. The test data are not used during model training. The data are normalised to eliminated outliers which are caused by high values of frequent words. For testing the models, the test data is used to extract the features. The features are normalised, and model prediction begins. The predicted labels and the original labels are used to evaluate the models' performance using the accuracy and the F measure.

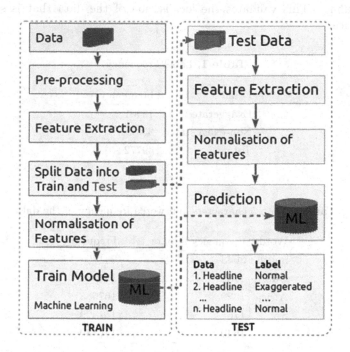

Fig. 1. The data flow diagram of the proposed method.

4   Data

This section discusses data, data processing, feature extraction, feature normalisation, and split of train and test data. The data consists of news headlines in the South African context. Data were acquired from Sefara et al. [19] in a comma separated value format (CSV). The data contains the following columns: publisher, article title, category, article content, link, publish date, label, and more. This paper focusses on the title, the content of the article, and the label. The label is the column that is used for the prediction based on the title and content of the article.

4.1 Data Exploration

Data exploration helps to understand the nature of the data. The data contain 49772 titles with their corresponding labels. The distribution of labels is shown in Table 1 with most titles labelled as not exaggerated. This shows that the data are unbalanced and that one needs to use the F score metric to evaluate the models' performance. Table 2 shows the word and character distributions of the data. The words *South, Cup, Africa, SA* and *World* appeared frequently in the news headlines. This validates the localisation of the data that is specific to South Africa.

Table 1. Label Frequency

Label	Frequency
Exaggerated	16401
Non-exaggerated	33371
Total	49772

Table 2. Top 5 characters and words frequency of the data

Words	Frequency	Characters	Frequency
South	3043	e	241856
Cup	1832	a	181982
Africa	1822	s	167685
SA	1765	i	151829
World	1649	r	147380

Figure 2 shows a graphical representation of the word distribution in the data. Larger words appeared more frequently in the data. The word cloud shows that the data are localised to South Africa and shows some of the companies and political organisations of South Africa.

Fig. 2. Word cloud of the data.

4.2 Data Pre-processing

Data pre-processing includes the cleaning of the news headlines to remove noise. Noise can be expressed in terms of characters that may confuse machine learning models. Such characters include numbers, links, emojis, and stopwords. Stopwords are function words that exist in the data and do not provide meaningful information. Removal of noise results in smaller data in terms of size, and then reduces the model training time due to a lower number of tokens in the training data. Table 3 shows some of the 145 stopwords identified in the data, and it shows the frequency of special characters identified in the data. Data preprocessing is an important step in NLP to help determine which type of token should be removed from the data without losing the quality of the data. The tokens shown in Table 3 do not help the models learn significant features for each label; therefore, we removed these tokens.

4.3 Feature Extraction

Feature extraction in NLP involves the transformation of natural language into feature representation that the models can understand. There are different types of feature extraction technique including:

- Word embeddings (BoW)
- TF-IDF

Word embedding is a learnt token representation in which each token is represented as a vector in a lower-dimensional space. Word embeddings maintain the context and relationship of words in such a way that it is simple to find word similarity. Word embeddings can be implemented in various ways, including Word2vec [10], GloVe [13], FastText [2] and more. There are two approaches to implement word embedding, namely, skip-gram, which predicts the context using target word, and continuous bag-of-word, which predicts the target word using the context.

Table 3. Frequency of stopwords and special characters.

Stopwords	Frequency	Special characters	Frequency
to	14194	:	12717
in	10870	'	11808
of	8784	,	9655
the	8662	...	4843
for	7704	'	4039
and	4867	'	3239
on	4623	-	1925
a	4326	?	1673
as	3029	[	1668
with	2870	]	1668

TF-IDF is a statistical measure that evaluates word relevance for a document in a set of documents. TF-IDF is used mostly in NLP applications such as:

- text summarisation - a process of summarising large text into summaries
- information retrieval - a process of identifying and obtaining relevant information
- text categorisation - a process of classifying text into relevant categories, and
- sentiment analysis - a process of analysing sentiments in a text.

TF-IDF highlights significant terms that are not too frequent in the data but are of great importance. This paper explores the use of the TF-IDF method to extract features using n-gram of size 3. A Python library scikit-learn is used to implement feature extraction [12].

4.4 Train and Test Data

After features are extracted for each title of the article, we divided the news headlines data into a train and test set to avoid overfitting. The train set contained 80% of the news headlines, while the test set has 20% of the news headlines. The news headlines in the test set are used only to evaluate the trained model to measure the performance and quality of the model. There are two main reasons to split the data.

- **To avoid overfitting**. Overfitting occurs when a machine learning model learns the training data too well and fails to generalise to new data. By splitting the data into a training set and a test set, we can train the model on the training set and evaluate its performance on the test set. This helps us to ensure that the model is generalising well and is not simply memorising the training data.

- **To get an unbiased estimate of the performance of the model.** If we train and evaluate the model on the same data, we will get an over-optimistic estimate of its performance. This is because the model will already have seen the data on which it is being evaluated. By dividing the data into a training set and a test set, we can ensure that the model is being evaluated on data that it has never seen before. This gives us a more unbiased estimate of the performance of the model on real-world data.

4.5 Feature Normalisation

The significance of normalising the features is to improve the performance of machine learning models. Feature normalisation is a method used to scale the independent variable or data features. This method is an important step in data preprocessing. Both the train and test data set features are normalised using a standard scaler which is defined by the following equation:

$$x' = \frac{x - \mu}{\sigma} \tag{1}$$

where x represents the feature, μ represents the mean, and σ represents the standard deviation of the data [17]. The feature normalisation step may have advantages in domains such as speech [17], but in this study the models were unable to improve performance. Therefore, we chose to keep the original features that exhibit the characteristics and context of the data.

5 Machine Learning Models

The following techniques are implemented and used to build the classification model on the acquired data set.

- **Logistic Regression** is a supervised machine learning classifier that is used for predicted binary or multiclass variables. The binary logistic regression model performs classification using two classes, which provides probabilities ranging from 0 to 1 using the cross-entropy loss function. In this paper, we used logistic regression to classify exaggerated news titles as the model performs better in a binary classification.
- **Decision Tree** is a non-parametric supervised machine learning method that is utilised for classification and regression applications. This model uses a tree-like model to make decisions. In this paper, we used the popular decision tree to classify exaggerated news titles, as it performs better on a small data set.
- **SVM** is a supervised machine learning method that can be used for classification and regression applications. In classification problems, SVM labels the data by finding the optimal decision boundary that divides the labels. SVM has a set of mathematical functions that are called kernels. Kernels are used to transform the input data into the required form. In this paper, we used different SVM kernels to choose the best kernel to classify exaggerated news titles.

– **XGBoost** is a machine learning technique that uses a group of decision trees with gradient boosting to make predictions [15]. It can be used in regression, binary classification, and multiclass classification. In this paper, it is used for binary classification.

6   Model Evaluation

Model evaluation helps to understand the performance, weaknesses, and strengths of the model. The models are tested on an unseen test data set that was not used during model creation. We used the following performance metrics to measure the quality of the models.

– **Accuracy**: is the proportion of the number of correctly predicted headlines out of all the headlines.
– **F1 score** is used to measure the accuracy of the model by calculating the harmonic mean of recall and precision. This metric helps to obtain accurate results under the conditions of imbalanced data.
– **Confusion matrix**: is an X x X matrix used to measure the performance of a classification task, where X is the number of labels.

7   Results and Discussion

The machine learning models were trained using 80% of the news headlines and evaluated on the unseen news headlines, which is 20%. The results are shown in Table 5. The models were fitted to the training data with default parameters except SVM. The SVM was fitted with the parameter *kernel* set to the following values: *polynomial, linear, radial basis function (RBF)*, and *sigmoid*. Table 4 shows the results of the test of the best SVM kernel for the classification of exaggerated news headlines. Both *polynomial* and *RBF* obtained an accuracy of 70% while *linear* and *sigmoid* obtained an accuracy of 69%. Since the data were unbalanced, we considered measuring the quality of the models using the F score. As shown in Table 4, *linear* SVM outperformed other SVM kernels, obtaining 64% followed by *RBF*, *sigmoid* and *polynomial* with 63%, 63% and 62%, respectively.

Table 4. SVM kernels.

SVM Kernel	Accuracy (%)	F1 score (%)
Linear	**69**	**64**
RBF	70	63
Sigmoid	69	63
Polynomial	70	62

Other models included XGBoost, LR, and decision trees as shown in Table 5. These machine learning models were built and tested using the proposed data. XGBoost obtained an accuracy of 70% followed by logistic regression and decision trees with an accuracy of 69%, 64%, respectively. Furthermore, we used the F1 score as one of the evaluation metrics for the models, as our dataset was unbalanced, the F1 score was calculated with linear SVM obtaining 64% followed by decision trees, XGBoost and LR with a score of 63%, 62% and 61%, respectively.

Table 5. Model prediction results in percentage based on the proposed data.

Model	Accuracy (%)	F1 score (%)
XGBoost [14]	70	62
Linear SVM [11]	69	64
LR [11]	69	61
Decision trees [5]	64	63

The performance of the models can be further improved by implementing data augmentation to overcome label imbalance. Data augmentation is a method to artificially increase the data size by introducing noise. Using the SMOTE oversampling method [3], we applied data oversampling on the minority label to have balanced labels. This increased the dataset by 33% to 66742 headlines. The results in Table 6 show the improved performance in all the models. SVM outperformed other models with accuracy and F score of 84% followed by LR, XGBoost and decision trees with accuracy of 78%, 72% and 71%, respectively. Data oversampling improved the performance of the models to reach state-of-the-art results. The models were fitted with the default parameters implemented on scikit-learn [12].

Table 6. Model prediction results after over sampling.

Model	Accuracy (%)	F1 score (%)
XGBoost	72	70
Linear SVM	84	84
LR	78	78
Decision Trees	71	71

To further evaluate the performance of the oversampled models, we computed confusion matrices in Fig. 3 for XGBoost, decision tree, logistic regression and linear SVM. The confusion matrices show that linear SVM outperformed other models in correctly predicting 84% of news headlines and missing only 16%.

The second-best model is LR which correctly predicted 78% of the new titles and missing 22%. The third-best model is XGBoost in Fig. 3a which correctly predicted 72% of the news headlines and missing 28%. The decision tree model correctly predicted 71% news headlines and missing 29%. All the models were able to predict the non-exaggerated news headlines. But linear SVM was able to predict exaggerated news headlines with an accuracy of 41% followed by LR with an accuracy of 38%. XGBoost confused 26% exaggerated news headlines with normal news, resulting in type I error, meaning that the model failed to predict the news headlines as exaggerated. In general, the four models performed better in predicting news headlines, as normal and linear models performed better in predicting exaggerated news headlines.

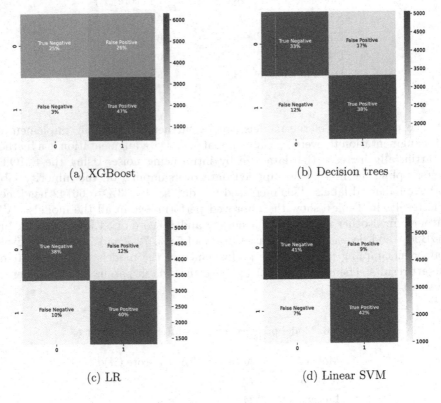

(a) XGBoost   (b) Decision trees

(c) LR   (d) Linear SVM

Fig. 3. The confusion matrices after oversampling

8   Conclusion and Future Work

This paper proposed the classification of exaggerated news headlines using machine learning algorithms applied to an exaggerated news dataset in the context of South Africa. The data contained news titles and labels. The data was preprocessed and cleaned. TFIDF was used to transform the data to generate feature sets. Several machine learning models were trained on the data and evaluated using the accuracy, F score, and confusion matrix. The models were trained on 80% of the data and tested on the rest. The models produced baseline results for the proposed dataset. Linear SVM obtained better results with an accuracy and an F score of 84% after oversampling the data. The classification of news headlines remains a challenging task when applying machine learning. Therefore, we recommend the use of linguistic techniques to properly label the data. Future work will focus on improving the results using deep learning methods.

References

1. Aggarwal, A., Chauhan, A., Kumar, D., Verma, S., Mittal, M.: Classification of fake news by fine-tuning deep bidirectional transformers based language model. EAI Endorsed Trans. Scalable Inf. Syst. **7**(27), e10–e10 (2020)
2. Bojanowski, P., Grave, E., Joulin, A., Mikolov, T.: Enriching word vectors with subword information. Trans. Assoc. Comput. Linguist. **5**, 135–146 (2017)
3. Chawla, N.V., Bowyer, K.W., Hall, L.O., Kegelmeyer, W.P.: SMOTE: synthetic minority over-sampling technique. J. Artif. Intell. Res. **16**, 321–357 (2002)
4. Ibrishimova, M.D., Li, K.F.: A machine learning approach to fake news detection using knowledge verification and natural language processing. In: Barolli, L., Nishino, H., Miwa, H. (eds.) INCoS 2019. AISC, vol. 1035, pp. 223–234. Springer, Cham (2020). https://doi.org/10.1007/978-3-030-29035-1_22
5. Jehad, R., Yousif, S.A.: Fake news classification using random forest and decision tree (j48). Al-Nahrain J. Sci. **23**(4), 49–55 (2020)
6. Jehad, R., Yousif, S.A.: Classification of fake news using multi-layer perceptron. In: AIP Conference Proceedings. AIP Publishing (2021)
7. Lai, C.M., Chen, M.H., Kristiani, E., Verma, V.K., Yang, C.T.: Fake news classification based on content level features. Appl. Sci. **12**(3), 1116 (2022)
8. Mao, K., Xiao, X., Zhu, J., Lu, B., Tang, R., He, X.: Item tagging for information retrieval: a tripartite graph neural network based approach. In: Proceedings of the 43rd International ACM SIGIR Conference on Research and Development in Information Retrieval, SIGIR 2020, pp. 2327–2336. Association for Computing Machinery, New York (2020). https://doi.org/10.1145/3397271.3401438
9. Mehta, D., Dwivedi, A., Patra, A., Anand Kumar, M.: A transformer-based architecture for fake news classification. Soc. Netw. Anal. Min. **11**, 1–12 (2021)
10. Mikolov, T., Chen, K., Corrado, G.S., Dean, J.: Efficient estimation of word representations in vector space (2013). http://arxiv.org/abs/1301.3781
11. Patel, A., Meehan, K.: Fake news detection on Reddit utilising countvectorizer and term frequency-inverse document frequency with logistic regression, multinominalnb and support vector machine. In: 2021 32nd Irish Signals and Systems Conference (ISSC), pp. 1–6 (2021). https://doi.org/10.1109/ISSC52156.2021.9467842
12. Pedregosa, F., et al.: Scikit-learn: machine learning in Python. J. Mach. Learn. Res. **12**, 2825–2830 (2011)
13. Pennington, J., Socher, R., Manning, C.D.: GloVe: global vectors for word representation. In: Proceedings of the 2014 Conference on Empirical Methods in Natural Language Processing (EMNLP), pp. 1532–1543 (2014)

14. Rao, V.C.S., Radhika, P., Polala, N., Kiran, S.: Logistic regression versus XGBoos: machine learning for counterfeit news detection. In: 2021 Second International Conference on Smart Technologies in Computing, Electrical and Electronics (ICSTCEE), pp. 1–6 (2021). https://doi.org/10.1109/ICSTCEE54422.2021.9708587

15. Sahin, E.K.: Assessing the predictive capability of ensemble tree methods for landslide susceptibility mapping using XGBoost, gradient boosting machine, and random forest. SN Appl. Sci. **2**(7), 1308 (2020)

16. Samadi, M., Momtazi, S.: Fake news detection: deep semantic representation with enhanced feature engineering. Int. J. Data Sci. Anal., 1–12 (2023). https://doi.org/10.1007/s41060-023-00387-8

17. Sefara, T.J.: The effects of normalisation methods on speech emotion recognition. In: 2019 International Multidisciplinary Information Technology and Engineering Conference (IMITEC), pp. 1–8 (2019). https://doi.org/10.1109/IMITEC45504.2019.9015895

18. Sefara, T.J., Rangata, M.R.: A natural language processing technique to identify exaggerated news titles. In: Ranganathan, G., Papakostas, G.A., Rocha, Á. (eds.) International Conference on Information, Communication and Computing Technology. pp. 951–962. Springer, Heidelberg (2023). https://doi.org/10.1007/978-981-99-5166-6_64

19. Sefara, T.J., Rangata, M.R.: A natural language processing technique to identify exaggerated news titles. In: Ranganathan, G., Papakostas, G.A., Rocha, Á. (eds.) Inventive Communication and Computational Technologies, ICICCT 2023. LNNS, vol. 757, pp. 951–962. Springer, Singapore (2023). https://doi.org/10.1007/978-981-99-5166-6_64

20. Shewalkar, A., Nyavanandi, D., Ludwig, S.A.: Performance evaluation of deep neural networks applied to speech recognition: RNN, LSTM and GRU. J. Artif. Intell. Soft Comput. Res. **9**(4), 235–245 (2019)

21. Thilagam, P.S., et al.: Multi-layer perceptron based fake news classification using knowledge base triples. Appl. Intell. **53**(6), 6276–6287 (2023)

22. de Wet, H., Marivate, V.: Is it fake? News disinformation detection on South African news websites. In: 2021 IEEE AFRICON, pp. 1–6 (2021). https://doi.org/10.1109/AFRICON51333.2021.9570905

23. Zhang, X., Ghorbani, A.A.: An overview of online fake news: characterization, detection, and discussion. Inf. Process. Manage. **57**(2), 102025 (2020). https://doi.org/10.1016/j.ipm.2019.03.004. https://www.sciencedirect.com/science/article/pii/S0306457318306794

Inappropriate Text Detection and Rephrasing Using NLP

Sanyam Jain[iD] and B. K. Tripathy[✉][iD]

School of Computer Science Engineering and Information Systems, Vellore Institute of
Technology, Vellore, India
sanyamjain3003@gmail.com, tripathybk@vit.ac.in

Abstract. The impact of offensive language on public and professional discourse
highlights the need for efficient mitigating measures. Cutting-edge computational
linguistic techniques were used to identify and treat such language in a novel way.
A two-pronged mechanism is used when hazardous content is found: offending
terminology is either removed or put through Natural Language Pre-processing,
producing rephrased information that maintains the original meaning of the text.
Additionally, this work uses two freely accessible datasets for text categorization. The technique is unique, because during the rephrasing stage, we consider
the incorrect words to get their synonyms, and we choose to fit for replacement
in the phrase. Classification best accuracy we have achieved of about 95%. The
method is comprehensive and aims to create a setting that encourages courteous
and peaceful discussion while maintaining semantic integrity. This research provides a sophisticated approach to fostering meaningful relationships in both public
and professional contexts by fully addressing incorrect language.

Keywords: Inappropriate Text · NLP · Toxic Classification · Rephrasing

1 Introduction

In the digital age, text-based communication has increased at an unprecedented rate
due to the quick spread of internet platforms. While technology has made it possible
to connect to the world and share knowledge, it has also created new problems due
to offensive language and material. Concerns about cyberbullying, hate speech, and
abusive language have grown as a result of the user-generated content boom and the
frequent anonymity of online interactions. As a result, there is an urgent need for effective
and efficient techniques for identifying and dealing with offensive material in online
communication channels.

The detection and moderation of offensive language is essential for upholding inclusive and respectful online communities as well as for safeguarding people from harm and
maintaining the integrity of digital platforms. Advanced Natural Language Processing
(NLP) methods and machine learning models that can analyze, categorize, and rewrite
text while keeping context and meaning are needed to solve this problem.

K. K. Patel et al. (Eds.): icSoftComp 2023, CCIS 2030, pp. 261–273, 2024.
https://doi.org/10.1007/978-3-031-53731-8_21

To improve the identification and rephrasing of improper content, this research article takes a thorough strategy that blends machine learning algorithms with NLP approaches. To create solutions that are compatible with the changing online communication scene, we put a lot of effort into leveraging the potential of textual data analysis.

How do we define Inappropriate in this research? The word is inappropriate might sound like the word that is mistakenly added to the text. But, we have another meaning for it i.e., the word which is used as a bad word in the text either to criticize or offend someone or a thing.

The main difficulty is developing a system that can automatically detect text that contains offensive words and then reword it while maintaining the original purpose of promoting polite conversation. A more advanced strategy utilizing NLP and machine learning is necessary to solve this difficult challenge as traditional rule-based systems are limited in their ability to interpret context.

The wide-ranging effects of improper content on people, communities, and digital platforms highlight the importance of this study. Online platforms may come to be seen negatively as a result of instances of cyberbullying, hate speech, and inappropriate language. Advanced NLP methods and machine learning models may be used to create strong solutions that proactively identify and reduce the negative effects of inappropriate text.

The main goal of this study is to develop, put into practice, and assess a thorough approach that can accurately identify offensive content in online communication and rewrite it to encourage civil conversation. Our strategy comprises investigating the efficacy of three machine learning models—XG Boost, Support Vector Classification, and Random Forest—for detecting improper content. Additionally, we look into the application of NLP methods to recognize portions of speech in improper text and apply the proper rephrasing procedures.

In the context of online communication platforms, this research focuses on tackling the issue of incorrect text identification and rephrasing. Since we understand that language is complex and context-specific, our research seeks to find a compromise between upholding communication integrity and encouraging polite language.

In the sections that follow, we discuss the background and theoretical underpinnings of our research, the datasets used for experimentation, the methodologies used, experimentation followed by the results and analyses, and finally, conclusions and possible directions for future research.

2 Background

Meaningful dialogue is frequently disrupted by user-generated conversations that veer into improper stuff like abusive language and insulting statements [1]. An innovative "Convolutional, Bi-Directional LSTM (C-BiLSTM)" model, developed in a recent study, combines convolutional neural networks (CNN) with bi-directional long short-term memory (BLSTM) to address this issue. This architecture outperforms baseline models in efficiently learning query word characteristics using a convolutional layer and encoding sequential patterns using BLSTM. Compared to standalone CNN, LSTM, and BLSTM models, the C-BiLSTM architecture exhibits improved accuracy, offering efficient unsuitable content detection and management.

A novel grammatical relation-based sentence-level semantic filtering method is presented that efficiently removes offensive content from online chats [2]. With this method, offending words are identified, their syntactic and semantic relationships are extracted, and then the relationships and patterns are modified using heuristic rules. This method improves the detection and elimination of offensive words, which helps with the better regulation of online content. In addition, conventional keyword censorship techniques are investigated, which include comparing text words to a blacklist of offending phrases. Then, offensive words are identified and either eliminated, partially replaced, or entirely replaced. By addressing language issues in general, these techniques promote a more civil and welcoming online community.

An important goal is to use natural language processing (NLP) to find unsuitable and sexual text on the internet [3]. In this case, three text encoders (Bag of Words, TF-IDF, and Word2Vec) and four classifiers (SVM, Logistic Regression, k-Nearest Neighbors, and Random Forests) were used to evaluate twelve model permutations. The best-performing model obtained an excellent accuracy of 0.97 and an F-score of 0.96 (precision 0.96/recall 0.95), using TF-IDF text encoding and an SVM classifier with a linear kernel. The erotic-sexual class served as the positive class and the neutral class served as the negative class in the model assessment process, which included accuracy, precision, recall, and F-score metrics.

EfficientNet-B7 is a framework that extracts video descriptors for bidirectional LSTM (BiLSTM) network input, increasing video representation and enabling multi-class classification [4]. It uses an ImageNet pre-trained CNN. EfficientNet-BiLSTM has been evaluated using a manually annotated dataset of 111,156 YouTube cartoon clips, and the results show that it is superior to attention-based competitors (95.30% accuracy) in this regard. Deep learning classifiers outperform conventional ones, with top-tier performance being achieved by the EfficientNet and BiLSTM (128 hidden units) architecture (f1 score = 0.9267).

For censoring films, an intelligent deep learning-based system is developed. It makes use of sophisticated convolutional neural networks (CNNs) for quick and precise unsuitable material recognition and localization [5]. To automate inappropriate speech detection, a collection of bad language is gathered, annotated, processed, and examined. It contains both isolated words and continuous speech. On innovative foul language dataset measures, the system outperforms baseline techniques, obtaining a high macro average AUC (93.85%) and weighted average AUC (94.58%), along with an improved F1-score. The suggested technique is also quicker than manual human screening for audible content suppression. Max pooling, Average pooling, and Sum pooling are pooling layers that aid in parameter reduction in image processing.

Traditional (logistic regression, SVM) and deep learning (CNNs, LSTMs) models were used in the experiments [6]. The inverse regularization strength C within a given range was optimized for hyperparameters. The weighted F1 score on validation data had to be maximized throughout optimization. The bidirectional LSTM network emerged as the best-performing model. Additionally, improvements were made by fusing BiLSTMs with CNNs and other logistic regression-SVM variations, acting as a meta-classifier and producing gains in a few instances.

The study uses iterative methods to grow the dataset by creating keyword-based seed videos from web sources [7]. The classification accuracy of a deep learning classifier is 84.3% when it is focused on toddler-friendly content. Using this classifier, a thorough study of YouTube videos for toddlers indicates that 1.05% of them are improper. Additionally, there is a worrisome 3.5% chance that infants watching appropriate content will see incorrect suggestions within the first ten suggestions.

Broader abusive language training datasets produce robust algorithms capable of accurately identifying certain abusive language kinds in cross-domain trials [8]. Even in out-of-domain conditions, joint-learning systems routinely outperform the competition. Cross-domain knowledge transfer is demonstrated by HurtLex, which improves true positive rates.

The study [9] uses binary classification and has an accuracy rate of 91.2%. When processing these three classes using a neutral category, the model produces an accuracy of 85.70%. Resampling techniques are used to address multiclass imbalance. Sentiment and semantic analysis after that assesses the gravity of immoral content. Insights into content polarity and ethical sensitivity are revealed by the research's exploration of Decision trees, SVM, and Naive Bayes for classification.

In the investigation of abusive user comment identification, distributed representations for text are examined [10]. The first step is classifying various distributed representation techniques that include word and sentence embedding models. An empirical evaluation of these approaches' performance in classifying abusive text follows. Significantly, multi-task sentence embedding outperforms alternatives that have been learned. Positive results are obtained when word embeddings are simply averaged with SVM classifiers. When compared to several distributed representation strategies, traditional n-grams perform robustly.

For classification purposes, researchers have successfully used machine learning methods, particularly Bag-of-Words (BoW) and N-grams [11]. Deep learning architectures have also shown interesting results in this area. However, because datasets differ, evaluating the effectiveness and performance of various features and classifiers is still difficult. Standardized annotation procedures and benchmark datasets are requirements for insightful comparison evaluations.

According to experimental results, bidirectional GRU networks containing LTC are the most effective at spotting offensive language [12]. The models perform well when analyzing "spam" and "hateful" tweets, with the RNN-LTC model earning a high F1 score of 0.551 for "spam" and CNN with context tweets achieving a score of 0.309 for "hateful." Notably, character-level features help classic machine learning classifiers perform better by utilizing TF-IDF values for important character elements.

To classify abusive content, the deep learning models CASE and KIPP [13] achieve average accuracy rates of 92.2% and 98.21%, respectively, with AUROC scores of 0.977 and 0.986. The Malang method puts data privacy first by not sending delicate information via the cloud. This emphasis on privacy is in line with a global trend, hence MaLang's anonymization features are essential for organizational adoption. Malang, which may be used for inter-departmental surveillance, can identify employee harmful content exchanges thanks to weight matrix updates that improve model performance through federated learning.

To find hate speech in brief text, particularly tweets, an ensemble classifier is presented. Standard word unigrams and attributes indicating users' prior propensity for abusive messages are used by the base classifiers [14]. The method addresses the issue of obfuscation since users tend to disguise objectionable statements using slang or inventive spelling. It is highlighted that for a language-independent solution, word frequency vectorization outperforms pre-trained word embeddings.

The study shows preliminary findings from tests on the identification of offensive content in various social media datasets [15]. It investigates the use of pre-built deep neural networks for feature representation and classification in two text classification problems. Findings show that averaging pre-trained word embeddings frequently beats ngrams representation when traditional feature representation and word embeddings are compared.

3 Dataset Used

We used two datasets toxic comment classification dataset [18] and the hate speech and offense language dataset [19]. They have 24,783 and 59,572 samples respectively. The second dataset also contains the hate speech classification which we don't need so we selected only toxic class samples. So, we got the final number of 80,192 samples by combining both datasets.

However, we will not be merging them as the essence of the place from where it is extracted might add confusion to the models. Because both of them were made for different reasons and the sections of comments/texts present in them are different. At last both the datasets were decreased to two features, one is the comment text and the other the class of the sample (toxic or non-toxic).

These datasets were specifically used because they contain major offensive language conversation texts and are publicly renowned for their accurate dataset. Moreover, these datasets contain real-world tweets and comments, which will support to preparation of our model to handle real-world inappropriate texts. Furthermore, we have used the dataset [20] because it was mainly designed to support the toxic comment classification dataset. This enhanced our model by identifying specific toxic words in the text.

4 Methodology

4.1 TF-IDF Vectorization

TF-IDF (Term Frequency-Inverse Document Frequency) serves as a widely employed method in natural language processing and information retrieval to transform text data into numerical vectors, rendering it suitable for diverse machine learning algorithms. Its purpose is to capture the significance of individual words within a document or a corpus of documents. The fundamental concept behind TF-IDF is to indicate how pertinent a word is to a document in a collection of documents.

The operational process of TF-IDF vectorization is outlined as follows:

- Calculation of Term Frequency (TF):

Term Frequency quantifies the frequency of a word in a specific document. It is computed as the ratio of the count of times a word (term) emerges in a document to the total count of terms in that document. A higher TF value suggests greater importance of a word within the document.

$$TF = (NOD)/(TOD) \tag{1}$$

Here, NOD = Number of occurrences of term in document, TOD = Total number of terms in document.

- Calculation of Inverse Document Frequency (IDF):

Inverse Document Frequency gauges the significance of a word across the entire corpus of documents. It is calculated as the logarithm of the ratio of the total number of documents to the number of documents containing the specific term. IDF assigns a higher weight to words appearing in fewer documents, which are thus more informative.

$$IDF = \log((TOD) / (\text{Number of documents containing the term})) \tag{2}$$

- TF-IDF Score Computation:

The TF-IDF score for a word in a given document is determined by the product of its TF and IDF values. This score reflects the relevance of the word to the document in comparison to its importance across the entire corpus.

$$TF - IDF = TF * IDF \tag{3}$$

- Vector Representation:

Following the calculation of TF-IDF values for all terms across all documents, each document is transformed into a vector. Each dimension of the vector corresponds to a unique term in the entire corpus, and the value within each dimension represents the TF-IDF score of the corresponding term in the document. This procedure yields a high-dimensional numerical representation of textual data.

TF-IDF vectorization offers a method to quantify the significance of words in a document relative to their occurrence in other documents. It is frequently employed in various NLP tasks like text classification, clustering, information retrieval, and more. Nevertheless, it's important to note that while TF-IDF is a potent technique, it might not capture intricate word relationships comprehensively and could encounter challenges in grasping semantics or word associations. In such instances, more advanced techniques such as word embeddings could prove more suitable.

4.2 Latent Semantic Analysis

Latent Semantic Analysis (LSA) is a technique employed in natural language processing and information retrieval to unveil concealed relationships between words and documents through analysis of their statistical patterns. It is based on the notion that words sharing similar contexts generally carry akin meanings. LSA finds application in tasks like text classification, document similarity assessment, and information retrieval.

The fundamental operation of Latent Semantic Analysis comprises several steps:

- Collection and Preprocessing of Text Corpus:

A substantial collection of text documents is amassed, followed by the application of preprocessing procedures to cleanse the data. This typically encompasses converting all words to lowercase, eliminating punctuation and stop words (common words such as "the," "and," and "is" that bear little significance), and executing stemming or lemmatization to reduce words to their root forms.

- Construction of Term-Document Matrix:

A term-document matrix is created, where rows denote distinct words (terms) within the corpus and columns represent documents. Each entry in the matrix contains the frequency or another measure of the term's presence in the corresponding document. As most words do not appear in most documents, this matrix often exhibits sparsity.

- Application of Singular Value Decomposition (SVD):

SVD, a matrix factorization technique, decomposes the term-document matrix into three matrices: U, Σ, and V^T. U signifies relationships between terms and concepts, Σ is a diagonal matrix conveying the significance of each concept, and V^T indicates relationships between documents and concepts.

Mathematically, for a term-document matrix A, SVD decomposition appears as

$$A = U\Sigma V \wedge T \qquad (4)$$

- Dimensionality Reduction:

LSA entails dimensionality reduction by truncating matrices U, Σ, and V^T. This involves retaining only the top k singular values and their corresponding columns from U and V^T, along with the top k values from the diagonal matrix Σ. This process diminishes noise and captures the foremost latent concepts in the data.

- Calculation of Semantic Similarity:

LSA can subsequently gauge similarity between terms or documents. Often, cosine similarity between rows (terms) of the reduced U matrix or columns (documents) of the pared-down V^T matrix is employed to measure this similarity. Higher cosine values signify heightened semantic resemblance.

The formula for cosine similarity between two vectors x and y:

$$\text{Cosine Similarity } (x,\ y) = (x.y) / (||x|| * ||y||) \qquad (5)$$

Here, "." signifies the dot product, and "|| ||" signifies the Euclidean norm.

Latent Semantic Analysis proves particularly valuable in apprehending the intrinsic semantic structure of a corpus and addressing synonymy and polysemy, where words possess multiple meanings or different words reference the same concept.

4.3 Wordnet

NLP (Natural Language Processing) WordNet is a lexical database designed to aid in the understanding of word relationships and meanings within natural language texts. It is a valuable resource for various NLP tasks such as text analysis, machine translation, information retrieval, and more. The core concept of WordNet revolves around organizing words into synsets (sets of synonyms) and establishing semantic relationships between these synsets. This facilitates the exploration of word meanings and connections beyond simple dictionary definitions.

WordNet's Structure and Elements:

- Synsets:

The central building blocks of WordNet are synsets, which group together words that share similar meanings. A synset consists of a group of synonymous words (lemmas) that can be interchanged without altering the meaning of the sentence. For instance, the synset for "car" might include lemmas like "automobile," "vehicle," and "motorcar."

- Lemmas:

Lemmas are the base forms of words, stripped of inflections and tenses. They are the fundamental units within synsets.

5 Experimentation

The texts in both of the datasets were actually like comments and were scrambled with many different types of characters. It was challenging to convert them into similar understandable strings which can be used by future processes. So, before using it for classification we had cleaned the dataset. The series of steps followed were:

1. Expanding all the concentration words (For example, can't is converted to cannot).
2. Converting all the characters to lowercase.
3. Removing brackets and the text present in them.
4. Removing any types of URL or HTTP/HTTPS link from the text.
5. Removing XML/HTML tags.
6. Removing the "@" (mention) and "#" (Hashtag) symbols from the text which are most commonly used in social media comments.
7. Getting all the alphabetic and whitespace characters only.
8. Removing punctuations in the text.

9. Make them in the same line by removing any new line characters.
10. Also, remove the words having words in them.
11. Converting all of the words to their closest ASCII value using Unicode.
12. Tokenizing the text sentences to remove the stop words from the text (using the NLP stopword library).
13. Combining the result array of words into a string for further processing.

Furthermore, we can't use strings directly to make the classification model. We have used TF-IDF Vectorization and Latent Semantic Analysis. Firstly, the TF-IDF vectorization considers the importance of words in documents relative to their frequency across the entire dataset. By which we will get the most 5000 frequent words for both the datasets. Following TF-IDF vectorization, the code employs Latent Semantic Analysis (LSA) to reduce the dimensionality of the data. LSA is a technique that captures the underlying semantic structure in a document-term matrix. This is useful for extracting latent patterns and relationships within the text data. The LSA involves applying Truncated Singular Value Decomposition (SVD), which reduces the dimensions of the TF-IDF matrix while preserving its important characteristics. This is followed by the normalization to ensure the normality of the numeric values.

For classification, we are using three different machine learning models namely, XG Boost, Support Vector Classification, and Random Forest. Moreover, the classification models' best parameters were established by using various combinations of parameters. Furthermore, they were used to make hybrid models which can outperform the classification. The hybrid models made were Support Vector Classification + XG Boost, Random Forest + XG Boost, and Support Vector Classification + Random Forest. These hybrid models were found to have higher accuracy than isolated models for both datasets.

After successful classification as inappropriate, each word of the text is checked as inappropriate in the bad word list [20], and then the subsequent part of the code defines a function that takes each inappropriate word of text as input and aims to generate synonyms for that word using WordNet. The function iterates through the synsets (sets of synonymous words) associated with the input word and retrieves the lemmas (synonymous word forms) for each synset. These lemmas are then appended to a list of synonyms, excluding the words present in the bad word or the input word itself. The function ultimately returns the list of synonyms generated for the given word.

The pseudocode for the complete process goes as shown below:

```
for sample in dataset:
       sample class = 1 if sample['toxic']
for text in dataset:
    text <- clean(text)
Split the dataset (test size = 0.2)
dataset text vectorization <- TF-IDF vectorization (max features = 5000)
dataset text dimension reduction <- Latent Semantic Analysis (component = 300)
dataset text normalized <- Normalize
Support Vector Classification <- SVC ('kernel', C=11)
Random Forest <- RF ('entropy', component = 200, max depth = 3)
XG Boost <- XGB (learning rate = 0.1, component = 500, max depth = 3)
Stacking Classifier:
    Support Vector Classification + XG Boost
             Random Forest + XG Boost
          Support Vector Classification + Random Forest
predict <- Best classification model
Input: input text
input text <- clean (input text)
input text vectorization <- TF_IDF vectorizer transform
input text dimension reduction <- Latent Semantic Analysis transform
prediction <- model.predict(input text dimension reduction)
if inappropriate:
    import bad words list
    for word in input text:
        if word is bad word:
            synsets <- wordnet.synsets(word)
            lemmas <- synsets lemmas
            synonym word <- non-bad word in lemmas
        if length synonym == 0:
            input text <- remove(word)
        else:
            input text <- replace (word, synonym word [-1])
```

6 Result Analysis

Accuracy measured by all of the models on both datasets is shown in Fig. 1.

The F1 score, R2 score, Precision, and Recall measures for each dataset by all the models are given in Table 1.

It was very challenging to process these collected datasets and convert them into a constant form that can be used for processing and classification. Moreover, the functions and methods used to clean the data have gone through many trials, as the format of data in both datasets was not similar to a larger extent, due to which it became challenging to make a combined cleaning method for both of them. The combined method was needed because it helps us to not restrict the scrambled future inputs format and makes the program handle extreme cases.

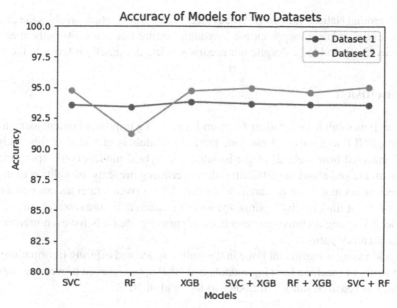

Fig. 1. Accuracy of the models on both datasets.

Table 1. F1 score, R2 score, Precision, and Recall of models.

Dataset 1	Model	F1 score	R2 score	Precision	Recall
	XG Boost	0.9678	−0.0031	0.9453	0.9914
	Support Vector Classification	0.9665	−0.0505	0.9371	0.9979
	Random Forest	0.9660	−0.0703	0.9342	1.0
	SVC + XGB	0.9671	−0.0268	0.9421	0.9935
	RF + XGB	0.9665	−0.0426	0.9429	0.9914
	SVC + RF	0.9666	−0.0466	0.9371	0.9981
Dataset 2	XG Boost	0.6638	0.4048	0.8880	0.5299
	Support Vector Classification	0.6590	0.4067	0.9082	0.5171
	Random Forest	0.1973	0.0118	0.9922	0.1095
	SVC + XGB	0.6926	0.4304	0.8622	0.5787
	RF + XGB	0.6610	0.3896	0.8601	0.5368
	SVC + RF	0.6922	0.4361	0.8766	0.5719

The utilization of hybrid models has resulted in a notable improvement in accuracy compared to the approaches presented in references [5, 7], and [9]. The aforementioned references employed singular methodologies, relying either on deep learning models or traditional classification-based machine learning models. In contrast, our approach not only combines the strengths of these paradigms but also introduces an innovative feature centered around Natural Language Processing (NLP). This feature involves the capacity to proficiently rephrase inappropriate language, setting our work distinctly apart from references [3, 4], and [13], despite our accuracy being marginally inferior to theirs.

7 Conclusion

A thorough methodology for identifying and rephrasing improper content using a combination of NLP methods and machine learning models is provided in this study. We looked analyzed how well all of the isolated and hybrid models could spot offensive material and talked about the difficulties that overfitting presents. In addition to increasing detection accuracy, our research shows how NLP-driven strategies can enhance the quality of online dialogue by rephrasing incorrect material. Future research might investigate deep learning architectures and hone rephrasing methods for even greater word sets and accuracy gains.

By addressing a significant issue in the online space and offering information about the capacities of machine learning models and NLP approaches in preventing improper material, this research study contributes to the field of NLP.

References

1. Yenala, H., Jhanwar, A., Chinnakotla, M.K., Goyal, J.: Deep learning for detecting inappropriate content in text. Inter. J. Data Sci. Anal. **6**, 273–286 (2018)
2. Xu, Z., Zhu, S.: Filtering offensive language in online communities using grammatical relations. In: Proceedings of the Seventh Annual CEAS 2010 (2010)
3. Parnell, A.C., González-Castro, V., Alaiz-Rodríguez, R., et al.: Machine Learning techniques for the detection of inappropriate erotic content in text. Inter. J. Comput. Intell. Syst. **13**(1), 591 (2020) ISSN 1875–6883
4. Yousaf, K., Nawaz, T.: A deep learning-based approach for inappropriate content detection and classification of youtube videos. IEEE Access **10**, 16283–16298 (2022). https://doi.org/10.1109/ACCESS.2022.3147519
5. Wazir, A.S.B., Karim, H.A., Lyn, H.S., Ahmad Fauzi, M.F., Mansor, S., Lye, M.H.: Deep learning-based detection of inappropriate speech content for film censorship. IEEE Access **10**, 101697–101715 (2022). doi: https://doi.org/10.1109/ACCESS.2022.3208921
6. Golem, V., Karan, M., Šnajder, J.: Combining shallow and deep learning for aggressive text detection. In: Proceedings of the First Workshop on Trolling, Aggression and Cyberbullying (TRAC-2018), pp. 188–198 (August 2018)
7. Papadamou, K., et al.: Disturbed youtube for kids: characterizing and detecting inappropriate videos targeting young children. In: Proceedings of the International AAAI Conference on Web and Social Media, vol. 14(1), pp. 522–533 (2020). https://doi.org/10.1609/icwsm.v14i1.7320
8. Endang, W.P., Patti, V.: Cross-domain and cross-lingual abusive language detection: a hybrid approach with deep learning and a multilingual lexicon. In: Proceedings of the 57th Annual Meeting of the Association for Computational Linguistics: Student Research Workshop (2019)
9. Shah, F., Anwar, A., ul haq, I., AlSalman, H., Hussain, S., Al-Hadhrami, S.: Artificial Intelligence as a Service for Immoral Content Detection and Eradication (2022)

10. Chen, H., McKeever, S., Delany, S.J.: The use of deep learning distributed representations in the identification of abusive text. In: Proceedings of the International AAAI Conference on Web and Social Media, vol. 13(01), pp. 125–133 (2019). https://doi.org/10.1609/icwsm.v13 i01.3215
11. Kaur, S., Singh, S., Kaushal, S.: Abusive content detection in online userGenerated data: a survey, Procedia Comput. Sci. **189**, 274- 281 (2021). ISSN 1877–0509,
12. Lee, Y., Yoon, S., Jung, K.: Comparative studies of detecting abusive language on twitter. arXiv preprint arXiv:1808.10245 (2018)
13. Kompally, P., Sethuraman, S.C., Walczak, S., Johnson, S., Cruz, M.V.: Malang: a decentralized deep learning approach for detecting abusive textual content. Appl. Sci. **11**(18), 8701 (2021)
14. Pitsilis, G.K., Ramampiaro, H., Langseth, H.:Detecting offensive language in tweets using deep learning. arXiv preprint arXiv:1801.04433 (2018)
15. Chen, H., McKeever, S., Delany, S.J.: Abusive text detection using neural networks. In: AICS (2017)
16. Urrutia Zubikarai, A.: Appled NLP and ML for the detection of inappropiarte text in a communications platform. MS thesis. Universitat Politècnica de Catalunya (2020)
17. Tripathy, B.K.: Audio to Indian sign language interpreter (AISLI) using machine translation and NLP techniques. In: Hybrid Computational Intelligent Systems. pp. 189–200. CRC Press (2023)
18. Cjadams, J.S., Elliott, J., Dixon, L., Mark McDonald, N., et al.: Toxic Comment Classification Challenge. Kaggle (2017). https://kaggle.com/competitions/jigsaw-toxic-comment-classific ation-challenge
19. Samoshyn, A.: Hate Speech and Offensive Language Dataset. Kaggle (2020). https://www. kaggle.com/datasets/mrmorj/hate-speech-and-offensive-language-dataset
20. Nicapotato Bad Bad Words. Kaggle (2017). https://www.kagglc.com/datasets/nicapotato/ bad-bad-words

Predicting Suicide Ideation from Social Media Text Using CNN-BiLSTM

Christianah T. Oyewale[1]([✉])[iD], Joseph D. Akinyemi[2][iD], Ayodeji O.J Ibitoye[3][iD],
and Olufade F.W Onifade[4][iD]

[1] Mohammed VI Polytechnic University, Rabat, Morocco
oyewalechristianahtitilope@gmail.com, christianah.oyewale@um6p.ma
[2] University of York, YO10 5GH York, UK
joseph.akinyemi@york.ac.uk
[3] University of Greenwich, London, UK
a.o.ibitoye@greenwich.ac.uk
[4] University of Ibadan, Ibadan, Nigeria
ofw.onifade@ui.edu.ng
https://um6p.ma, https://www.york.ac.uk/, https://www.gre.ac.uk/,
https://ui.edu.ng/

Abstract. Predicting suicide ideation is crucial for mental health assessment, especially as clinical methods have been unsuccessful due to victims' reluctance to seek help. Also, due to the time-consuming nature of clinicians having to review patient case notes individually, there is a risk that victims might commit suicide before being detected. Deep learning models such as the Convolutional Neural Network (CNN) and Long Short-Term Memory (LSTM) models have shown promise in improving suicide risk assessment. A key challenge, however, is the need to find the right combination of word embeddings for vectorizing texts for these Deep-learning methods. This work uses a deep learning network composed of CNN and Bidirectional LSTM layers with two different word embedding techniques, Word2Vec and FastText. Using Word2Vec as the baseline word embedding. Experiments on a Reddit dataset of 232,074 posts gave test set F1-scores of 94% using FastText and 90% using Word2Vec. It was observed that FastText could give better performance with less overfitting than Word2Vec.

Keywords: Suicide ideation · Machine learning · FastText embedding · Mental health · CNN · LSTM · Overfitting

1 Introduction

Suicide is harm carried out on oneself to end one's life, known as a suicide attempt [14]. Suicide which can be categorized into individual, social, and situational factors, is the twelfth cause of death in the United States and takes its ranking depending on age group [6,21]. Individual factors are those that result from psychological or demographic issues. The social factors relate to the victim's lifestyle through media use, societal beliefs, stigma, and barriers. The last category, situational factors, is caused by unexpected circumstances in which the victim finds himself or herself [21]. The prevention of suicide requires different levels of measurement to combat it [21]. One way to prevent this suicide is to identify it when the victim is in the process of finding or thinking of a solution to harm himself/herself. This task is known as suicide ideation prediction (SIP), which is crucial in public health to combat suicide attempts due to the negative effect this brings to the victim, the public, and the fact that it claims about 8,000 people's lives yearly [17].

SIP tasks can be complex to identify, especially when the victims refuse to visit the clinician, who started researching suicide ideation identification. The refusal can be due to the stigma associated with it [2]. SIP can't be carried out by taking any of the victim's blood or samples. Additionally, it is time-consuming because healthcare professionals make predictions based on the victim's responses during consultation and/or displaying the behavioral traits [7]. The victim's abstinence from the clinic could also mean they can leverage easy access to the internet to express their emotions. This gave rise to using Machine Learning models to identify patterns that could be used for SIP.

This research is dedicated to optimizing SIP by exploring the right combination of word embeddings, classifiers, and datasets. In addition to traditional datasets, the study employs Word2Vec [20] and FastText [3] word embeddings, coupled with a CNN-BiLSTM classifier, to enhance prediction accuracy. The research's significance lies in its contribution to improving the accuracy of SIP models and their ability to detect subtle linguistic cues indicative of suicidal thoughts. By effectively identifying linguistic patterns associated with suicidal ideation, these models offer a valuable resource for mental health professionals and support systems. This potential for early intervention aligns with the overarching goal of reducing suicide rates and providing timely assistance to individuals in distress.

The remaining parts of this research work are as follows: Section Two discusses the related works to this research work. The methodology employed in this research work was discussed in Section Three. Section Four discussed and analyzed the results obtained after the methodology implementation. Finally, Section Five concludes with summaries and recommendations.

2   Related Works

In mental health research, a collective of studies has embarked on a profound journey to unravel the enigmatic threads of suicide ideation, weaving a comprehensive tapestry that sheds light on the intricate landscape of human emotions and behaviors [1,11]. Similarly, studies have emphasized the need for early suicide detection [10,13]. To detect suicide ideation, the clinicians invested a significant amount of time in this process. In addition to the efforts of the clinicians and the availability of big data on the internet, other techniques like natural language processing [4,16], thematic analysis [5], machine learning, and deep learning [18] which rely more on the use of social media [22] were discovered. These techniques help in the identification and also factors that contribute to suicide ideation in the victim [15,19].

The three essential factors to consider in parts of the mentioned technique(Machine Learning and Deep Learning) are dataset, word embedding, and classifiers. The model selection can, at times be influenced by the size of the dataset [5,8,10]. Accessing the internet quickly made vast datasets exist; this strengthens deep learning research [2,12], which requires a more significant amount of data than traditional machine learning models. Another factor, as mentioned, is word embedding like Word2Vec, TF-IDF, RoBERTa, XLM-R, SpanBERT, mBERT, Bi-directional Long Short-Term Memory (LSTM), Convolutional Neural Networks (CNN), Transformer Model which helps in capturing subtle linguistic nuances in datasets [2,12]. The selection of the third factor, Machine Learning, is very important and must be accurately combined with dataset and, most significantly, word embedding.

It is very important to be skillful in the combination of these three factors to improve the ability of classifiers to quickly identify suicide ideation in social media text and avoid unexpected situations like overfitting of the model [2]

3   Methodology

The methodology involves data preprocessing, feature selection, dataset splitting, model training and fine-tuning, and model prediction and evaluation (Fig 1). The preprocessing of the Reddit dataset undergoes several preprocessing techniques, including accent removal, contraction expansion, case normalization, special character removal, and word lengthening correction. Stopwords are handled, ensuring uniformity across datasets. The datasets are split for model training and evaluation, maintaining consistency. Word embeddings, such as Word2Vec and FastText, enrich textual representations for robust feature extraction. A neural network architecture, including CNNs and BiLSTM layers, captures patterns. Fine-tuning with a validation set optimizes hyperparameters. The model was tested on a separate set, and predictions were compared to ground truth labels, generating a classification report and confusion matrix. Evaluation is done separately for each dataset, covering clinical and social contexts.

3.1   Dataset

The Reddit suicide detection dataset, obtained from Aldhyani et al. [2], contains 232,074 posts from the "SuicideWatch" subreddit, with a balanced distribution of suicidal and non-suicidal posts (Fig. 2). Statistically, the suicidal post contains more negative words and data-points, and more uniform distribution while non-suicidal post consists of more neutral and positive words and data-points, and skewed to the right. The dataset was divided into 70% for training, 10% for validation, and 20% for testing. Class distribution analysis reveals a balanced

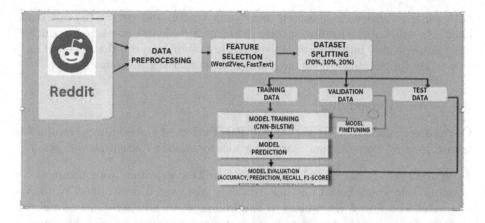

Fig. 1. Generic Methodology

dataset. The text data preprocessing steps include lowercase transformation, special character removal, and stop word removal. Text length analysis was conducted to understand message length. Word frequency analysis was performed for suicidal and non-suicidal texts, creating a vocabulary of unique words and their frequencies. Word clouds were generated to visualize frequently occurring words in both datasets, highlighting prominent terms based on frequency. These steps laid the groundwork for feature engineering and provided insights into the dataset's content and characteristics.

	Unnamed: 0	text	class
0	2	Ex Wife Threatening SuicideRecently I left my ...	suicide
1	3	Am I weird I don't get affected by compliments...	non-suicide
2	4	Finally 2020 is almost over... So I can never ...	non-suicide
3	8	i need helpjust help me im crying so hard	suicide
4	9	I'm so lostHello, my name is Adam (16) and I'v...	suicide

Fig. 2. Snapshot of the Dataset

3.2 Data Preprocessing

Data preprocessing in this research involved multiple steps to prepare text data for suicide ideation detection. The data cleaning included converting text to lowercase, removing special characters, eliminating stop words, lemmatization, tokenization, and calculating sentence lengths. A Tokenizer object was used to convert preprocessed text into numerical sequences, and we specified the maximum token to be 30,000 words. Sequences with varying lengths were padded with zeros to achieve a consistent maximum length, and we set the maximum length to 430. Label encoding transformed categorical class labels (e.g., "suicide" and "non-suicide") into numerical values. These steps ensured that the text data was ready for training a machine-learning model for text classification.

3.3   Word Embedding

Our research employed Word2Vec and FastText word embedding techniques to transform text into compact feature vectors for suicide content classification.

1. **Word2Vec**: Using neural networks, Word2Vec captures word meanings by predicting words based on their contexts. We used a pre-trained Word2Vec model with a vast vocabulary of two billion words and 300-dimensional vectors. This allowed us to encode syntactic and semantic relationships among words, making it suitable for various natural language processing tasks, including classifying suicidal and non-suicidal content.
2. **FastText**: FastText is similar to Word2Vec but excels at handling out-of-vocabulary (OOV) words and sub-word information. It breaks words into sub-words, enabling it to represent even rare or domain-specific words. Fast-Text combines words and sub-words in training, resulting in 300-dimensional vectors that capture finer-grained semantic and morphological information. Our FastText implementation used a vocabulary of about one million words, making it a valuable tool for tasks like text classification. These word embedding techniques provided us with the foundation to distinguish between suicidal and non-suicidal content effectively.

3.4   CNN-BiLSTM Model

Convolutional Neural Networks (CNNs) efficiently process text data by identifying local patterns. This lets them capture crucial features within sentences, even in lengthy texts. The model combines CNN with Bidirectional Long Short-Term Memory (BiLSTM) to address the challenge of capturing long-range dependencies in sequential data. While CNNs excel at identifying local patterns, BiLSTM processes text sequences both forward and backward, comprehensively understanding dependencies across various text parts. This combination capitalizes on the strengths of both architectures.

The CNN-BiLSTM architecture (Fig. 3) implemented comprises key components:

1. **Embedding Layer and Random Initialization**: Input text was encoded into a numeric format using pre-trained word embeddings like Word2Vec or FastText. The input layer, the dataset, was passed into the embedding layer, which converted the dataset into the numerical form.

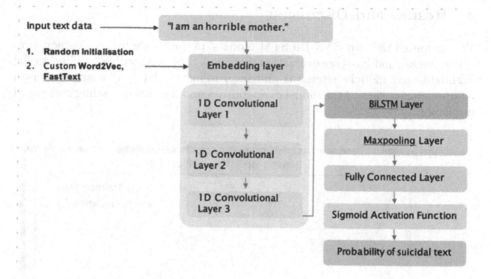

Fig. 3. CNN-BiLSTM Model

2. **Three Stacked One-Dimensional CNNs**: These layers detect specific patterns and features in text data, progressively learning hierarchical representations. The output of the embedding layer was passed into a One-dimensional convolutional layer (1D CNN). Since three 1D CNNs were employed, the output of the first 1D CNN was passed as the input of the second 1D CNN, and the output of the second was passed as the input of the third.

3. **Bidirectional Long Short-Term Memory (BiLSTM)**: This layer captures long-range dependencies and temporal context in the text. The output of the third 1D CNN was passed as the input of the BiLSTM layer.

4. **MaxPooling for Dimensionality Reduction**: Max-pooling extracted the most essential information from BiLSTM outputs. It reduced the nodes from the BiLSTM's output. The output was then passed into the fully connected layer.

5. **Sigmoid Activation for Classification**: The final layer shows the text's probability of expressing suicidal sentiments. This helped in classifying the text into suicide or non-suicide.

We tracked training, validation loss, and accuracy to gauge learning and potential overfitting. The test loss and accuracy were calculated to evaluate how well the model predicts unseen data. The model made predictions on the test dataset, enabling further analysis. Detailed metrics, including precision, recall, F1 score, and a visual confusion matrix, offered insights into the model's performance and errors.

4   Results and Discussion

We evaluated the two CNN-BiLSTM models' performance using accuracy, precision, recall, and F1-score evaluation metrics. Comparing with a baseline model highlights our model's strengths and improvements [2]. The chapter offers a comprehensive view of the model's capabilities and limitations, aiding real-world applicability.

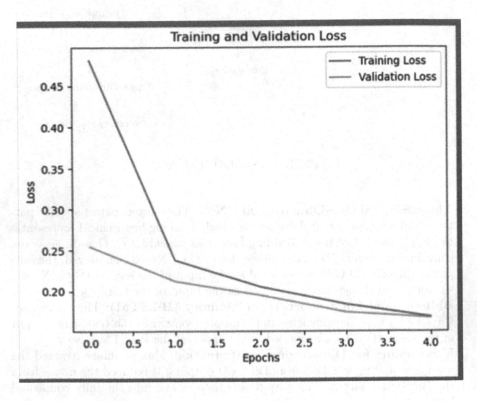

Fig. 4. CNN-BiLSTM with FastText Loss Curve

4.1   CNN-BiLSTM Model with FastText Word Embedding

The first model that was developed is CNN-BiLSTM using FastText as word embeddings. The initial model exhibited overfitting issues with an F1-score of 93%, indicating a need for improvements. Varying embedding dimensions and introducing dropout layers did not substantially enhance performance. Incorporating a 1-dimensional (1D) Convolutional layer and max-pooling showed minimal improvement. Reducing the training epoch to 3 showed a trend towards improvement, although the F1-score dropped slightly to 91%. Initial attempts with early stopping and regularization resulted in disappointing outcomes. Refining regularization techniques by removing L1 regularization stabilized training but caused some oscillations. Adding an extra BiLSTM layer increased complexity and led to oscillations in the accuracy curve, with a 1% accuracy boost. Aligning with insights from prior work led to a significant F1-score increase of 93%. By closely following the literature's model architecture, we replicated their reported 95% F1 score [2]. Experimenting with stacked 1D CNNs and a single Bi-LSTM layer resulted in a 92% F1 score. Fine-tuning the learning rate and eliminating early stopping eventually led to convergence and an impressive F1 score of 94% (Fig. 4).

```
Model: "sequential"

 Layer (type)                  Output Shape              Param #
=================================================================
 embedding (Embedding)         (None, 430, 300)          9000000

 conv1d (Conv1D)               (None, 426, 128)          192128

 conv1d_1 (Conv1D)             (None, 422, 64)           41024

 conv1d_2 (Conv1D)             (None, 418, 32)           10272

 bidirectional (Bidirectiona   (None, 418, 256)          164864
 l)

 global_max_pooling1d (Globa   (None, 256)               0
 lMaxPooling1D)

 dense (Dense)                 (None, 64)                16448

 dropout (Dropout)             (None, 64)                0

 dense_1 (Dense)               (None, 1)                 65

=================================================================
Total params: 9,424,801
Trainable params: 424,801
Non-trainable params: 9,000,000
```

Fig. 5. CNN-BiLSTM with FastText model summary

Our final model consists of an embedding layer, three stacked 1D CNN layers, a BiLSTM layer, max-pooling, and a fully connected layer, with sigmoid activation (Fig. 5). This optimization journey highlights the importance of architectural adjustments, regularization, and parameter tuning. Ultimately, we achieved a remarkable F1-score of 94%, affirming the model's effectiveness in identifying suicidal content.

4.2   CNN-BiLSTM Model with Word2Vec Word Embedding

The second model developed is CNN-BiLSTM using Word2Vec as word embeddings highlighted the iterative journey of fine-tuning and optimization to achieve optimal performance. We initially applied the FastText fine-tuned architecture to Word2Vec, but overfitting became evident early on, indicating the need for adjustments. The attempt to address overfitting by reducing or increasing epochs did not lead to ideal convergence, posing a challenge. Lowering the learning rate improved convergence tendencies, but simplifying the architecture led to unusual validation curve behavior. Reverting to an appropriate learning rate yielded more promising convergence behavior (Fig. 6).

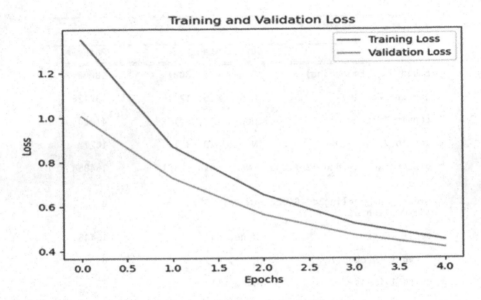

Fig. 6. CNN-BiLSTM with Word2Vec Loss Curve

Our evaluation of the CNN-BiLSTM model with Word2Vec embeddings emphasized the importance of model adaptability and the intricate relationship between architectural configurations and hyperparameters. Through iterative adjustments, we navigated overfitting and convergence challenges, ultimately reaching a model configuration with promising performance and convergence tendencies. This process underscores the iterative nature of deep learning model development, highlighting the need for thorough experimentation and optimization.

4.3   Model Performance and Evaluation Metrics

The model performance analysis revealed differences in accuracy, precision, recall, and F1 scores among Word2Vec and FastText. The FastText model exhibited balanced precision, recall, and F1 scores, making it consistent across metrics.

The Word2Vec model, while competitive in accuracy, lags in precision, recall, and F1 scores. The choice of embedding technique impacts the model's performance, with each model having strengths in different aspects. FastText offers a balanced performance, and Word2Vec requires further refinement to achieve balance (Table 1).

Table 1. Model Evaluation Result.

Model	Accuracy	Precision	Recall	F1 Score
CNN-BiLSTM (Word2Vec)	0.9356	0.9045	0.9031	0.9030
CNN-BiLSTM (FastText)	0.9356	0.9356	0.9356	0.9356

In the confusion matrix analysis: the Word2Vec model performed well in true negatives and true positives but needs improvement in minimizing false negatives and false positives (Fig. 8). The FastText model demonstrates strong performance in true negatives and true positives, with fewer false negatives and false positives (Fig. 9). FastText showcases robust and balanced performance in the confusion matrix, making it reliable, while Word2Vec requires optimization for better balance.

Choosing the most suitable model ultimately depends on task priorities, whether balanced performance or a specific focus on precision or recall.

4.4 Comparison with an Existing Model

The existing model achieved an impressive F1 score of 95% but exhibited overfitting, indicated by a rising training accuracy, while the validation accuracy plateaued or declined [2]. The second experiment carried out by the same researchers showed an F1 score of 87%, and the curve converged, which showed that there was no overfitting. This formed the baseline for our comparison. Our model showed convergence in the accuracy curve, reaching a peak F1 score of 94%. Our model outperformed the existing model [2]. Unlike these researchers who used SoftMax for binary classification, we used the sigmoid activation function, more suitable for binary classification, and fine-tuned hyperparameters.

Addressing overfitting is crucial, as the existing model's accuracy curve highlights. With a higher minimum F1 score, our model demonstrated efforts to optimize and create a robust model for identifying suicidal content in text data (Fig. 7).

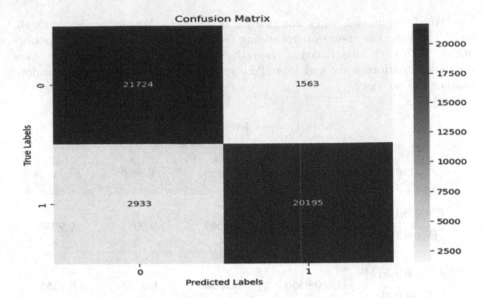

Fig. 7. Word2Vec Confusion Matrix

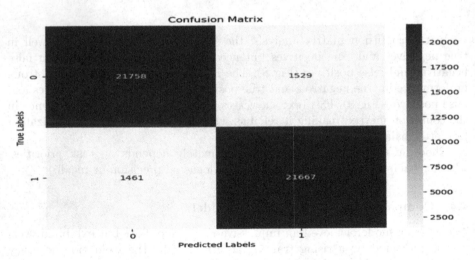

Fig. 8. FastText Confusion Matrix

5   Conclusion

Suicide is a pressing public health concern, necessitating continuous efforts from medical professionals and beyond. Word embeddings are pivotal in text classification, though they confront challenges when identifying out-of-vocabulary words. Our journey culminated in developing an exceptionally effective text classification model, boasting an impressive F1 score of 94%. Through comparative analysis, we have underscored its superiority over existing models. Our future endeavors will involve enhancing data cleaning and potentially employing advanced techniques such as data augmentation. Furthermore, we will explore more versatile word embeddings and address memory constraints associated with transformer models like BERT. This research underscores the significance of creating precise suicide detection systems and encourages collaboration between mental health organizations and technology companies. The ethical considerations and user privacy protection must remain central in this technology's responsible and effective implementation.

References

1. Ahn, S.Y., Yu, S., Kim, J.E., Song, I.H.: The relationship between suicide bereavement and suicide ideation: analysis of the mediating effect of complicated grief. J. Affect. Disord. **331**, 43–49 (2023). https://doi.org/10.1016/J.JAD.2023.03.008

2. Aldhyani, T.H.H., Alsubari, S.N., Alshebami, A.S., Alkahtani, H., Ahmed, Z.A.T.: Detecting and analyzing suicidal ideation on social media using deep learning and machine learning models. Int. J. Environ. Res. Public Health **19**(19), 1–16 (2022). https://doi.org/10.3390/ijerph191912635

3. Bojanowski, P., Edouard, G., Armand, J., Tomas, M.: Enriching word vectors with subword information. Trans. Associat. Comput. Ling. **5**, 135–146 (2017). https://doi.org/10.1162/tacl_a_00051

4. Cusick, M., et al.: Portability of natural language processing methods to detect suicidality from clinical text in US and UK electronic health records. J. Affect. Disorders Reports **10**, 100430 (2022). https://doi.org/10.1016/j.jadr.2022.100430

5. Davidson, J.E., et al.: Job-related problems prior to nurse suicide, 2003–2017: a mixed methods analysis using natural language processing and thematic analysis. J. Nurs. Regul. **12**(1), 28–39 (2021). https://doi.org/10.1016/S2155-8256(21)00017-X

6. Garnett, M.F., Curtin, S.C., Stone, D.M.D.: Suicide Mortality in the United States, 2000–2020 Key findings Data from the National Vital Statistics System, Mortality (2022). https://www.cdc.gov/nchs/products/index.htm

7. Huang, Y., et al.: Comparison of three machine learning models to predict suicidal ideation and depression among Chinese adolescents: A cross-sectional study. J. Affect. Disord. **319**, 221–228 (2022). https://doi.org/10.1016/J.JAD.2022.08.123

8. Ji, S., Pan, S., Li, X., Cambria, E., Long, G., Huang, Z.: Suicidal ideation detection: a review of machine learning methods and applications. IEEE Trans. Comput. Soc. Syst. **8**(1), 214–226 (2021). https://doi.org/10.1109/TCSS.2020.3021467

9. Jones, N.C.: Prediction and Analysis of Degree of Suicidal Ideation in Online Content (2020)

10. Kang, C., et al.: Prevalence, risk factors and clinical correlates of suicidal ideation in adolescent patients with depression in a large sample of Chinese. J. Affect. Disord. **290**, 272–278 (2021). https://doi.org/10.1016/j.jad.2021.04.073

11. Lauderdale, S.A., Martin, K.J., Oakes, K.R., Moore, J.M., Balotti, R.J.: Pragmatic screening of anxiety, depression, suicidal ideation, and substance misuse in older adults. Cogn. Behav. Pract. **29**(1), 105–127 (2022). https://doi.org/10.1016/J.CBPRA.2021.06.003

12. Malviya, K., Roy, B., Saritha, S.: A transformers approach to detect depression in social media. In: 2021 International Conference on Artificial Intelligence and Smart Systems (ICAIS), pp. 718–723 (2021). https://doi.org/10.1109/ICAIS50930.2021. 9395943

13. Miller, N.E., North, F., Duval, M., Tieben, J., Pecina, J.L.: Comparison of screening for suicidal ideation in the year prior to death by suicide. J. Affect. Disorders Reports **10**, 100446 (2022). https://doi.org/10.1016/j.jadr.2022.100446

14. NIH. Frequently Asked Questions About Suicide (2023)

15. Ni, Y., Barzman, D., Bachtel, A., Griffey, M., Osborn, A., Sorter, M.: Finding warning markers: Leveraging natural language processing and machine learning technologies to detect risk of school violence. Int. J. Med. Informat. **139**, 104137 (2020). https://doi.org/10.1016/j.ijmedinf.2020.104137

16. Palmon, N., Momen, S., Leavy, M., Curhan, G., Boussios, C., Gliklich, R.: PMH52 use of a natural language processing-based approach to extract suicide ideation and behavior from clinical notes to support depression research. Value Health **24**, S137 (2021). https://doi.org/10.1016/J.JVAL.2021.04.674

17. Sinha, P.P., Mahata, D., Mishra, R., Shah, R. R., Sawhney, R., Liu, H.: Suicidal - a multipronged approach to identify and explore suicidal ideation in twitter. In: International Conference on Information and Knowledge Management, Proceedings, pp. 941–950 (2019). https://doi.org/10.1145/3357384.3358060

18. Tadesse, M.M., Lin, H., Xu, B., Yang, L.: Detection of suicide ideation in social media forums using deep learning. Algorithms **13**(1), 7 (2019). https://doi.org/10. 3390/a13010007

19. Tull, M.T., et al.: The roles of borderline personality disorder symptoms and dispositional capability for suicide in suicidal ideation and suicide attempts: Examination of the COMT Val158Met polymorphism. Psychiatry Res. **302**, 114011 (2021). https://doi.org/10.1016/j.psychres.2021.114011

20. Tomas, M., Kai, C., Greg, C., Jeffrey, D.: Efficient Estimation of Word Representations in Vector Space. arXiv preprint arXiv:1301.3781

21. WHO., Saxena, S., Saxena, S., Krug, E.G., Krug, E.G., Chestnov, O., Chestnov, O.: World Health Organization. Department of Mental Health and Substance Abuse. Preventing Suicide: a Global Imperative. World Health Organization (2014)

22. Zhang, D., Wang, R., Tian, Y., Qi, C., Zhao, F., Su, Y.: Exploring life and help-seeking experiences regarding suicidal ideations among nursing home residents. Geriatr. Nurs. **45**, 69–76 (2022). https://doi.org/10.1016/j.gerinurse.2022.03.004

Enhanced Multi-step Breast Cancer Prediction Through Integrated Dimensionality Reduction and Support Vector Classification

Ritika Wason[✉] [iD], Parul Arora, M. N. Hoda, Navneet Kaur, Bhawana, and Shweta

Bharati Vidyapeeth's Institute of Computer Applications and Management (BVICAM), New Delhi, India
ritika.wason@bvicam.in

Abstract. Cancer is a dreaded disease which can affect any body part. Breast cancer is a significant health alarm worldwide, especially distressing millions of women each year. It is a notable pathological condition with high mortality rates especially in the developing nations. Traditional machine learning models have been working for breast cancer prediction, but they face challenges when dealing with high-dimensional and noisy data. Notably, early detection can lead to timely detection and eradication of the associated mortality risk. This approach can help battle the associated concern and stigma with breast cancer in a big way. This manuscript proposes an enhanced multi-step approach for breast cancer prediction by integrating dimensionality reduction with the SVM classifier. This integrated approach not only improves the prediction accuracy but also enhances interpretability by identifying the most relevant features associated with breast cancer. When applied to the WDBC dataset the approach gave 96.50% accuracy.

Keywords: Breast cancer · Dimensionality Reduction · Logistic Regression · Diagnosis · Supervised Classification

1 Introduction

Breast cancer is the recorded most diagnosed cancer types among women globally [1]. In India it is one of the most prominent cancer types especially among women [1]. It should further be noted that it is associated with high mortality rate [1]. The burden of the same is expected to increase each year at least till 2025 [2]. According to the World Health Organization (WHO) recent Global Breast Cancer Initiative Framework (GBCIF), 2023 a roadmap should be enabled to save at least 2.5 million people from breast cancer by 2040 [2]. This is keeping in mind that over 2.3 million cases of breast cancer are detected annually, making it the utmost diagnosed cancer among adults [2]. Survival from this cancer is again inequitable within nations, with over 80% of deaths due to breast cancer being recorded in low- and middle-income countries [2, 3].

Extensive research has shown that early diagnosis and prompt treatment significantly increase a patient's chance of survival [4]. WHO also recommends nations to concentrate on early breast cancer detection programs so that at least 60% of the breast cancers can be diagnosed and treated at early-stage [2]. Treatment if started within 3 months of initial appearance will increase chances of survival [3]. Worldwide, the disease now demands early prediction and management of breast cancer before it metastasizes to significantly reduce its resulting morbidity [5]. Breast cancer usually proliferates in the ductal region and to some extent in the lobules of the breast [5]. It is demands effective predictive models to aid in early detection and treatment planning [4].

Mammography has been the primary reference technique for breast cancer screening [5]. The breast consists of lobules, connective tissue, and ducts, with breast cancer naturally originating in either the ducts or the lobules [1]. Indicators of breast cancer include the presence of a breast lump or enlargement, alterations in size, shape, dimpling, redness, pitting, alterations in the presence of the nipple, and aberrant nipple discharge [7].

Tumors caused by breast cancer can be classified as either benign or malignant [8]. For breast cancer detection and prediction, numerous machine learning (ML) algorithms, ensemble ML (EML) algorithms, and deep learning techniques are available [9]. These algorithms utilize the effectiveness of each classifier to determine the optimal outcome. However, they often struggle with the curse of dimensionality and fail to give best results when working on high-dimensional data sets containing irrelevant or redundant features [10]. This manuscript is an effort to augment the accuracy of breast cancer prediction through a multi-step framework. First, we decrease the number of features while conserving the essential information, to prevent overfitting and enhance the result interpretability [11]. The experimental details and evidence of the work has been reported here. It can serve as an important landmark in improving the early breast cancer prediction efforts using existing, available machine learning classifiers.

The rest of this manuscript is structured as follows: Sect. 2 analyzes the current work in this domain. Section 3 elaborates the proposed methodology. Section 4 justifies the need of dimensionality reduction techniques. Section 5 details the machine learning classifiers used in breast cancer prediction. Section 6 discusses the results of the experiment. Section 7 concludes the findings of the study. Section 8 details the future enhancements possible.

2 Literature Review and Gap Analysis

Many studies and research related to cancer diagnosis and prediction have already been conducted in the past. Here we try to analyse a few considerable ones in Table 1 below:

Most of the research listed in Table 1 above have investigated the application of dimensionality reduction techniques for breast cancer prediction. Many of the previous works have focused on either the dimensionality reduction aspect without deeply exploring the impact on subsequent predictive models. None of the models first tried to find the best correlation between the selected parameters. However, we believe that if the best correlated parameters are chosen then breast cancer prediction can be accurately defined. Hence, the key goal of our learning was to find the best correlated dimensions that could accurately help in detecting breast cancer. We have detailed our methodology for the same in the next section.

Table 1. State of Art of Breast Cancer Research.

S.No	Ref	Achievements	Limitations
1.	[12]	PCA, and its variants used to decrease the dimensionality of the RNA sequencing data, tested on neural network and Support Vector Machines. **Reported Accuracy -96.4% (SVM)**	Dimensionality reduction could improve performance of machine learning models to detect prostate cancer. However, impact on similar datasets not evaluated
2.	[13]	Factors leading to increasing burden of breast cancer in India evaluated	Importance of early detection in breast cancer management highlighted
3.	[5]	The maximum accuracy achieved in the blood's database during the baseline models generation 93% with XGboosting regressor,94% in the EIT database	The SBRA project aims to make a full system (end-to-end) for early breast cancer detection, despite being limited by databases
4.	[1]	Presents a novel technique in which deep learning applied to categorize breast cancer	Tested with a single data set. The model also lacked interpretability and constraints
5.	[14]	To advance the accuracy of machine learning models from the present ones, PCA and LDA applied	Though high accuracy was obtained, 100% accuracy remains a future scope
6.	[15]	Varied association and classification approaches were evaluated to understand breast cancer forms	Proved data mining as a viable tool for breast cancer diagnosis in a single geographical region
7.	[16]	Explores the different selection strategies for creating ensembles of classification models from a population. The objective is to enhance the accuracy of decision support systems used for the early detection and diagnosis of breast cancer	The process of model selection involves the identification of the "single best" method, which is determined by linking the relative accuracy of a restricted set of models in a cross-validation study
8.	[17]	Diagnostic evaluation of breast cancer was performed through several approaches	Usage of PCA and backpropagation neural networks have resulted in maximum accuracy
9.	[18]	Ensemble methods employed a nested ensemble approach that incorporated the Stacking and Vote (Voting) techniques for combining classifiers	The existing models exhibit limitations due to their fixed loop structure, which restricts the potential for further refinement and enhancement of the algorithm's precision
10.	[19]	The proposed method for feature selection aims to recognize the most appropriate set of features for breast cancer classification	The efficacy of the planned method has been assessed using a restricted set of authentic datasets sourced from the UCI machine learning source

(continued)

Table 1. (*continued*)

S.No	Ref	Achievements	Limitations
11.	[20]	The classification process involves the introduction of a novel ensemble classifier that utilizes a Group Method of Data Handling (GMDH) neural network	The findings indicate that the proposed methodology exhibits an accuracy range of merely 7% to 26%
12.	[21]	Ensemble classifiers used for enhancing the predictive capabilities of classification models in order to attain superior accuracy levels	The proposed algorithm exhibits a 17% improvement in performance when compared to conventional algorithms
13.	[22]	The utilisation of clinical and imaging characteristics to aid in the prioritisation of care for individuals particularly those who are at an elevated risk of developing severe illness	Although clinical features alone did not yield as accurate results as imaging features, the grouping of clinical and imaging structures proved to be beneficial only for the LR model
14.	[23]	This study introduces a novel approach that combines principal component analysis (PCA) with the boosted C5.0 decision tree algorithm, incorporating a penalty factor	Improved diagnostic accuracy has been achieved
15.	[24]	Study assesses the efficacy of F-LDA as a method for reducing dimensionality in the classification of cancer data	The findings indicate that F-LDA exhibits notable efficacy in the context of cancer diagnosis when applied to small datasets. Application to large datasets not yet explored
16.	[25]	Principal Component Analysis (PCA) utilised as a method for dropping the dimensionality of the dataset prior to the application of algorithms for the purpose of diagnosing breast cancer	Although the results were encouraging, the application of these findings to real-time datasets has not yet been accomplished
17.	[26]	Matures two prognostic models for the prediction of survival in breast cancer patients. The aim is to determine whether the Hybrid Bayesian Network (HBN) model demonstrates superior performance compared to the logistics regression (LR) model	The data utilised for both internal and external validation were obtained from the Surveillance, Epidemiology, and End Results (SEER) database

(*continued*)

Table 1. (*continued*)

S.No	Ref	Achievements	Limitations
18.	[27]	The LR-PCA model, consisting of 8 components, demonstrates consistent and effective performance in achieving outcomes with an accuracy rate of 99.1%	This study offers a comprehensive evaluation of several feature selection and dimensionality reduction techniques, ensuring a fair and unbiased comparison
19.	[28]	Employs feature selection techniques allowing for the identification and selection of significant features, which can then be utilized in the subsequent classification process	The records containing incomplete data were excluded. Many variables excluded due to their unavailability. This omission may have potentially impacted the efficacy of the models
20.	[29]	The efficacy and precision of prominent supervised and semi-supervised machine learning algorithms undertaken	In the context of a limited dataset, it is plausible for semi-supervised learning to serve as a viable alternative to supervised learning algorithms
21.	[30]	The accuracy of classifying benign cases is high, while the accuracy of classifying malignant cases is reduced	Methods may exhibit high accuracy in classifying images, but this comes at the cost of increased computational complexity
22.	[31]	A regression model developed to predict clonal growth and cancer outcomes by incorporating manually traced histologic variables	Limitation of this model is its constrained diagnostic range, as it is only applicable to endometrial tissues within the normal-in-cancer spectrum
23.	[10]	Anticipates and diagnoses breast cancer, through machine-learning algorithms	The support vector machine (SVM) algorithm demonstrated superior performance
24.	[6]	Focuses on the utilisation of machine learning techniques for the detection of breast cancer	The proposed deep neural network with support vector integration has demonstrated promising outcomes. The quality of images can pose a significant challenge
25.	[32]	Evaluates the effectiveness of Multi-Layer Perceptron Neural Network (MLP) and Convolutional Neural Network (CNN), in the detection of breast malignancies	The proposed model yielded precise outcomes

3 Proposed Methodology

We aim to develop a predictive model that can predict the likelihood of an individual having breast cancer based on most correlated features. Figure 1 depicts the methodology followed by us. The same is explained below:

To detect breast cancer, it is significant to first have input data to train the system. The input dataset should first be pre-processed followed by feature extraction, feature selection and finally classification to detect cancerous tumors. Machine Learning techniques are basically employed to develop a classification model on a dataset that contains labelled classes. The steps carried out in our approach are summarized below:

1. **Data Collection**: The WDBC (Wisconsin Breast Cancer Dataset) dataset was used [33]. It consists of 10 features which further contain three features each namely mean, standard error and worst. Hence in total the dataset consists of 33 features (10*3 + 3) of 569 patients. Of these 357 are benign and 212 are malignant.
2. **Pre-processing**-Pre-processing is important to ensure that the input data is free of noise [34]. This involves handling missing values and scaling [1]. Data scaling may have a huge impact on model performance.
3. **Dimensionality Reduction and Feature Extraction**-PCA dimensionality reduction algorithm was applied to the pre-processed data. They helped reduce the dimensionality of the feature space while preserving the most relevant information.
4. **Model Construction**- Following feature extraction, we compared the logistic regression, decision tree, SVC and random forest models to find the best one.
5. **Model Evaluation**-The performance of the above models was evaluated using metrics like accuracy, precision, recall and F1-score. We compared the results of each to find the most accurate classifier model.

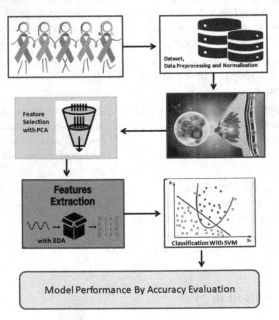

Fig. 1. Proposed Methodology.

4 Need of Dimensionality Reduction

Prediction quality shall always vary according to the number of variables or features considered [12]. One should note that it requires n x d memory to store data, where data can be represented as Eqn. 1 below:

$$\{x_i\}_{i=1}^n \text{with} x_i \in R^d \tag{1}$$

However, many real-time data have repeated patterns. Dimensionality reduction [28] is the mapping of the data to a lower dimensional space to discard the uninformative variance in the same [11].It helps detect the subspace in which the data lives. This is crucial to help us suggest better models for inference while overcoming the following:

a. **Curse of Dimensionality:** In medical datasets, including those used in breast cancer diagnosis, the number of features (dimensions) can be large, leading to the curse of dimensionality [13]. This phenomenon can make data analysis, visualization, and modeling challenging. Dimensionality reduction helps mitigate this problem by dropping the quantity of features while retaining the most relevant information [1].

b. **Feature Extraction**: Not all features are equally relevant for diagnosing breast cancer [28]. Some features may be redundant or noisy, and others may not contribute significantly to the diagnosis [14]. Dimensionality reduction helps recognize the most important features, improving the efficiency of the diagnostic process and reducing the risk of overfitting. There are two main categories of dimensionality reduction methods namely: Feature Selection and Extraction. Feature Extraction helps construct a new feature subspace with the goal of maintaining the most relevant information. It helps in improving the predictive performance as well as the computational efficiency.

c. **Visualization**: Reducing high-dimensional data to a lower number of dimensions also allows for effective visualization. Dimensionality reduction techniques like Principal Component Analysis (PCA) can project the data into a lower-dimensional space while preserving the structure and relationships between data points, making it easier for clinicians and researchers to explore the data [15].

d. **Computational Efficiency**: High dimensionality often leads to increased computational complexity [17]. Dimensionality reduction can significantly speed up the analysis process, making it more feasible to process large datasets and enabling more efficient machine learning algorithms for classification and prediction.

e. **Reducing Overfitting**: When dealing with high-dimensional data, there's a higher risk of overfitting the model to the noise in the data. By reducing the dimensionality, the model is less likely to memorize noise and is more likely to generalize well to new, unseen data.

f. **Improved Model Performance**: High-dimensional data can cause issues such as the curse of dimensionality and multicollinearity. These issues can negatively impact the performance of machine learning models. Dimensionality reduction can lead to simpler and more interpretable models, potentially improving diagnostic accuracy.

Dimensionality Reduction thus helps in transforming high dimensional data into low dimensional data while preserving data integrity. The criteria for reduction is removing the features that are less significant in outcome prediction [17]. Feature extraction methods like principal component analysis have proven to work well when there are several correlated features and dimensionality reduction is more crucial than interpretability. Dimensionality reduction also helps simplify the data format, increases model generalization, and makes visualization and exploratory analysis easier by decreasing the number of features.

To improve the accurateness of machine learning models, we have used the PCA followed by EDA to reduce the dimensionality of the features, which helps to increase the efficiency of models with faster rate.

4.1 Principal Component Analysis (PCA)

PCA or Principal Component Analysis or Orthogonal Linear Transformation decreases the dimension of input data by projecting it onto a smaller space through linear operations including the assessment of covariance matrix [12]. It can be computed as stated in Eqn. 2.

Given a dataset,

$$X \in R^{NXd} \tag{2}$$

where,

N= each of the rows that represent a unique data point
d= each of the d columns that gives a particular feature
The goal here is to find an orthogonal weight matrix, as given in Eqn. 3

$$W \in R^{dXd} \tag{3}$$

This should result in maximizing the variance and minimizing the reconstruction error as given in Eqn. 4 below.

$$Z = XW \tag{4}$$

where Z= score matrix, columns of which are the principal components of X.

Many varied algorithms can be used to maximize the variance. However, it should be noted that the maximum number of principal components cannot surpass the original dimension [35]. One notable point related to PCA is that it can only be applied to continuous data as it only relies on the linear relationship between each dimension [34]. To standardize the data Standard Scalar was used in our experiment.

4.2 EDA

Exploratory Data Analysis (EDA) is a critical step in any research analysis to examine the data for distribution outliers and anomalies [36]. It is an important step after data collection and pre-processing, where data can be graphically visualized. It is also critical to analyze the data for an experiment for detection of mistakes, assumption checking, preliminary selection of appropriate methods, determining relationship among explanatory variables etc. In this study we have used correlation along with scatterplot, heatmap etc.

for performing EDA. The general formula for sample correlation as well as covariance have been listed in Eqn. 5 and 6 below.

$$Cor(X, Y) = \frac{Cov(X, Y)}{s_x s_y} \tag{5}$$

and

$$Cov(X, Y) = \frac{\sum_{i=1}^{n}(x_i - \bar{x})(y_i - \bar{y})}{n - 1} \tag{6}$$

5 Machine Learning in Breast Cancer Diagnosis

The dimensionality reduction process when combined with the machine learning algorithms provides a mechanism to improve the overall performance of the predictive model. The goal is to create a more accurate and efficient model that can potentially lead to accurate, early prediction leading to early intervention.

5.1 Support Vector Machines (SVM)

Support Vector Machines (SVM) are a powerful machine learning algorithm commonly used for classification tasks, and they can indeed be applied to breast cancer prediction [25]. SVMs work by finding a hyperplane that best separates data points of different classes in a high-dimensional space [25]. In the context of breast cancer prediction, SVMs can be used to classify whether a given patient's data corresponds to benign or malignant tumors. To learn a classifier using SVM, we can mathematically represent the same as follows:

We have N training vectors $\{(x_i, y_i)\}$, where

$$x \in R^D, y \in \{-1, 1\} \tag{7}$$

A classifier can be learnt for a new x, such that:

$$f(x) = w^T \varnothing(x) + b \tag{8}$$

where f(x) is the classifier output

$\varnothing(x)$ is the feature space such that the training exemplars are well separated in the feature space into a 2D or 3D feature plane. However, to impose boundary (margin for error) on the above feature plane different mechanisms may be used. However, we do not discuss them further as that is not our goal.

It's significant to note that while SVMs can be effective, they are not the only approach for breast cancer prediction. Other machine learning algorithms, such as random forests, gradient boosting, deep learning (e.g., convolutional neural networks), and logistic regression, can also be used and might perform well depending on the precise characteristics of the dataset.

6 Results and Discussions

We have outlined our strategy in the sections above. We started with the WDBC dataset which had 569 rows and 33 columns. It should be noted however that there were only 10 nuclear features [7]. These features are listed in Table 2 below:

Table 2. 10 nuclear features with their standard deviation

S. No	Feature	Standard Deviation (std)
1.	Radius	3.524049
2.	Perimeter	24.298981
3.	Area	351.914129
4.	Compactness	0.052813
5.	Smoothness	0.014064
6.	Concavity	0.079720
7.	Concave points	0.038803
8.	Symmetry	0.027414
9.	Fractal Dimension	0.018061
10.	Texture	4.301036

Table 3. Accuracy of various Classifiers

S.No	Classifier	Accuracy
1	Logistic Regression	86.84%
2	Decision Tree	87.4%
3	SVC	96.5%
4	Random Forest	95.61%

Each of the above feature was computed for each nucleus and then the mean value, largest or worst value and standard error of each feature were found over the range of isolated cells. As the dataset was supervised hence the diagnosis dimension was also given along with two other dimensions, thus making it 33 dimensions in total. These 33 dimensions call for dimensionality reduction which we have done using PCA. We found the correlation of each dimension. After reducing 33 dimensions to 13 after PCA, we performed EDA on the same. During EDA, a heatmap was generated which is shown in Fig. 2 below:

Through analysis of the heatmap we realised that the two features radius_mean and texture_mean was the most correlated and could be used for prediction. We plotted their relation into a scatter plot as given in Fig. 3 below:

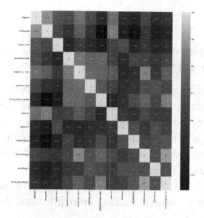

Fig. 2. Heatmap for WDBC dataset

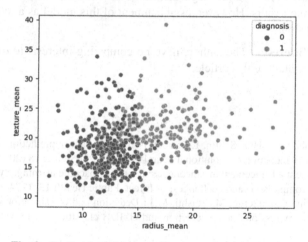

Fig. 3. Scatterplot between radius_mean and texture_mean

With this dimensionality reduction of the dataset, we now applied different ML algorithms namely: Logistic Regression, Decision Tree, Random Forest and the SVC to find the most accurate classifier. The results of the same are listed in the table below:

From Table 3 above it is clear that SVC gives the highest accuracy of 96.5%.

7 Conclusion

From the thorough analysis of the available literature, we finally concluded that early, accurate breast cancer prediction could be achieved by integrating a multi-step dimensionality reduction involving PCA followed by EDA. The final dimensions obtained could then be coupled with the SVC classifier to obtain accurate breast cancer prediction model. The results obtained prove the same.

8 Future Scope

Though the results of the proposed framework are promising, however, room for potential growth and development still remain. The major areas being:

- Integrating Data: The proposed approach has only been applied to standardised WDBC dataset. Application to some real-time dataset may further result in identifying subtle patterns that may not be evident presently.
- Personalized Medicine: The enhanced predictive capabilities of this approach can contribute to tailoring treatment strategies based on individual patient profiles, optimizing the chances of successful outcomes.
- Translational Research: The integrated model can serve as a decision support tool for healthcare providers, aiding in making more informed decisions regarding patient care and follow-up.

Thus, an early breast cancer diagnosis model has been reported in this work which promises accurate results. However, establishment of this model as a tool is subject to the above extensions.

Disclosure of Interests. The authors have no competing interests to declare that are relevant to the content of this article.

References

1. Xu, H.Q., Shao, X,, Hui, S., Jin, L.: Supervised breast cancer prediction using integrated dimensionality reduction convolutional neural network. PLoS One. **18** (2023)
2. Saba, T.: Recent advancement in cancer detection using machine learning: Systematic survey of decades, comparisons and challenges. J. Infect. Public Health **13**, 1274–89 (2020)
3. Khandezamin, Z., Naderan, M., Rashti, M.J.: Detection and classification of breast cancer using logistic regression feature selection and GMDH classifier. J. Biomed. Inform. **1**, 111 (2020)
4. Manikandan, P., Durga, U., Ponnuraja C.: An integrative machine learning framework for classifying SEER breast cancer. Sci Rep. **13** (2023)
5. Zuluaga-Gomez, J.: Breast Cancer Diagnosis Using Machine Learning Techniques (3 May 2023). http://arxiv.org/abs/2305.02482
6. Vaka, A.R., Soni, B.K.: SR. Breast cancer detection by leveraging Machine Learning. ICT Express. **6**(4), 320–324 (2020)
7. Street, W.N., Wolberg, W.H., Mangasarian, O.L.: Nuclear feature extraction for breast tumor diagnosis, vol. 1905, pp. 861–670. [Internet]. (29 Jul 1993). https://www.spiedigitallibr ary.org/conference-proceedings-of-spie/1905/0000/Nuclear-feature-extraction-for-breast-tumor-diagnosis/, https://doi.org/10.1117/12.148698.full (Accessed 9 Aug 2023)
8. Hosni, M., Abnane, I., Idri, A., Carrillo de Gea, J.M., Fernández Alemán, J.L.: Reviewing ensemble classification methods in breast cancer. Comput Methods Programs Biomed. **177**, 89–112 (2019)
9. Breast Cancer Facts & Statistics 2023 [Internet]. (Accessed 11 Jul 2023). https://www.bre astcancer.org/facts-statistics
10. Naji, M.A., Filali, S.E., Aarika, K., Benlahmar, E.H., Abdelouhahid, R.A., Debauche, O.: Machine learning algorithms for breast cancer prediction and diagnosis. Proc. Comput. Sci., 487–92 (2021)

11. Burges, C.J.C.: Dimension reduction: A guided tour. Foundat. Trends Mach. Learn. [Internet], **92**(4), 275–365 (2009). https://www.researchgate.net/publication/220416606_Dimension_Reduction_A_Guided_Tour (Accessed 16 Aug 2023)
12. Kabir, M.F., Chen, T., Ludwig, S.A.: A performance analysis of dimensionality reduction algorithms in machine learning models for cancer prediction. Healthcare Anal. **1**(3), 100125 (2023)
13. Mehrotra, R., Yadav, K.: Breast cancer in India: Present scenario and the challenges ahead. World J. Clin. Oncol. [Internet] **13**(3), 209 (2022) (Accessed 11 Jul 2023). /pmc/articles/PMC8966510/
14. Subramanian, D.: Breast Cancer Prediction using Some Machine Learning Models by Dimensionality Reduction of Various Features [Internet]. https://www.researchgate.net/publication/358521975
15. Pendharkar, P.C., Rodger, J.A., Yaverbaum, G.J., Herman, N., Benner, M.: Association, statistical, mathematical and neural approaches for mining breast cancer patterns. Expert Syst. Appl. **17**(3), 223–232 (1999)
16. West, D., Mangiameli, P., Rampal, R., West, V.: Ensemble strategies for a medical diagnostic decision support system: a breast cancer diagnosis application. Eur. J. Oper. Res. **162**(2), 532–551 (2005)
17. Gupta, K., Janghel, R.R.: Dimensionality reduction-based breast cancer classification using machine learning. Adv. Intell. Syst. Comput. [Internet], **798**, 133–46 (2019). https://link.springer.com/chapter/, https://doi.org/10.1007/978-981-13-1132-1_11 (Accessed 11 Jul 2023)
18. Abdar, M., Zomorodi-Moghadam, M., Zhou, X., Gururajan, R., Tao, X., Barua, P.D., et al.: A new nested ensemble technique for automated diagnosis of breast cancer. Pattern Recognit Lett. **1**(132), 123–131 (2020)
19. Sun, X., Qourbani, A.: Combining ensemble classification and integrated filter-evolutionary search for breast cancer diagnosis. J. Cancer Res. Clin. Oncol. (2023)
20. Tuerhong, A., Silamujiang, M., Xianmuxiding, Y., Wu, L., Mojarad, M.: An ensemble classifier method based on teaching–learning-based optimization for breast cancer diagnosis. J. Cancer Res. Clin. Oncol. (2023)
21. Li, X., Chen, X., Rezaeipanah, A.: Automatic breast cancer diagnosis based on hybrid dimensionality reduction technique and ensemble classification. J. Cancer Res. Clin. Oncol. [Internet] (2023). https://pubmed.ncbi.nlm.nih.gov/36995408/ (11 Jul 2023)
22. Krawczyk, B., Schaefer, G., Woźniak, M.: A hybrid cost-sensitive ensemble for imbalanced breast thermogram classification. Artif. Intell. Med. **65**(3), 219–227 (2015)
23. Tian, J.X., Zhang, J.: Breast cancer diagnosis using feature extraction and boosted C5.0 decision tree algorithm with penalty factor. Math. Biosci. Eng. **19**(3), 2193–2205 (2022)
24. Fabiyi, S.D., Ezechukwu, D.N.: Feature extraction and dimensionality reduction of cancer data using folded LDA. In: 3rd International Informatics and Software Engineering Conference, IISEC 2022. Institute of Electrical and Electronics Engineers Inc. (2022)
25. Yadav, A., Jamir, I., Jain, R.R., Sohani, M.: Breast cancer prediction using SVM with pca feature selection method. Inter. J. Sci. Res. Comput. Sci. Eng. Inform. Technol. **5**, 969–978 (2019)
26. Taghizadeh, E,, Heydarheydari, S., Saberi, A., JafarpoorNesheli, S., Rezaeijo, S.M.: Breast cancer prediction with transcriptome profiling using feature selection and machine learning methods. BMC Bioinformatics **23**(1) (2022)
27. Bahrami, M., Vali, M.: Wise feature selection for breast cancer detection from a clinical dataset. In: 2021 28th National and 6th International Iranian Conference on Biomedical Engineering, ICBME 2021, pp. 160–164 (2021)
28. Ibrahim, S., Nazir, S., Velastin, S.A.: Feature selection using correlation analysis and principal component analysis for accurate breast cancer diagnosis. J. Imaging. **7**(11) (2021)

29. Al-Azzam, N., Shatnawi, I.: Comparing supervised and semi-supervised Machine Learning models on diagnosing breast cancer. Annals Med. Surgery. **1**(62), 53–64 (2021)

30. Sathyavathi, S.: Breast cancer identification using logistic regression. Biosci Biotechnol Res Commun. **13**(11), 34–36 (2020)

31. Sánchez-Cauce, R., Pérez-Martín, J., Luque, M.: Multi-input convolutional neural network for breast cancer detection using thermal images and clinical data. Comput. Methods Programs Biomed. **1**, 204 (2021)

32. Desai, M., Shah, M.: An anatomization on breast cancer detection and diagnosis employing multi-layer perceptron neural network (MLP) and Convolutional neural network (CNN), vol. 4, pp. 1–11. Clinical eHealth. KeAi Communications Co. (2021)

33. UCI Machine Learning Repository [Internet]. http://archive.ics.uci.edu/dataset/17/breast+cancer+wisconsin+diagnostic (Accessed 9 Aug 2023)

34. Deisenroth, M,P., Faisal, A.A, Ong, C.S:. Mathematics for machine learning, p. 371

35. Gallier, J.H., Quaintance, J.: Linear algebra and optimization with applications to machine learning

36. Secondary Analysis of Electronic Health Records. Secondary Analysis of Electronic Health Records. Springer International Publishing, pp. 1–427 (2016)

Designing AI-Based Non-invasive Method for Automatic Detection of Bovine Mastitis

S. L. Lakshitha(✉) ⓘ and Priti Srinivas Sajja

PG Department of Computer Science and Technology, Sardar Patel University,
Vallabh Vidyanagar 388 120, Gujarat, India
lakshithas163@gmail.com

Abstract. Mastitis is a major disease in dairy animals as a consequence of udder inflammation. The occurrence of disease in dairy animals impacts farmers economic status and hinders dairy industry growth. Many conventional and rapid methods are used by the dairy industry to detect mastitis by considering the moderate to severe symptoms of infected animals and their milk. The limitations of such methods are that they are expensive, arduous, and require samples. In the precision dairying era, rapid, low-cost, automated alternative techniques for early prediction of diseases in animals are in demand by stakeholders, especially farmers. Infrared Thermography (IRT) is an emerging, non-invasive tool to predict diseases in humans and animals. IRT, coupled with machine learning algorithms, has the potential to detect bovine mastitis. Researchers explored algorithms such as K-nearest neighbourhood (KNN), Support vector machine (SVM), Random Forest (RF), and Convolutional Neural Network (CNN) for automatic, real-time detection of mastitis using thermographic images. This paper discusses the prevalence of mastitis, related works using IRT with machine learning models, designing a model by embedding domain heuristics to fine-tune the decision, and details about the experiment carried out by employing KNN and SVM for thermal and demographic data. Both SVM and KNN classifiers classified the disease with an accuracy of 60%; this low accuracy is due to limited data. However, the results provide new insights to develop a non-invasive method for the detection of mastitis. The future work envisaged by using large field data for the detection of mastitis more accurately.

Keywords: Mastitis · Machine Learning · IRT · Thermal Images · Artificial intelligence-in-Animal Health

1 Introduction

Agriculture, along with dairy farming, remains an integral part of human life. India has become the world leader in dairy production, accounting for 24 percent of global milk production. Dairy and livestock contribute 6.2% to national GDP and 31% to agricultural GDP [1]. Milk production in the country is estimated at Rs 9.3 lakh crore. It is the largest agricultural product in terms of value and even exceeds the total value of cereals, pulses, and sugarcane. Dairying in India has significantly contributed to the socio-economic

K. K. Patel et al. (Eds.): icSoftComp 2023, CCIS 2030, pp. 301–313, 2024.
https://doi.org/10.1007/978-3-031-53731-8_24

development of rural areas by providing opportunities for self-reliance to millions of small and marginal farmers and landless farmers. Even though India has achieved the top position in dairy production, it still faces several challenges, such as animal health, nutrition, and low productivity, that create bottlenecks for development. Among the many challenges dairy farmers face, recurrent animal diseases and disease outbreaks are a threat to the growth of the dairy industry. Mastitis in milking animals poses a global concern and has severe financial bearing to the dairy industry. In India, an average of 50% of dairy animals are affected by mastitis [2]. An estimated 2.65 lakh crore (USD 32.2 billion) is lost financially each year worldwide, and 7165.51 crore rupees is lost in India [3, 4]. Milk from mastitis-infected animals is unfit for human consumption and may contain zoonotic bacteria and cause infections in humans [5].

Mastitis can be classified into subclinical mastitis (SCM), clinical mastitis (CM), and chronic or contagious mastitis, depending on the severity of the symptoms. The incidence of SCM (42%) was higher than that of CM (15%) globally, similarly in India SCM (45%) and CM (18%) observed in a meta-analysis of mastitis across six continents reported between 1967 and 2019, demonstrating the significance of SCM in cattle's [6]. Subclinical mastitis is difficult to recognize because there are no detectable clinical signs in milk. Detectable physical signs such as changes in colour of milk, watery appearance of the milk, the presence of clots or flakes in the milk, swelling of the udder, and rapid temperature change appear in clinical mastitis. Persistent inflammation of the mammary gland leads to chronic or contagious mastitis in dairy animals [3, 7]. These visible symptoms on the udder of the infected animal and the physicochemical changes in the milk produced by the animal help in the diagnosis of mastitis. Electric Conductivity (EC), California Mastitis Test (CMT) and Somatic Cell Counts (SCC) are considered as better predictors of mastitis. Some of the diagnostic methods, such as bacterial culture tests, biomarkers, proteomic techniques, immunoassay (ELISA), and polymerase chain reaction (PCR), are subjective, arduous, and require the collection of a milk sample from the infected animal. Some of the aforementioned methods require state-of-the-art equipment and a trained analyst, making them impractical at farm level. Therefore, the detection of mastitis by a non-invasive method in real-time on the farm is a preferred method. In the era of digitization, rapid, robust, low-cost, and handy alternative techniques are in demand by stakeholders, especially dairy farmers. Hence, automated methods for early detection of animal diseases using Artificial Intelligence (AI) and Machine Learning (ML) platforms are currently in focus under precision dairying [8].

Infrared Thermography (IRT) is an emerging, non-invasive tool to predict diseases in humans and animals. IRT, coupled with machine learning algorithms, has the potential to detect bovine mastitis [9]. The basis of infrared thermography is that all objects emit infrared radiation in proportion to their temperature, as stated by the Stefan-Boltzmann law [10]. IRT is a helpful technique for assessing biometric changes brought on by stress, inflammation-related pain in the animal or infected area, and blood flow variations brought on by physiological or environmental factors [11]. IRT is a simple, efficient, non-invasive, on-site technique that produces visual images without exposing users to radiation while detecting surface heat. Thermographic images of animals, in conjunction with machine learning algorithms, are extensively used for predicting or classifying diseases in humans as well as animals. Many researchers have used ML algorithms as potential tools in the detection of mastitis. ML algorithms such as SVM, KNN, CNN, Stochastic Gradient Descent, Naive Bayes Classifier, Random Forest, and Multilayer Perceptron are used to classify the thermal images of the udder and eye of the cow to diagnose different stages of mastitis based on temperature threshold values. The target detection algorithms such as YOLO, R-CNN, and single-shot multi-box detector (SSD) are explored for the identification of key parts of animals like the udder, eye, and head for temperature.

The accuracy of the methods using IRT relies on environmental factors like solar radiation and the velocity of the wind [12]. Researchers used single-breed images or videos at large farm levels for training the ML models which may not be suitable for Indian dairy where farmers have small herd sizes and different breeds. Thermal camera features, operator position and placement of animals, and types of milking also affect the reading [13, 14]. Non-availability of freely accessible datasets of udder thermal images for developing machine learning models and the lack of a standardized thermal image acquisition procedure pose another challenge [15]. Requirement of large data sets considering different breeds to be compared against reference methods to train the ML model to detect mastitis precisely. The model is designed to overcome all these challenges, to embed domain heuristics to fine-tune the decision for initiating the remedial measures by the farmers.

There are seven sections in this article. The relevant research on the automatic identification of mastitis using thermal images is discussed in Sect. 2. In Sect. 3, the architecture of the designed model and its various phases are described in detail. The classification of mastitis using thermal and demographic data by ML algorithms, its evaluation, and results are discussed in Sects. 4, 5 and 6, respectively. Section 7 conclude the paper and also presents future work by using large field data for the accurate detection of mastitis.

2 Related Work in the Detection of Mastitis

The features of IRT are explored in the detection of bovine diseases [16]. The udder becomes a crucial body part for the detection of any febrile disease by a non-invasive method [17]. The elevation of 1–1.5 °C in the surface temperature of the udder is due to slight variation in the blood flow during infection before visible symptoms against other controlled parts [18]. The variable temperature of infected udder has positive correlation with SCC of milk, temperature extracted from thermal images is a useful index for detection of mastitis with high precision [19, 20]. The potentials of IRT in aiding diagnosis of mastitis can be further explored by combining with machine learning techniques. Research carried out by using thermal images coupled with ML shows the foundation for diagnosis of mastitis accurately in early stages. Target detection algorithms were used to detect the variable temperature between control parts, such as the udder and the eyes. These algorithms were trained to detect the disease, and their accuracy was checked against reference methods. The ocular surface temperatures from thermal images can automatically detected by using algorithms like YOLO [21]. The factors like left and right Udder Skin Surface Temperature difference (USST), Ocular Surface Temperature and USST difference considered for automatic mastitis detection with deep learning methods. Deep learning network like You Only Look Once v5 (YOLOv5) [21],

EFMYOLOv3 (Enhanced Fusion Mobile Net V3 You Only Look Once v3) [22] based on the bilateral filtering enhancement of thermal image is used to detect mastitis. YOLO is an efficient object detection algorithm which identified key parts with accuracy of 96.8% [22]. Hoshen-Kopelman algorithm used for classifying the thermal images of udder into healthy, subclinical or clinical mastitis [13]. Python programming was used to create distinct linear regression models that were intended to determine whether there was a dependency between the groups of healthy and mastitis-infected animals. It was observed that the milk yield and the temperature of their udder surface were significantly correlated in mastitis-infected cows, but there was no correlation in healthy cows [23]. A significant relationship was observed between temperature extracted by IRT, somatic cell count, and the California mastitis test in the detection of mastitis by the Classification and Regression Tree (CART) algorithm [24]. IRT in conjunction with machine learning tools is a promising technique for predictive and precision farming.

3 Architecture of the System

To make the model robust and by considering the above limitations in the field, we have conceptualized the model in three major phases as described in Fig. 1. The key components of the model in order to detect mastitis using thermal images with machine learning algorithms is discussed under this section.

3.1 Data Collection and Pre-processing

Acquiring images is the initial step in the first phase. Good-resolution thermal images of an animal's udder are obtained through a thermal camera. Any object's image is captured by a thermal camera using the heat radiation it emits. Thermal imaging is the process of converting infrared (IR) radiation, or heat, into visual representations that demonstrate how temperature variations are distributed throughout a scene as observed by a thermal camera [4]. The temperature readings of a certain region are represented by the pixels in the thermal image. Pre-processing is the important step for removing undesirable features because the captured image may be noisy due to some factors like atmospheric conditions, mud or grass, or hairs present on the udder surface. Pre-processing of images is necessary to get clean images with appropriate features. The captured thermal images are pre-processed to reduce the noise and converted into gray-scale images for simplification. This preliminary step enhances the image quality for analysis. This can be achieved by using various available linear and non-linear filters [25]. Filtering of an image involves smoothing, sharpening, and edge enhancement. Smoothing deals with reducing noise in the image by using a low-pass filter, while sharpening deals with enhancing the edges.

These enhanced thermal images contain parts other than the region of interest (ROI). Extracting the ROI is a crucial step in thermal image processing, as this affects the result of the machine learning model while classifying the disease. The temperature matrix of the udder, minimum, maximum, and average temperatures can be collected for further processing. Along with these thermal data, demographic information about the animal, such as breed, age, stage of lactation, history of the animal, milk yield, type of milking, and image before or after milking, is collected from the field for predicting mastitis. These factors are correlated with disease and are very important in classifying diseases. For example, a reduction in milk yield was observed in mastitis-infected animals due to parity and the time of mastitis occurrence [5]. Parallelly, milk samples are collected for measuring SCC or CMT to train and assess the model output. After pre-processing, the dataset can be cross-validated to make two subsets, the training set and the testing set, which are used to train and test the model, respectively. The test data is used to evaluate the efficiency of the model on unseen data. The training set should be larger compared to the testing set because machine learning model performance depends on the past information given to it.

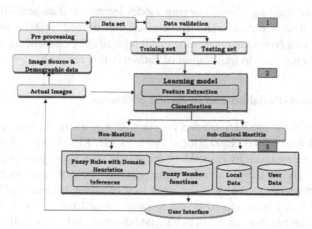

Fig. 1. Architecture of the System

3.2 Learning Model-Automatic Disease Diagnosis

The second phase is to train the model using the good quality, balanced, and validated training data set obtained in the previous phase using machine learning techniques. Both the data and the learning algorithms' performance are necessary for a machine learning model to be successful. Before making a decision, the robust model must be trained using real-world data that has been gathered and knowledge about the intended application [26]. Removing irrelevant data from the data sets aids in the accuracy of the decision. This can be achieved by various feature extraction and feature selection methods. The feature extraction is embedded in the model to simplify by reducing the number of features, decreasing the training duration, reducing overfitting by enhancing generalization, and avoiding the curse of dimensionality [27]. The information from the original data set can be summarized by eliminating redundancy using feature extraction techniques in machine learning like Feature Pyramid Networks and Path Aggregation Networks [21].

The data obtained in the previous stages used for image classification by supervised machine learning algorithms. This algorithm is classified into regression and classification methods; it is appropriate for classification problems and is also easy to implement as compared to unsupervised learning algorithms [28]. Regression is used when we are dealing with continuous types of data. Classification is used when we are dealing with categorical data. The role of the classification model is to match the given input with the respective predefined output label. Classification can be based on binary or multiclass. In this case, we have a binary classification with two classes, 0 and 1, indicating whether the animal is affected by disease or not, respectively. The classification algorithm is considered as good learner if it can generalize the classes on unseen data [29]. We can use single classifiers or hybrid classifiers; the available classifiers are provided in Fig. 2. [30]. During the training phase, learning model learns the data and observations for classifying the data into their predefined classes. The trained model is used to classify thermal image into Non-mastitis or subclinical mastitis. The classification model, SVM and KNN are being used to train subset of dataset in this research.

3.3 Fuzzy Rules–Decisions and Remedial Measures

The third phase of the model fine-tunes the classification using domain heuristics stored in the form of fuzzy rules. Fuzzy logic can be applied to classify alerts for mastitis [31]. This can be achieved by embedding the system with fuzzy logic. Fuzzy rule base, fuzzification, and defuzzification are the major components of fuzzy logic. Fuzzy logic mimics human reasoning by defining rules in a common language [32]. Equation 1 represents the fuzzy rules stored in the form of "if and then". An inference system makes use of membership functions to translate the provided input features to a specific output. The membership function shows each input's level of participation in a system graphically [33]. Figure 3. Illustrates the graphical representation of fuzzy membership functions.

The different forms of knowledge about the problem are represented by using fuzzy sets and fuzzy logic. The general format of fuzzy rule is as follows.

$$\text{if condition (P1, P2, P3, ..., Pn) then action (A1, A2, A3, ..., An)} \quad (1)$$

where, P1, P2, P3,......, Pn are conditions based on local data & stage of mastitis and A1, A2, A3,......, An are actions to be taken by the farmer.

The foremost aim of the designed system is to detect mastitis precisely along with advice on specific disease remedial procedures to be initiated by farmers. Some of the rules and actions initiated by farmers are provided in Table 1.

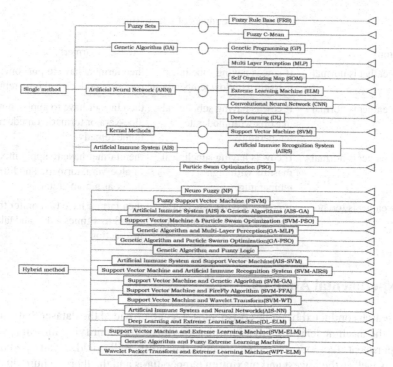

Fig. 2. Single classifier and Hybrid classifier Methods [30]

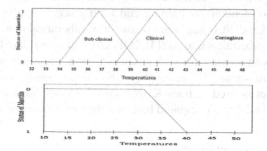

Fig. 3. Triangular and Trapezoidal functions

Table 1. Examples of Fuzzy Rules

Conditions		Advise to farmers
If the weather is normal	If cow is having subclinical mastitis with low temperature	then farmers have to put some ice packs on udder as a home remedy
If the weather is normal	If cow is having subclinical mastitis with moderate temperature	then farmer have to apply some aloe-vera or turmeric on udder as a home remedy
If the weather is normal	If cow is having subclinical mastitis with high temperature	then farmer have to apply mixture of aloe-vera, turmeric and lime on udder as a home remedy
If the weather is too hot	if cow is having subclinical mastitis with low temperature	then farmer have to monitor the cow 3 to 4 times a day and take the next action

4 Experimental Analysis

The research conducted by Velasco-Bolanos, J. et al. provided the dataset that was used to train the machine learning model in this experiment for identifying mastitis [14]. For the experiment, factors such as temperature, humidity, milk yield, mean, median, and mode, as well as the lowest and maximum temperatures and the thermal humidity index (THI), were considered. The model is built using the Python programming language. Pre-processing is performed on the dataset to eliminate null values. The datasets are divided in the ratio of 60:40 as training and testing sets, respectively, by using the Python sklearn package. SCC is a better predictor of subclinical mastitis; a SCC of > 2 lakhs is considered subclinical mastitis [4]. In the attributes considered, SCC greater than threshold is used as an output label with a value of 0 or 1, indicating whether the animal is affected by mastitis or not. The remaining attributes are considered input parameters. The training dataset is used to train KNN and SVM machine learning models. The accuracy of these model was compared based on their results on the testing dataset.

4.1 K-nearest Neighbour (KNN)

KNN is a supervised machine learning algorithm that classifies the data based on the neighbour datapoints. Since the Euclidean distance is used to measure resemblances between training and testing data points without requiring prior knowledge about data distribution, hence, the implementation of KNN is simple [34]. Based on the Euclidean distance of the training set point, the nearest neighbor (k) for each row in the test set is observed, and classification is accomplished based on the majority vote [35, 36].

Euclidian distance is calculated using the Eq. 2:

$$d = \sqrt{[(x2 - x1)^2 + (y2 - y1)^2]} \tag{2}$$

where, x1, y1 and x2, y2 are new data point and its neighbour data point respectively. The KNN model is illustrated in Fig. 4a [37].

4.2 Support Vector Machine (SVM)

SVM is a popular binary classification-supervised machine learning algorithm technique. SVM classifies into potential categories by analysing historical data [38]. SVM classifies the data by placing a margin among the classes. By drawing the largest gap between the margin and the classes, the margins are created in a way that minimizes the classification error [39, 40]. As a result, it is also referred to as an optimal margin classifier. The SVM algorithm finds the hyperplane with the maximum margin among all feasible options, which divides an initial set of data into two smaller sets (the margin is the distance between the hyperplane and the closest points). The hyperplane is defined by support vectors, which are usually the points closest to the hyperplane and those that define it [41]. The SVM model is illustrated in Fig. 4b [42]. When new data comes in, it is assigned to its class by identifying a hyperplane that effectively divides the two classes based on the closest point of the categories separated by the hyperplane.

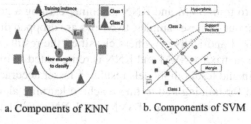

a. Components of KNN   b. Components of SVM

Fig. 4. Components of KNN [37] and SVM [42]

5 Evaluation of Classification

The efficiency of classification models can be evaluated based on their performance using evaluation metrics. The most popular evaluation metrics used are accuracy, sensitivity, specificity, and precision [43]. These are calculated using the formula given in Fig. 5.

Metric	Definition	Formula
Accuracy (Acc)	It is the ratio of number of correct predictions to the total number of samples.	$\dfrac{TP + TN}{TP + FN + TN + FP}$
Sensitivity (Se)	It measures the proportion of actual positives that are correctly identified as positives.	$\dfrac{TP}{FN + TP}$
Specificity (Sp)	Measures the proportion of actual negatives that are correctly identified negatives.	$\dfrac{TN}{FP + TN}$
Precision (Pr)	The proportions of positive observations that are true positives.	$\dfrac{TP}{TP + FP}$
F1 Score	Harmonic mean of the precision and recall	$\dfrac{2TP}{2TP + FN + FP}$

TP - True Positive, FP - False Positive, TN – True Negative, FN – False Negative

Fig. 5. Evaluation Classification Formula

Confusion matrix to represent these values are shown in Fig. 6.

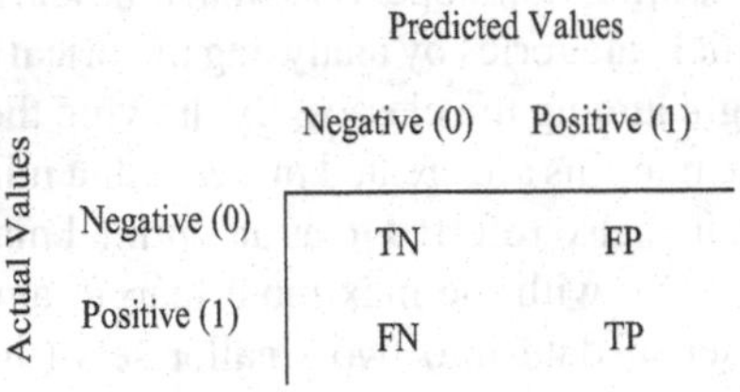

Predicted Values
Negative (0)
Positive (1)
Actual Values
Negative (0)
Positive (1)
TN
FP
FN
TP

Fig. 6. Confusion matrix

6 Results

The dataset is used to train SVM and KNN machine learning algorithms. The dataset contains 55 samples; out of these samples, two rows are dropped due to the presence of null values. Both SVM and KNN classifiers classified the results with an accuracy of 60%. The confusion matrix for SVM and KNN is represented in Fig. 7. The small data size available to train and test both models results in lesser accuracy. Future work with larger data sets used for training and testing machine learning algorithms would yield better results. The classification report of SVM and KNN is tabulated in Table 2.

Confusion Matrix(SVM)
0
1
8
7
1
4
Actuals
Predictions
Confusion Matrix(KNN)
0
1
10
5
3
2
Actuals
Predictions

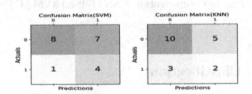

Fig. 7. Confusion Matrix -SVM & KNN

Table 2. Classification report of SVM and KNN

Algorithm	Profile feature	Precision	Recall	F1-Score	Support
SVM	NM	0.89	0.53	0.67	15
	M	0.36	0.80	0.50	5
	Avg./total	0.63	0.67	0.58	20
KNN	NM	0.77	0.67	0.71	15
	M	0.29	0.40	0.33	5
	Avg./total	0.53	0.53	0.52	20

NM: Non-mastitis; M: Mastitis

7   Conclusion and Future Work

The thermographic data captured from thermal images of the udder and other data like parity, milk yield, THI, SCC, etc. collected from the field are very useful in training the machine learning model against the reference methods for detection of mastitis. The SVM and KNN classifiers used in the experiment with a 53 data set are able to classify the data for predicting mastitis in dairy animals. Both SVM and KNN classifiers classified the results with an accuracy of 60%. Due to the small sample dataset the accuracy of the both models are not good. Machine learning algorithms yields better results when they are trained with large amount of data. This experiment is nascent step, the future work envisages collecting field-level data along with other details to be trained as per the proposed model to detect mastitis along with advice or alerts on specific disease remedial procedures to be initiated by farmers by fine-tuning the classification using domain heuristics stored in the form of fuzzy rules. The major focus of the conceptualized model is to build a non-invasive method to detect mastitis precisely using thermal images coupled with a machine learning algorithm, along with advice on specific disease remedial procedures to be initiated by farmers by embedding fuzzy rules.

Acknowledgment. The authors are thankful to Velasco-Bolanos, J. et al. [14] for providing the data.

References

1. Department of Animal Husbandry and Dairying. https://dahd.nic.in/. (Accessed 20 July 2023)
2. Saini G, Yadav V, Sharma M, Bisla A. Mastitis: A challenge in doubling the farmers income by 2022. Indian Farming 70 (7) (2020)
3. Martins, S.A.M., et al.: Biosensors for on-farm diagnosis of mastitis. Front. Bioeng. Biotechnol. **7**, 186 (2019)
4. Sinha, R., et al.: Infrared thermography as non-invasive technique for early detection of mastitis in dairy animals-a review. Asian J. Dairy Food Res. **37**(1) (2018)
5. Harjanti, D.W., Sambodho, P.: Effects of mastitis on milk production and composition in dairy cows. IOP Conf. Ser.: Earth Environ. Sci. **518**, 012032 (2019)
6. Krishnamoorthy, P., Goudar, A.L., Suresh, K.P., Roy, P.: Global and countrywide prevalence of subclinical and clinical mastitis in dairy cattle and buffaloes by systematic review and meta-analysis. Res. Vet. Sci. **136**, 561–586 (2021)
7. Vyas, S., Shukla, V., Doshi, N.: FMD and mastitis disease detection in cows using internet of things (IOT). Proc. Comput. Sci. **160**, 728–733 (2019)
8. Zhang, Y., Zhang, Q., Zhang, L., Li, J., Li, M., Liu, Y., et al.: Progress of machine vision technologies in intelligent dairy farming. Appl Sci (Basel) **13**(12), 7052 (2023)
9. Wang, Y., Li, Q., Chu, M., Kang, X., Liu, G.: Application of infrared thermography and machine learning techniques in cattle health assessments: A review. Biosyst. Eng. **230**, 361–387 (2023)
10. Poikalainen, V., Praks, J., Veermäe, I., Kokin, E.: Infrared temperature patterns of cow's body as an indicator for health control at precision cattle farming, pp. 187–194 (2012)
11. McManus, C., Tanure, C.B., Peripolli, V., Seixas, L., Fischer, V., Gabbi, A.M., et al.: Infrared thermography in animal production: An overview. Comput. Electron. Agric. **123**, 10–16 (2016)

12. Mota-Rojas, D., et al.: Clinical applications and factors involved in validating thermal windows used in infrared thermography in cattle and river buffalo to assess health and productivity. Animals (Basel) **11**(8), 2247 (2021)
13. Angelo, R., et al.: Thermal image thresholding for automatic detection of bovine mastitis. International J. Comput. Appli. **183**(14), 0975–8887 (2021)
14. Velasco-Bolaños, J., et al.: Application of udder surface temperature by infrared thermography for diagnosis of subclinical mastitis in Holstein cows located in tropical highlands. J. Dairy Sci. **104**(9),: 10310–23 (2021)
15. Cockburn, M.: Review: application and prospective discussion of machine learning for the management of dairy farms. Animals **10**(9), 1690 (2020)
16. Stelletta, C., Gianesella, M., Vencato, J., Fiore, E., Morgante M.: Thermographic Applications in Veterinary Medicine [Internet]. Infrared Thermography. InTech (2012)
17. Hovinen, M., Siivonen, J., Taponen, S., Hänninen, L., Pastell, M., Aisla, A.-M., et al.: Detection of clinical mastitis with the help of a thermal camera. J. Dairy Sci. **91**(12), 4592–4598 (2008)
18. Colak, A., Polat, B., Okumus, Z., Kaya, M., Yanmaz, L.E., Hayirli, A.: Short communication: early detection of mastitis using infrared thermography in dairy cows. J. Dairy Sci. **91**(11), 4244–4248 (2008)
19. Zaninelli, M., et al.: First evaluation of infrared thermography as a tool for the monitoring of udder health status in farms of dairy cows. Sensors **18**(3), 862 (2018)
20. Machado, N.A.F., Da Costa, L.B.S., Barbosa-Filho, J.A.D., De Oliveira, K.P.L., De Sampaio, L.C., Peixoto, M.S.M., et al.: Using infrared thermography to detect subclinical mastitis in dairy cows in compost barn systems. J. Therm. Biol. **97**(102881), 102881 (2021)
21. Wang, Y., Kang, X., Chu, M., Liu, G.: Deep learning-based automatic dairy cow ocular surface temperature detection from thermal images. Comput. Electron. Agric. **202**(107429), 107429 (2022)
22. Xudong, Z., Xi, K., Ningning, F., Gang, L.: Automatic recognition of dairy cow mastitis from thermal images by a deep learning detector. Comput. Electron. Agric. **178**(105754), 105754 (2020)
23. Khakimov, A.R., Pavkin, D.Y., Yurochka, S.S., Astashev, M.E., Dovlatov, I.M.: Development of an algorithm for rapid herd evaluation and predicting milk yield of mastitis cows based on infrared thermography. Appli. Sci. **12**(13), 6621 (2022)
24. Coskun, G., Aytekin, İ: Early detection of mastitis by using infrared thermography in Holstein-Friesian dairy cows via classification and regression tree (CART) analysis. Selcuk J. Agricult. Food Sci. **35**(2), 115–124 (2021)
25. Kim, H.J., Shrestha, A., Sapkota, E., Pokharel, A., Pandey, S., Kim, C.S., et al.: A study on the effectiveness of spatial filters on thermal image pre-processing and correlation technique for quantifying defect size. Sensors (Basel) **22**(22), 8965 (2022)
26. Sarker, I.H.: Machine learning: Algorithms, real-world applications and research directions. SN Comput Sci. **2**(3) (2021)
27. Chen, R.-C., Dewi, C., Huang, S.-W., Caraka, R.E.: Selecting critical features for data classification based on machine learning methods. J Big Data **7**(1) (2020)
28. International Business Machine. https://www.ibm.com/blog/supervised-vs-unsupervised-learning/ (Accessed 28 July 2023)
29. Mohamed, A.: Comparative study of four supervised machine learning techniques for classification. J. Appli. Sci. Technol. (2017)
30. Kalantari, A., Kamsin, A., Shamshirband, S., Gani, A., Alinejad-Rokny, H., Chronopoulos, A.T.: Computational intelligence approaches for classification of medical data: State-of-the-art, future challenges and research directions. Neurocomputing **276**, 2–22 (2018)
31. De Mol, R.M., Woldt, W.E.: Application of fuzzy logic in automated cow status monitoring. J. Dairy Sci. **84**(2), 400–410 (2001)

32. Matiko, J.W., Beeby, S.P., Tudor, J.: Fuzzy logic based emotion classification. In: 2014 IEEE International Conference on Acoustics, Speech and Signal Processing (ICASSP). IEEE (2014)
33. Adil, O., Ali, M., Ali, A., Sumait, B.S.: Comparison between the Effects of Different Types of Membership Functions on Fuzzy Logic Controller Performance (2015)
34. Adeniyi, D.A., Wei, Z., Yongquan, Y.: Automated web usage data mining and recommendation system using K-Nearest Neighbor (KNN) classification method. Appl. Comput. Inform. **12**(1), 90–108 (2016)
35. Sharma, M., Kumar Singh, S., Agrawal, P., Madaan. V.: Classification of clinical dataset of cervical cancer using KNN. Indian J. Sci. Technol. **9**(28) (2016)
36. Bhole, V., Kumar, A., Bhatnagar, D.: A texture-based analysis and classification of fruits using digital and thermal images. In: ICT Analysis and Applications, p. 333–343. Springer, Singapore (2020)
37. Taunk, K., De, S., Verma, S., Swetapadma, A.: A brief review of nearest neighbor algorithm for learning and classification. In: 2019 International Conference on Intelligent Computing and Control Systems (ICCS). IEEE (2019)
38. Vardasca, R., Vaz, L., Mendes, J.: Classification and decision making of medical infrared thermal images. In: Dey, N., Ashour, A., Borra, S. (eds.) Classification in BioApps. LNCVB. vol. 26. Springer, Cham. (2018). https://doi.org/10.1007/978-3-319-65981-7_4
39. Mahesh, B.: Machine learning algorithms - a review. Inter. J. Sci. Res. **99**(1) (2020)
40. Chauhan, V.K., Dahiya, K., Sharma, A.: Problem formulations and solvers in linear SVM: a review. Artif. Intell. Rev. **52**(2), 803–855 (2019)
41. Romero, P.E., Rodriguez-Alabanda, O., Molero, E., Guerrero-Vaca, G.: Use of the support vector machine (SVM) algorithm to predict geometrical accuracy in the manufacture of molds via single point incremental forming (SPIF) using aluminized steel sheets. J. Mater. Res. Technol. **15**, 1562–1571 (2021)
42. Rani, A., Kumar, N., Kumar, J., Kumar, J., Sinha, N.K.: Machine learning for soil moisture assessment. In: Deep Learning for Sustainable Agriculture, pp. 143–68. Elsevier (2022)
43. Naser, M.Z., Alavi, A.H.: Error metrics and performance fitness indicators for artificial intelligence and machine learning in engineering and sciences. Archit. Struct. Constr. (2021)

Author Index

K. K. Patel et al. (Eds.): icSoftComp 2023, CCIS 2030, pp. 315–316, 2024.
https://doi.org/10.1007/978-3-031-53731-8

Printed in the United States
by Baker & Taylor Publisher Services